BUILDING
WASHINGTON

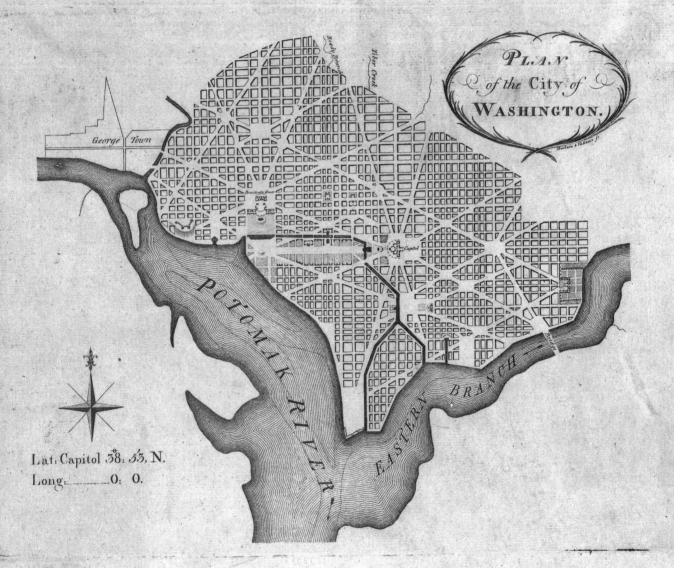

George Town

PLAN
of the City of
WASHINGTON.

Thackara & Vallance sc.

POTOMAK RIVER

EASTERN BRANCH

Lat: Capitol 38: 53, N.
Long: _____ 0: 0.

BUILDING
WASHINGTON

Engineering and Construction of the New Federal City, 1790–1840

ROBERT J. KAPSCH

JOHNS HOPKINS UNIVERSITY PRESS
Baltimore

This book was brought to publication through the generous assistance of the Kiplinger Foundation and the Charles E. Peterson Fellowship program of the Athenaeum of Philadelphia.

Johns Hopkins University Press
2715 North Charles Street
Baltimore, Maryland 21218-4363
www.press.jhu.edu

Library of Congress Cataloging-in-Publication Data

Names: Kapsch, Robert James, 1942–, author.
Title: Building Washington : engineering and construction of the new Federal City, 1790–1840 / Robert J. Kapsch.
Description: Baltimore : Johns Hopkins University Press, 2018. | Includes bibliographical references and index.
Identifiers: LCCN 2017030368| ISBN 9781421424873 (hardcover : alk. paper) | ISBN 9781421424880 (electronic) | ISBN 1421424878 (hardcover : alk. paper) | ISBN 1421424886 (electronic)
Subjects: LCSH: City planning—Washington (D.C.) | Public works—Washington (D.C.) | Washington (D.C.)—History. | Washington (D.C.)—Social conditions.
Classification: LCC HT168.W3 K37 2018 | DDC 307.1/21609753—dc23
LC record available at https://lccn.loc.gov/2017030368

A catalog record for this book is available from the British Library.

Frontispiece: Courtesy of the Library of Congress.

Special discounts are available for bulk purchases of this book. For more information, please contact Special Sales at 410-516-6936 or specialsales@press.jhu.edu.

Johns Hopkins University Press uses environmentally friendly book materials, including recycled text paper that is composed of at least 30 percent post-consumer waste, whenever possible.

To my wife, Perry

CONTENTS

July 16, 1790   Residence Act of 1790 is enacted, authorizing construction of a new federal city.

THE FIRST PUBLIC BUILDING CAMPAIGN (1791–1802)

January 21, 1791   The first group of commissioners for the new city is appointed.

March 14, 1792   The commissioners inform L'Enfant that he is no longer involved in their business.

July 18, 1792   James Hoban's design for the President's House is selected.

October 13, 1792   The cornerstone for the President's House is laid.

April 5, 1793   William Thornton's design for the Capitol is selected.

June 1793   Architect Stephen Hallet submits his objections to Thornton's design.

September 18, 1793   The cornerstone of the Capitol is laid.

June 28, 1794   Architect Stephen Hallet is dismissed by the commissioners.

June 24, 1796   Architect George Hadfield submits his resignation, withdrawn later.

October 3, 1797   The commissioners cannot meet payrolls, starting a period of labor conflict.

November 17, 1800   The first planned day for Congress to meet in the new federal city.

December 14, 1801   The House of Representatives temporary building is completed.

June 1, 1802   Congress abolishes the Board of the Commissioners of Public Buildings.

THE SECOND PUBLIC BUILDING CAMPAIGN (1803–1811)

March 6, 1803   President Thomas Jefferson offers the position of surveyor of the public buildings to architect-engineer Benjamin Latrobe.

October 17, 1807   A banquet is held to celebrate the completion of the south wing of the Capitol for the House of Representatives.

March 23, 1808   Benjamin Latrobe reports a $52,000 cost overrun in his fifth report to Congress.

April 26, 1808   William Thornton attacks Benjamin Latrobe in a letter to the newspaper.

September 19, 1808	John Lenthall, clerk of the works, is crushed to death by falling masonry in the north wing of the Capitol.
June 24, 1813	Latrobe's lawsuit against Thornton is decided in favor of Latrobe.
August 24, 1814	British forces defeat the American militia at Bladensburg and continue to Washington to burn the Capitol, the President's House, and other buildings.

THE THIRD PUBLIC BUILDING CAMPAIGN (1815–1824)

February 13, 1815	Congress passes an appropriation to rebuild public buildings on their present sites in the city of Washington.
March 14, 1815	The commissioners for rebuilding Washington ask Latrobe to come to Washington to interview.
April 29, 1816	President Madison signs legislation abolishing the board of commissioners and establishing a single commissioner of Washington, Samuel Lane.
November 20, 1817	Latrobe resigns, to be replaced later by architect Charles Bulfinch.

My interest in the history of the early engineering and construction of Washington, DC, began more than forty years ago, when I enrolled in the first graduate-level historic preservation program in the Washington metropolitan area. The class was taught by Frederick Gutheim, whose enthusiasm for the history of the city was contagious. My studies led eventually to my appointment as Chief of the National Park Service's venerable Historic American Buildings Survey / Historic American Engineering Record (HABS/HAER), where I met Charles Peterson, the founder of HABS. Peterson became my mentor and friend, and over the years he generously shared with me his extensive knowledge of the early American building industry. As Chief of HABS/HAER for fifteen years, I also learned from the knowledgeable and talented engineers, architects, and historians of that program, particularly my deputies Sally Tompkins and John Burns.

Much of this book is about the construction of the President's House, now called the White House. My interest in that topic began twenty-five years ago, when I was approached by White House Curator Rex Scouten and White House Historical Association Director Bernie Myers to undertake the first comprehensive drawings and photographs of the White House, in time for the 200th anniversary of laying its cornerstone. The recordation project was made possible through the support of the American Institute of Architects (AIA), particularly through the efforts of the institute's president, Ben Brewer, and its executive director, James Cramer. This multiyear HABS delineation and documentation project, followed by the 200th anniversary exhibit and symposium, provided me with substantial information on the construction of the building and inspired me to pursue a second doctorate at the University of Maryland. For my dissertation, under the competent direction of Mary Sies, I researched and described the construction and labor history of the White House. Stewart Kaufman provided valuable insights into early labor issues.

Another result of the White House recordation project was that through Rex Scouten I became involved in the study of the quarries on Government Island in Aquia Creek, Stafford County, Virginia. I was editor of the final report prepared for County Executive C. M. Williams and the Stafford County Board of Supervisors, which led to the establishment of Government Island as a public park. On this picturesque island, visitors can now observe the eighteenth-century method of quarrying Aquia limestone for the White House and for the Capitol.

I am appreciative of many other work experiences that enriched my understanding of the engineering and construction history of the federal city, among them working on the rehabilitation of the 516-foot-long Monocacy Aqueduct on the C&O Canal under Denis Galvin, Doug Faris, and Kevin Brandt.

It is thanks to my friendship with William Allen, architectural historian of the US Capitol and author of the definitive history of the Capitol, that I learned so much of the convoluted history of that building. He was also a game investigator, dodging snakes with me during our explorations of the building-stone quarries of the Potomac valley.

Special mention should be made of the debt I owe to the researchers and scholars who preceded me. Fiske Kimball's and Wells Bennett's articles published in 1919 are still important today. I owe gratitude to the many anonymous scholars who compiled and published original records, such as the congressional staffers who compiled *Documentary History of the Construction and Development of the United States Capitol Building and Grounds* (1904) and the unnamed archivists at the National Archives who organized and microfilmed *Records of the District of Columbia Commissioners and of the Offices Concerned with Public Proceedings, 1797–1867* (1964). I am grateful for the significant contributions of John C. Van Horne, Lee W. Formwalt, Jeffrey A. Cohen, Darwin H. Stapleton, William B. Forbush III, Tina H. Sheller, the Maryland Historical Society, and Yale University Press in publishing *The Correspondence and Miscellaneous Papers of Benjamin Henry Latrobe* (3 volumes: 1984, 1986, 1988) and related works. Also important are Saul K. Padover's compilation *Thomas Jefferson and the National Capital* (1946) and Elizabeth S. Kite's work *L'Enfant and Washington, 1791–1792* (1929).

Much of the research that I undertook for this book was done at two great research institutions in Washington: the National Archives and the Library of Congress. The staffs of both institutions were gracious and generous in extending assistance, especially C. Ford Peatross, Curator of Architectural and Engineering Drawings, and Mary Ison, Librarian, Prints and Photographs Division, Library of Congress, who have greatly assisted me over the past four decades.

The history of the new federal city was determined in part by our nation's ties to the United Kingdom's engineering practices and construction developments.

Through the efforts of National Park Service Regional Director Terry Carlstrom and Associate Director Richard Powers, and while I was National Park Service Senior Scholar for Engineering and Architecture, I was able to undertake extensive research in London. There I had access to invaluable materials in the British Library, the library of the Institution of Civil Engineers (ICE), and the library of the Imperial College of Science, Technology and Medicine, London University. Michael W. Chrimes, Librarian for ICE, and Roland Paxton, Chairman of the Panel for Historical Engineering, ICE, were especially helpful.

I have been assisted in my research, both directly and indirectly, by many colleagues in the fields of engineering and industrial history. Neal Fitzsimons, with Herbert R. Hands, more than fifty years ago established the American Society of Civil Engineers History and Heritage program, within which I was able to share ideas with other engineering historians. That unique program continues today under the capable direction of Bernie Dennis, with Steve Pennington in charge of the National Capital Region. About as many years ago, Robert Vogel was named the first curator of civil and mechanical engineering at the Smithsonian Institution, another vital source. Fitzsimons and Vogel, together with James C. Massey, established the brilliant Historic American Engineering Record in 1969. Research by Don Myer, Emory Kemp, Francis E. Griggs Jr., and Donald Sayenga has been important in the discussion of key subjects, such as the development and construction of the chain bridge at Little Falls. Lance Metz, who for twenty-five years organized and managed the annual Canal and Technology Symposium held at Lafayette College, provided a forum for many of the ideas and subjects contained in this book. Dr. Patrick Martin at Michigan Technological University, who guided the work of the Society for Industrial Archeology and is now Chairman of the International Committee for the Conservation of the Industrial Heritage, was enormously supportive of this publication.

I am grateful also to Carrie Mullen, former Director of West Virginia University Press, who first suggested to me writing a book such as this. I thank multiple reviewers of the manuscript, among them William Allen, William Lermond, and James Goode. I am also grateful to Catherine Goldstead and Deborah Bors of Johns Hopkins University Press for sound publication decisions and to Lois Crum for her knowledge and hard work in copyediting.

Books on engineering history benefit from extensive illustrations, particularly those in color, and presenting them often benefits from subventions to subsidize publication. I thank Knight Kiplinger, President and Editor-in-Chief of the Kiplinger Washington Editors Inc., for his encouragement and support and thank the Kiplinger Foundation for funding it has provided for this publication. I am honored to have been awarded the Charles E. Peterson Fellowship, which has also helped fund the book. The Charles E. Peterson Fellowship, under the direction of Peter Conn and Bruce Laverty of the Athenaeum of Philadelphia, was established by Charles Peterson in conjunction with Roger Moss, the former director of the Athenaeum of Philadelphia.

In preparing the manuscript for this book, I received substantial and well-organized assistance from a hardworking research team: my stepdaughter, Sarah Frazer Prestemon, and my niece, Amanda Greenberry Brown. Finally, I thank my wife, Elizabeth Perry Kephart Kapsch, for her ongoing involvement ever since the editor of my first book noted that I occasionally write like an engineer, which I suspect may not have been meant as a compliment.

BUILDING
WASHINGTON

George Washington, Thomas Jefferson, and the Vision
for a New Federal City on the Potomac

*The country intended for the Permanent Residence of Congress, bears
no more proportion to the country about Philadelphia and German-
Town, for either wealth or fertility, than a crane does to a stall-fed Ox!*

MAJOR ANDREW ELLICOTT,
surveyor to the Commissioners of Public Buildings, to his wife, 1791

*For all the reasons I have detailed as succinctly as I could, it is
impossible to believe that, as projected and begun, the Federal City will
ever develop to the point where it will become a pleasant place to live
for the kind of people who are destined to inhabit it.*

DUC DE LA ROCHEFOUCAULD-LIANCOURT, 1797

With the signing of the Declaration of Independence in 1776, the thirteen
American colonies affirmed that they would secede from the Crown and
form a single sovereign power. This epic act of rebellion was followed by
years of military struggle that exhausted the patience, manpower, and
wealth of the colonials but led to England's relinquishing control of the
colonial territory in 1783 with the Treaty of Paris. Then, for all their differences, the thirteen colonies agreed to form an uneasy partnership: a
federal union constituted as a tripartite central government. That central
government was in need of a home, however, and it was necessary to decide whether its location should be in one of the established metropolitan areas, such as New York or Philadelphia, or in a new federal district
with no ties to any particular state—a new city for a new nation.

For decades, the nation's first president had envisioned a great city, a
commercial and political metropolis established along the banks of the
Potomac River. Now George Washington hoped to realize that vision with
a new city located at the symbolic and geographic center of the original
colonies. Strategically, President Washington wished to avoid founding
the central government in New England, where states were threatening
to secede from the new republic. The Potomac, situated far south of New
England, seemed a safer choice for fostering a young government. At the
same time, it would provide balance between northern and southern interests, with proximity also to those agricultural states whose cultural
and economic interests still gravitated toward the Caribbean.

President Washington's vision was a spectacularly ambitious one, without precedence and with little practical chance of success. The great European commercial and political centers of London and Paris had taken centuries to develop, their beginnings buried deep in the past. To bring an equivalent seat of power into existence in a short time was implausible. To create such a city in that place—the tidal reach of the Potomac River, a largely unsettled region lacking many of the amenities of other colonial towns—seemed impossible. It would have been far more pragmatic to designate a well-established metropolitan center such as Philadelphia or New York, both of which were competing for the honor, as the capital of the new republic.

Washington had for years been formulating his idea for the commercial future of the new city. Since his first excursions west, he had seen the potential for the agricultural wealth of the Northwest Territory of the Ohio to pour down the river via a system of navigable waterways to a great harbor city at the upper reaches of the tidal Potomac River, whence it would be transshipped throughout the world.

Other visionaries, most notably Thomas Jefferson, shared Washington's plan to develop the Potomac region as the new federal city. As fellow patriots, Whigs, agrarians, and Virginians, they had much in common. More importantly, it was not the first time these two men had undertaken a great enterprise against long odds. They had joined with others to create a new model for government in the form of a democratic republic, had fought to victory against the greatest military power on the planet, and had established a stable central administration amid conflicting interests, competing regions, and diverse peoples. Given these successes, perhaps planning and building a city from scratch in the middle of nowhere seemed a not quite insurmountable challenge.

In 1789 Washington was unanimously elected by the electoral college as the first president of the newly constituted United States. He immediately named Thomas Jefferson his secretary of state. Unfortunately, their positions of leadership did not ensure that their shared desire to locate a new capital city on the Potomac would be realized. Most of the newly elected congressmen and senators, especially those from the northern states, did not support the plan. And even if Washington and Jefferson did convince Congress that a new federal city should be located in the south, it was clear that congressional funds for such an enterprise would not be forthcoming.

Washington and Jefferson's plans for the new city extended far beyond its designation as the national capital. At that time, the federal government did not have enough employees to fully populate a crossroads village, and serving as a government center alone would not justify the construction of an entirely new city. To develop the great metropolitan hub they envisioned, Washington and Jefferson required a much more extensive economic base. The city's location on the Potomac River was key to its economic viability. Washington rightly saw this as the shortest way to develop communications and trade with the Northwest Territory, which had just been granted to the new nation by the Treaty of Paris. Beginning in 1785, Washington served as president of the Potomac Canal Company, formed to develop a navigable river route west. The tidal reach of the Potomac River—the area near Alexandria, Virginia, and Georgetown, Maryland—was within easy access of the Chesapeake Bay, and farther down, the bay provided access to the Caribbean and Europe. The Potomac had the potential to serve a new mercantile center as the conduit for transshipment of the immense production expected to come from the west, thus creating commercial ties through the nation's capital with the vast Northwest Territory and politically binding the western lands as they became settled to the thirteen founding states.

CHALLENGES TO BUILDING THE NEW CITY

Both Washington and Jefferson fully intended that the new city be the physical embodiment of the new nation. In addition to its commercial success, they hoped the federal city would surpass in scope and grandeur the long-established European capitals. They also knew that to achieve such a goal would require overcoming a daunting list of obstacles. They must locate and hire specialists in planning, design, and construction to build the largest and grandest stone building (the Capitol) and the largest and grandest mansion (the President's House) in the nation. They must also construct bridges over the broad Potomac River and the Eastern Branch, dig a canal through the center of the city, and put in place military defenses adequate to protect the new city. Their decision to build federal buildings out of stone created a conundrum of its own, as there were few stone buildings in the republic to serve as models, few trained stone masons on this side of the Atlantic, and few established quarries in the Potomac region. The defenses for the new capital—Fort Warburton and Fort Washington—were also some of the earliest built in the United States. In addition, it would be necessary to construct wharves, roads, warehouses, residences, commercial structures, and all the other buildings found in any city.

The new city was Washington and Jefferson's grand vision. They imagined it would be built in the manner of the eighteenth century, with gentleman planters and other gentry overseeing master builders. But society had been changing. Large construction projects such as this could no longer be adequately supervised by gentleman planters and lawyers. Now such an undertaking required experienced architect-engineers with special training in their field. Washington and Jefferson's grand vision soon met the hard realities of finance, design, and construction that were emerging by the end of the eighteenth century.

The scope of the construction plans went beyond the scope of any city building project in North America in the eighteenth century. Situating the capital on both banks of the Potomac would require a bridge more than a mile long simply to connect the halves of the city. The Washington Bridge became the longest ever built, and it was aptly nicknamed "the Long Bridge." Other crossings over the Potomac River eventually required the tallest abutments in the country, more than forty feet above water over Little Falls, and the Aqueduct Bridge crossing the Alexandria Canal needed one of the earliest and deepest cofferdams yet constructed. The projected canal through the city, later called the Washington City Canal, was built at a time when most Americans had never before seen a canal. President Jefferson initially proposed the Washington City Canal to water the dry dock at the Navy Yard, where 12 frigates of the new navy would be stored under the longest roof spans in the country. The Navy Yard and other military centers were the first in the nation.

When Washington and Jefferson formulated the Residence Act of 1790 to develop the new capital city and authorize the construction of a federal capital on the banks of the Potomac River, they had access to few architects or engineers with design and construction expertise. America had master builders, but virtually no one in the country knew how to build a city of such scope. Most of the few men who practiced architecture-engineering in the United States at the time had been educated and trained in Europe, under an older (pre-Enlightenment) model, and the new republic had no schools of architecture-engineering of its own. Following the French Revolution, Europeans' understanding of architecture and engineering had greatly expanded, from a military-engineering focus to one of Enlightenment thinking with an emphasis on scientific learning. Newly created schools for architects and engineers in England, France, and Germany began to teach those advances in construction knowledge, but no schools in the United States had yet been established to do the same.

Dedicated funding was another vast obstacle to building the new federal city. From the beginning, Washington and Jefferson knew Congress would not appropriate funds for the project. They expected that any efforts to obtain congressional funding would be blocked by the Philadelphia and New York interests desiring the federal city designation for their own cities. Borrowing the necessary funds seemed an unlikely prospect, because the United States had few financial reserves, and in a country with almost no liquid net worth, it was not possible to borrow enough money to cover the cost of construction. To overcome these problems, Jefferson proposed that local landowners donate one-half of their land to the new city. He proposed to use some of this land for public building sites, streets, and parks and then subdivide the remaining donated property into lots to be listed for sale in order to provide liquidity for the construction of the public buildings. By designating private lands as part of the new federal city, Jefferson expected to increase prices substantially, rewarding both the government and the landowners with an appreciation in the value of their holdings. He believed that the Virginia and Maryland legislatures would provide grants for a new federal city on the Potomac but did not expect other states to contribute, either because they preferred to found the new city elsewhere, or because their distance from the Potomac made them think the new city would not benefit them. However, Jefferson did not develop detailed cost estimates for the government buildings, and so it was never known whether the proposed funding would be adequate.

With neither adequate funding nor experienced architect-engineers (a term used in this book to encompass eighteenth- and early-nineteenth-century architects, engineers, and architect-engineers, such as Benjamin Latrobe), Washington and Jefferson appointed three commissioners from among the nearby landed gentry to oversee initial work on the construction project. Although these commissioners lacked design and construction expertise, they had plenty of zeal for promoting the new federal city. But it was unfortunately not enough. Their lack of knowledge led to numerous problems plaguing the construction of the new city, including excessive costs, construction delays, work stoppages, and workforce slowdowns. The original three commissioners were eventually replaced by two lawyers and a physician, who were equally ignorant of the demands of large construction projects.

The commissioners supervised a workforce comprising both skilled and unskilled workers. Many of the unskilled laborers were hired from within the immediate vicinity of the new city, but few skilled workers lived in the Potomac region. So the commissioners were forced to enlist carpenters and brick masons from northern cities and stone cutters and stone carvers from Europe. The widespread European wars at the time complicated this effort, since many countries prohibited emigration of their citizens.

Time, also, was in extremely short supply. Washington and Jefferson envisioned raising the capital city from bare ground in only ten years. This required addressing a series of basic urban-planning issues quickly. Adequate bridges and roads were a top priority for a new city located on a peninsula, with the Potomac River to the west and the Eastern Branch to the south. The completion of the new bridges and roads was complicated by the two men's vision for a city canal, which they wanted so that goods could be shipped throughout the city.

Though they were already pressed for time, the commissioners needed to quickly establish a series of building rules for private developments within the city limits. Their rules included height limitations to ensure that significant buildings would be constructed without hindering vistas and views, restrictions on wooden structures for fire prevention, and provisions enabling a landowner to make use of a neighboring landowner's structure when building next to it. Whereas most cities developed similar rules and regulations over a long span of time and in response to specific problems, the commissioners were under pressure to develop their rules and regulations almost instantaneously. The deadline only added to existing pressures. If the rules they established were too strict, city development would suffer. If they were too lenient, public health and safety would be affected and land values would plummet.

The commissioners also faced the task of building military and defensive works to protect the new city from attack. Given the colonies' relationship to Great Britain, these measures included fortifications on the Potomac, a navy yard on the Eastern Branch (later called the Anacostia River), and an arsenal at Greenleaf's Point, where the Potomac River and the Eastern Branch converge. As with the public buildings, there were few Americans who could plan, design, or build these projects.

In planning the federal city as a commercial center, the commissioners needed to develop a shipping route for goods and materials. The two stages of the route led first down the Potomac from the Piedmont by routes around the falls and, second, from the city down the tidal Potomac to the Chesapeake Bay. The Potomac Canal Company had completed considerable work on the system by 1790, when Washington and Jefferson initiated their efforts for a new federal city, and the company expected to soon complete bypass canals around Little Falls and Great Falls.

Washington and Jefferson did not anticipate the subtly changing environment of the region that would undermine plans to use the Potomac River for shipping. As increased sedimentation choked existing river channels, new river channels were created by momentous ice floods. Only special expertise in hydrology could keep shipping channels open to the city. But the few specialists with this expertise bitterly contested suggested remedies.

Acquiring building materials posed another challenge to the creation of the new city. Since the Potomac region was an underdeveloped backwater, there were few quarries and no mason yards, lumberyards, sawmills, limekilns, brickyards, or other services one would expect in cities like Philadelphia and New York. Before construction could begin, Washington, Jefferson, and the commissioners would need to develop a list of sources for the necessary building materials. They must open and expand stone quarries, such as Aquia; find and cut stands of oak; pit saw wood; recruit brick masters from urban areas to establish brick kilns; and build furnaces to burn limestone for lime.

CONGRESS AGREES TO AN AUTONOMOUS TERRITORY

The congressional need for an autonomous territory finally became abundantly clear to the new country's legislators following the Pennsylvania Mutiny of 1783. On June 17, the Congress of the Confederation, then sitting in the Pennsylvania State House (Independence Hall) in Philadelphia, received a message from Continental Army soldiers stationed in Lancaster, Pennsylvania, demanding their back pay and threatening action if their demands were ignored. Congress took no action, and two days later, 80 Continental Army soldiers left their post in Lancaster and marched the eighty miles to Philadelphia, where they joined Continental Army troops stationed in the city. With the troops stationed in Philadelphia, the force grew to approximately 500 men, who had effective control of arms and munitions within the city. The following morning, June 20, a mob of approximately 400 soldiers surrounded the Pennsylvania State House, refusing to allow the representatives to leave until they had addressed the soldiers' concerns for back pay. A small committee of delegates, led by Alexander Hamilton, met in secret and sent a message to the Supreme Executive Council of the Commonwealth of Pennsylvania, asking for the state's militia to provide them protection from the mob. Under the Articles of Confederation then in force, the federal government had no control of the Commonwealth of Pennsylvania's militia except in time of war. John Dickinson, the president of the commonwealth's Executive Council, agreed to consult with the militia commanders. In sympathy with the members of the Continental Army then in the city, Dickinson and the Executive Council refused to call up the state militia to protect the Congress. In response, Congress voted to relocate to Princeton, New Jersey. Clearly, Congress needed a federal city under its own control.

Passage of the Residence Act of 1790 was not automatic. It was a result of compromise: southern states agreed to support Hamilton's Assumption Bill, the authorization for the federal government to assume the war debts of the states, primarily northern, in exchange for the northern states' support for the Residence Bill, the authorization of a national capital below the Mason-Dixon Line. Jefferson was the principal engineer of the compromise, and Congress passed it into law on July 16, 1790. In recounting how the compromise came about, Jefferson explains that he had come upon Alexander Hamilton in New York City in late June 1790. "Hamilton was in despair. As I was going to the President's [residence] one day, I met him in the street. He walked me backwards and forwards before the President's door for half an hour. He painted pathetically the temper into which the legislature had been wrought, the disgust of those who were called the creditor States, the danger of the secession of their members, and the separation of the States." Jefferson responded to Hamilton that he was "a stranger to the whole subject (of system of finances for the new country)" but "that undoubtedly if its rejection endangered a dissolution of our Union . . ., I should deem that the most unfortunate of all consequences." Despite these comments, it was Jefferson who brought together the political accommodation leading to the Residence Act. Jefferson invited Hamilton to dinner the next day to discuss the matter with several of his friends. As a result of that dinner, the southern states agreed to support Hamilton's Assumption Bill in exchange for the northern states' support for the authorization of a national capital on the Potomac River. Jefferson explained:

> It was observed that this pill [the assumption of the war debts of the northern states] would be peculiarly bitter to the Southern States, and that some concomitant measure should be adopted to sweeten it a little to them. There had before been propositions to fix the seat of government either at Philadelphia or at Georgetown on the Potomac; and it was thought that by giving it to Philadelphia for ten years and to Georgetown permanently afterwards this might, as an

anodyne, calm in some degree the ferment which might be excited by the other measure alone. So two of the Potomac members (White and Lee, but White with a revulsion of stomach almost convulsive) agreed to change their votes, and Hamilton undertook to carry the other point.[1]

To implement the Residence Act, Washington turned to his secretary of state, Thomas Jefferson, for advice on how to proceed. The first task was to determine the location of the new city within the constraints of the act. Jefferson recommended that Washington find out who the rival contenders were for its location and what they might bid against each other in land or money to have the new city located on their land.[2]

FUNDING THE NEW FEDERAL CITY

A crucial factor was that Congress had not appropriated funds to implement the Residence Act. Nor did Washington or Jefferson expect that they would do so in the foreseeable future. Jefferson had recommended a land-for-money approach, supplemented by state grants from Maryland and Virginia, to build the new federal city.[3] No one knew how much the new city would cost, and therefore no one could know whether this plan was adequate. To explore the feasibility of a land-for-money deal, Jefferson met with the leading landowners in the vicinity of the future location of the new city. He reported to Washington that this approach was feasible.

Jefferson also provided Washington an overall estimate of how much land and money was required:

1,500 acres needed by the federal government
300 acres needed for public buildings, walks, etc.
1,200 acres to be divided into quarter-acre lots
2,000 lots then available for sale
At £100 per lot, to yield £200,000 Maryland currency ($520,000).[4]

The half of the landowners' lands ceded to the US government would be subdivided into lots and sold to provide money for the construction of the federal buildings. In addition, wrote Jefferson, it was likely that the legislatures of both Virginia and Maryland would pass grants of $120,000 each to assist in building the new federal city.

This land-for-money approach for funding never supplied enough money to support the construction costs of the public buildings. Initially few lots were sold, primarily because the city's planner, Pierre L'Enfant, refused to release his plan for the city, and few prospective buyers were willing to purchase lots without knowing where the lots were located. Later, the commissioners sold half of the available publicly owned lots, at bargain prices and under favorable terms, to the land speculation syndicate of James Greenleaf, Robert Morris, and John Nicholson. When these three defaulted on their contractual payments, title to the lots became clouded and the lots could not be sold. The inability to raise money through the sale of lots was the major reason for the continuing financial crisis of the commissioners during the first public building campaign (1791–1802).

With the failure of the land-for-money approach, money for construction of the new federal city would have to come from elsewhere. The area in which the city was to be built had little disposable income available for loan. The continuing European wars ensured that no money would be forthcoming from Europe. This meant that the commissioners would have difficulty borrowing money and that capital was also not available to building-materials suppliers for lumberyards, sawmills, stone quarries, or stone yards.

EXECUTING A VISION FOR THE NEW CITY

The three original commissioners were expected to have "sufficient respectability" and "good will to the general object," as well as "some taste of Architecture." "Sufficient respectability" was a key phrase indicating that the commissioners were to be individuals of the gentleman class, that is, wealthy landowners. There was no suggestion by Jefferson that these men should have knowledge of or experience in supervising a large engineering and construction project. The world was changing, and large, complicated projects like the new federal city needed the services of the newly emerging specialists in architecture-engineering. It was this lack of experience and knowledge on the part of the commissioners that led to many of the problems experienced from their establishment in 1791 until the commission was abolished in 1802.

The first group of commissioners were appointed on January 21, 1791: Daniel Carroll, Thomas Johnson, and Dr. David Stuart. All were friends of Washington and all were large-scale landowners of the region. They supervised the development of the new federal city until the last meeting of all three, on July 27–31, 1794. They were widely criticized for naivety for allowing themselves to be taken advantage of in the sale of six thousand lots to Greenleaf, Morris, and Nicholson's syndicate. They were also criticized for the manner in which they undertook their first construction project, the bridge over Rock Creek at what was later K Street. Instead of hiring a bridge engineer, they hired a general builder, Leonard Harbaugh. The structure he built was not large enough to span the floodwaters and tidal surge of Rock Creek and had to be expanded by the construction of side arches. In addition, the bridge developed a major sag and had to be rebuilt. The commissioners also had other difficulties, such as their continuing problems with Pierre L'Enfant, the first architect-engineer they hired to develop the plan and oversee its implementation.

Benjamin Henry Latrobe, by Charles Willson Peale, ca. 1804. On March 6, 1803, President Thomas Jefferson offered the position of surveyor of the public buildings, formerly held by James Hoban under the dissolved Commissioners of Public Buildings, to architect-engineer Benjamin Latrobe. The position paid $1,700 per year. Jefferson made it known to Latrobe that he intended to take a direct approach to the completion of the public buildings and not work through a commission or any other third party. Latrobe's principal responsibility was the completion of the south wing of the US Capitol, to house the House of Representatives. He also had to correct design and construction deficiencies in the north wing of the Capitol and the President's House. In doing so, he came into conflict with William Thornton and James Hoban. White House Collection / White House Historical Association.

From August 23, 1794, through May 18, 1795, the first group of commissioners resigned and was replaced by a second group of commissioners: Gustavus Scott, Dr. William Thornton, and Alexander White. The three gentleman landowners were replaced by two lawyers and a physician, who were equally devoid of knowledge and training in building. It was this second group of commissioners that oversaw most of the first public building campaign. They also received considerable criticism, for the high costs, recurring construction delays, continuing labor disputes, and other problems that resulted from their lack of oversight and construction understanding. The difficulties that the first group of commissioners had with architect-engineer Pierre L'Enfant were replicated in the great difficulties the second group of commissioners had with architects Stephen Hallet and George Hadfield and with the workers in general. Architect-engineer Benjamin Latrobe characterized this relationship as the commissioners waging war on their workers.

The conflict between the commissioners and the architect-engineers and workers were an outgrowth of Washington and Jefferson's eighteenth-century view of building. They wanted "respectable individuals," that is, persons of the gentleman class, to oversee construction. Because the world was changing, large and difficult construction efforts could no longer be adequately overseen by gentleman

William Thornton, by Charles Balthazar Julien Fevret de Saint Mémin. Thornton was the winner of the design competition for the US Capitol and became a commissioner in the second group appointed, along with Gustavus Scott and Alexander White. It was primarily Scott and Thornton who engendered difficulties with the labor force. The commissioners were criticized for not completing the needed federal buildings by the time the federal government moved to Washington in 1800, and for spending large sums of money. Thornton engaged in intense conflicts with professionally trained architects such as Stephen Hallet, George Hadfield, and Benjamin Latrobe. Most of these conflicts centered on deficiencies in Thornton's design of the US Capitol. Courtesy of the Library of Congress.

planters. The new specialist, architect-engineer, was required. Insistence on using the old ways failed to mesh with the new realities. The result was continuous conflict between the "respectable individuals" appointed as commissioners and the professional architect-engineers hired for their expertise. The conflict continued in the second and third public building campaigns. During the second building campaign (1803–1811), the differences between the amateur "respectable individual" and the professional architect-engineer were personified in the bitter battle between the nominal designer of the Capitol, physician William Thornton, and professional architect-engineer Benjamin Latrobe. And again during the third public building campaign (1815–1824), the rebuilding of the British-burned buildings, conflict arose between commissioner Samuel Lane, a war veteran, and architect-engineer Benjamin Latrobe. It was only after the arrival of architect-engineer Charles Bulfinch in January 1818—twenty-eight years after the beginning of the engineering and construction effort to build the new federal city—that the conflicts between amateur "respectable individuals" and professional architect-engineers receded.

One of the principal decisions made by Washington and Jefferson that led to high costs and construction delays was that the two principal buildings, the President's

House and the Capitol, would be faced with stone. For that purpose, a supply of adequate amounts of stone needed to be developed. There were several small existing quarries that might have sufficed if expanded, such as those along Aquia Creek in Stafford County, Virginia, approximately forty miles south of the new federal city, and the Seneca brownstone quarries north of Great Falls. The latter could not be used because the Potomac Canal had not yet been cut around the Great Falls, and therefore transporting stone from these quarries was impractical. L'Enfant turned to the Aquia quarries and negotiated with George Brent, the owner of Hissington Island (also called Wigginton Island and later renamed Government Island), and bought the entire island, excluding a one-acre quarry on the island owned by Baltimore stone cutter Robert Stewart, for six thousand dollars. The commissioners ratified this purchase on December 23, 1791.[5]

Although the President's House and the Capitol were to be stone-faced brick structures, wood would be needed for the floor joists (later changed by Latrobe and Jefferson to masonry vaulting in the House of Representatives wing) and for the rafters. The timber in the area surrounding the new federal city had been cut over, so crews had to be sent sixty or more miles south to White Oak Swamp for suitable, first-growth timber.[6] The cut wood was rafted up the Potomac. Wharves had to be built in the new city to receive these and other building materials. Clay needed to be dug locally and burned to make the bricks for the buildings. Also needed were foundation stone, firewood, lime, and other building materials.

Although Washington and Jefferson intended the new federal city to be built in ten years, some thirty-three years were actually required to complete the Capitol. Building the city took three public building campaigns: 1791–1802, 1803–1811, and 1815–1824, the third for rebuilding the public buildings after they were burned by the British on August 24, 1815.

No detailed time schedule had been developed for building the new city. The time specified in the Residence Act, ten years, was not adequate. Toward the end of the first public building campaign in 1800, the President's House was mostly finished, as was the Senate wing of the Capitol. The House wing had been barely begun. Nevertheless, the government moved to Washington in November 1800, as required by the Residence Act of 1790. Beginning in 1803, the second public building campaign concentrated on finishing the House of Representatives wing and correcting construction problems in the Senate wing and the President's House.

While the commissioners were wrestling with the problems of building grand public buildings, other men were addressing the problems of developing Washington as a mercantile center. The immediate problem was to keep the Potomac River ports of Georgetown and Alexandria open and available to oceangoing vessels. At the time, the Potomac River valley was being denuded of trees and vegetation by farmers, charcoal burners, and others; as a result, an increased volume of soil was washing into the Potomac River. The deposit of this river-borne sedimentation into shoals greatly increased at the point where the fresh water met tidewater at George-town. The deep port that General Edward Braddock had found at Georgetown in 1755 became less deep. In addition, the denudation of vegetation in the Potomac River valley increased runoff and thereby increased the intensity of floods. One of the most memorable was the ice flood of March 1783, which cut a new channel between Mason's Island (present-day Theodore Roosevelt Island) and the Virginia shore, reducing the utility of the Georgetown port on the other side of Mason's Island. These changes threatened Georgetown's viability as a commercial port and therefore threatened the vision that the Potomac would become a commercial avenue from the interior. Engineers such as Latrobe and self-taught engineers, such as Thomas Moore, were consulted, and various remedies were attempted to keep the port open. In the end nothing proved effective; the siltation of the Potomac River remained as one of the major barriers to developing the new city into a major mercantile center.

The new federal city straddled a very broad Potomac River. Ferries crossed the river, but the development of communications and commerce with the hinterland would require bridges. Early on, the commissioners decided they did not have money to build these needed bridges, and companies for that purpose were organized. The first of the bridges was built at Little Falls and came to be known as Chain Bridge. A group of merchants from Georgetown moved to form the Georgetown Bridge Company to construct it. On December 29, 1791, the Maryland Assembly granted the new company a charter.[7] Four years later, the Maryland Assembly granted the same company a charter to build a sixty-foot-wide road, now called Chain Bridge Road, from Georgetown to the new bridge at Little Falls.[8]

The first bridge at Little Falls was completed and opened for toll traffic on July 3, 1797.[9] The total cost was eighty-four thousand dollars.[10] The site selected for crossing the Potomac was at the head of Little Falls, where high bluffs on both the Virginia and Maryland sides of the river severely constricted floodwaters. During flood events, the river could rise vertically forty or more feet in twenty-four hours. Bridges built here would have to be above this flood line. No fewer than eight bridges were built at this place, most of them to replace flood-damaged structures.

The second crossing of the Potomac was at tidewater, just above the junction of the Potomac and the Eastern Branch. Here the river was wide but shallow. On February 5, 1808, Congress authorized the Washington Bridge Company to build a bridge at this location.[11] The construction of the mile-long Washington Bridge (also called the Long Bridge, the Potomac Bridge, and, later, the Fourteenth Street Bridge) began in June 1808 and was completed the next year at a cost of nearly one hundred thousand dollars. At the time it was the longest bridge in the country.[12] It was described as built on "a number of timber piles, driven into the bed of the river, [with] beams . . . laid thereon to form the carriage road, which is planked from end to end."[13]

In addition to these two Potomac River bridges, several bridges were built across the Eastern Branch on the same plan as the Long Bridge.

With construction of the new federal city proceeding, and with the uncertainty of the European war weighing on their minds, leaders of the new republic turned to the construction of seacoast fortifications to protect their coastal cities. To build the needed fortifications, the Department of War enlisted European trained military engineers. On May 12, 1794, Secretary of War Henry Knox sent instructions to French engineer Jean Arthur de Vermonnet to build a coastal battery to protect Alexandria and Washington. The original instructions were somewhat muddled, but once straightened out, they resulted in a rudimentary shore battery at Jones Point, immediately south of Alexandria. Although built, it was probably never garrisoned, and several years later it was a ruin.

In 1808 plans were prepared to replace the ruins of the coastal battery at Jones Point with a more substantial battery, to be constructed at Digges Point, on the Maryland shore of the Potomac River, 10 miles below the federal city. By December 21, 1809, the new coastal battery, sometimes called Fort Warburton and sometimes called Fort Washington, was completed. This new battery, essentially a coastal battery, was indefensible from attack from the land side. This deficiency led to its abandonment on August 27, 1814, in the days immediately after the defeat of American forces by the British at the Battle of Bladensburg. The coastal battery was blown up by its commander and garrison without firing a shot at the British ships then ascending the Potomac. It was replaced beginning in 1816 with a much larger fortification, called Fort Washington.

Other fortifications, such as the arsenal established at Greenleaf's Point (now Fort McNair), were limited to a single gun but were later increased. The navy yard on the Eastern Branch, the first navy yard in the new country, was substantial.

WASHINGTON, A TRIBUTE

The new city was named for George Washington not only for honorific reasons but because he was the one man who did the most, politically and economically, to bring the city into existence. Politically, it was Washington, as president, who oversaw the passage of the Residence Act of 1790. It was also Washington who selected the site of the new city along the Potomac River. And it was Washington who believed in the future commercial prosperity of the city. Washington and Jefferson would have been aware of the siltation of the Potomac and the changes wrought by the ice flood of 1784. They thought these environmental changes could be overcome.

Washington never saw his new city become the federal center, because he died in 1799. Even after his death, it was his prestige that provided the impetus to continue building the new federal city, and it was his friend Jefferson who directed much of this work. Jefferson learned from the problems of the commissioners of the first public building campaign. When he became president in 1801, he eschewed the appointment of a commission and hired the most experienced architect-engineer then in the United States, Benjamin Latrobe, to oversee the completion of the US Capitol and other projects. Latrobe reported directly to the president, as had originally been recommended by L'Enfant. The two wings of the Capitol (but not the center section) were completed by 1811.

Following the burning of the Capitol and the President's House after the Battle of Bladensburg in 1814, the vision of Washington and Jefferson was influential in the decision to rebuild the public buildings in Washington rather than elsewhere. This third public building campaign was overseen by many of the same men who initially built these buildings: James Hoban was hired to rebuild the President's House, and Benjamin Latrobe was hired to rebuild the Capitol. Latrobe was later replaced by Charles Bulfinch, who oversaw the completion of the Capitol dome in 1824—more than thirty years after Washington and Jefferson's initial vision for the federal city.

One is struck with the boldness of Washington and Jefferson's vision to build a great capital and commercial city on the Potomac River. This enterprise was undertaken by the two men without any clear source of funding; in the face of bitter hostility from the representatives of populous states like Pennsylvania and New York who wanted their own principal city designated the capital; in a semisettled region; and with what turned out to be a difficult time limit of ten years.

Certainly Washington and Jefferson's actions were bold, but they were also somewhat reckless. Not only was there no clear source of funding at the outset for building the new city, but in addition no detailed cost estimates were undertaken to determine how much money might be required for the principal buildings. Actual expenses of the first public building campaign came to several times the original rough estimates of Jefferson. For example, it was decided to give the Capitol a stone facing before there was any understanding of how much a stone facing might cost. The determination that the new federal city would be ready for the federal government in ten years was arrived at without any investigation of fact. When asked to assess Thornton's design of the Capitol, architect Stephen Hallet estimated that it would take thirty years to build—remarkably close to the actual construction time. Washington and Jefferson's belief that the Potomac River would become a major highway for the commerce of the Potomac River valley came to naught. Both men at the time were aware of the environmental changes that were choking navigation channels at Georgetown.

A key belief of both Washington and Jefferson was the eighteenth-century view that well-intended gentlemen of the landed gentry or professions were equipped to manage such a project and that professional architect-engineers were somewhat ancillary to the effort. Therefore, the overall management and supervisory organization for the construction of the new city was a commission made up of gentleman planters or town professionals. Professionally trained and experienced architect-engineers, such as Pierre L'Enfant, Stephen Hallet, and George Hadfield, were considered useful; but once they were hired by the commissioners, their advice was not necessarily followed.

An example of this bias is the adoption of William Thornton's design for the Capitol. Thornton was a physician, never trained or experienced in architecture. His design was severely criticized by French architect Stephen Hallet. But even though Hallet pointed out the shortcomings of Thornton's design, the commissioners, backed by Washington and Jefferson, went forward with the building.

The inclination to place men from the gentleman class in overall charge of building the new city resulted in improper work that had to be torn down and redone, labor disputes with the skilled workmen that led to work stoppages and slowdowns, delays in completing the needed buildings in time for the federal government to relocate from Philadelphia, and the expenditure of much more money than should have been required.

One man, Thomas Jefferson, eventually grasped the fundamental wrongness of placing socially positioned men in charge of a large construction effort for which they had no training or experience. On becoming president in 1801, Jefferson refused to reinvigorate the commission abolished by Congress. Instead, he sought out the best-trained and most experienced architect-engineer in the country to assume direction of the effort—Benjamin Latrobe. Jefferson maintained overall supervision of the project through the remaining years of his presidency. This was the original recommendation that Pierre L'Enfant made to George Washington on the management of the effort before he was dismissed.

Under Jefferson and Latrobe, deficiencies in the President's House and the Senate wing of the Capitol were corrected and, by 1811, the House of Representatives wing of the Capitol was largely completed.

Faced with the destruction of many of the public buildings of Washington by the British in 1814, Congress decided to rebuild at the same location. Latrobe again was hired to supervise the rebuilding of the Capitol, but this time he reported not to the president but, rather, first to a commission and later to a single commissioner, Samuel Lane, a soldier with little construction experience who had been wounded in the War of 1812.

It was a rocky start for the new federal city. Financing the construction of the new city was a continuing problem, and costs greatly exceeded original estimates. Recruiting the needed skilled workers and assembling the needed building materials also were challenges. The government buildings for the new city had not been completed when the government moved down from Philadelphia. Nevertheless, eventually the public buildings of the new city were completed—only to be burned by the British in 1814. Washington never became the commercial center envisioned by Washington and Jefferson, but it did become a great administrative center for the new country.

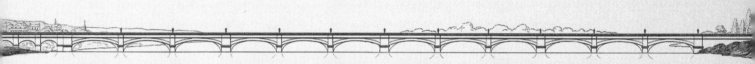

PART ONE   PROJECT PLANNING AND
CONSTRUCTION EFFORTS

1

Pierre L'Enfant's Two Plans for Executing
the President's Vision

*Wednesday evening arrived in this town Major
Longfont, a French gentleman employed by the
President of the United States to survey the lands
contiguous to Georgetown where the Federal city is
to be built. His skill in matters of this kind is justly
extolled by all disposed to give merit its proper
tribute of praise. He is earnest in the business and
hopes to be able to lay a plat of that parcel of land
before the President upon his arrival in this town.*

Georgetown (DC) Weekly Ledger, March 12, 1791

Washington and Jefferson required a detailed physical plan for the new city, not only to lay out the location of public buildings, streets, and avenues but also to define the boundaries of private lots that could be sold to finance construction efforts. Only a professionally executed plan would entice wealthy individuals to buy city lots and invest in the new city. To accomplish this, Washington and Jefferson sought an experienced and talented individual from the newly emerging architect-engineer profession. They turned to their war associate Pierre L'Enfant.

L'Enfant was a French aristocrat educated at the Royal Academy of Painting and Sculpture in Paris. Though today he is often described as an architect and a civil and military engineer, L'Enfant likely acquired his early knowledge of civil and military engineering largely through observation — France was then the most advanced country in these fields — and his later experience as a military engineer in the Continental Army.

NOT ONLY A PHYSICAL PLAN FOR THE CITY

L'Enfant arrived in Georgetown in March 1791 and quickly set to work developing the physical plan for the new city. His plan had two significant aspects: first, it was the grand depiction of Washington's concept for a national capital that would serve as both a government center and a large mercantile and economic port for the nation, and second, it defined a

Pierre L'Enfant, by Bryan Leister. While in America, planner and architect L'Enfant referred to himself as Peter L'Enfant; it was only after a campaign by the French ambassador in the late nineteenth century that he again was called by his given name, Pierre L'Enfant. L'Enfant was the author of two separate and different plans of Washington. He is best known for the physical plan of the new federal city (see plate 3). He also created a construction plan for the city that, had it been followed, would probably have completed the public buildings several years before 1800, the date when the federal government moved from Philadelphia to Washington. L'Enfant had no known formal portrait prepared during his lifetime. This is a 1993 painting based on existing sketches and other information. The artist well captured L'Enfant's proud manner (some say arrogance). Courtesy of Bryan Leister and the Washington Historical Society.

large number of city lots that, following Jefferson's financial plan, when sold could produce substantial revenue for the commissioners. His physical plan for the new federal city was grand. He overlaid the usual grid pattern for the streets with broad radial avenues. He selected the most prominent physical feature of the new city, Jenkins Hill, as the site for the Capitol. A mile away he chose the site for the President's House, connected to the Capitol by the broad avenue later named Pennsylvania. L'Enfant used a series of bridges to connect the new federal city with the surrounding area. Through the middle of the city, he laid out the placement of a new canal.

A second plan that L'Enfant produced for constructing the new city set forth an alternative approach to Jefferson's suggestions for the financing, organization, and completion of the public buildings. In a twenty-two-page memo dated January 17, 1792, detailing his own recommendations to Washington, L'Enfant outlined the development for early construction in 1792 and beyond. He believed the city lots should not be sold in the initial phases to finance construction, but only after the federal government had moved to the new city and the price of the lots had appreciated. He believed the commissioners should borrow money to support their initial construction operations and that the loans could be repaid through the sale of those building lots after the government had moved from Philadelphia to Washington. L'Enfant also disagreed with Jefferson's recommendation that construction should be under the supervision of a commission. Instead, he recom-

PROJECT PLANNING AND CONSTRUCTION EFFORTS

mended that a single individual, a director general, should supervise the entire project. The remainder of L'Enfant's construction plan described the sixteen steps necessary for obtaining materials and developing infrastructure. He estimated that the labor of 1,070 men over the course of four years (1792–1796) would bring the task to completion.[1]

It was this second plan, the construction plan, which brought L'Enfant into direct conflict with the commissioners. L'Enfant envisioned himself in the role of director general, reporting to Washington and leaving little or nothing for the commissioners to do. Washington, Jefferson, and their appointed commissioners, however, held the view that members of the gentleman class could and should oversee the planning and construction of the new city; thus, they were opposed to L'Enfant and his innovative ideas. L'Enfant, for his part, was a product of the changes taking place in the building industry at that time in France, and he held that an educated and experienced architect-engineer reporting to the highest administrative officer should be in charge. This basic difference in views concerning supervisory structure continued to plague the construction of the new city in years to come.

In what was perhaps a misguided effort to secure his leadership position and advance his own plan for financing construction of the city, L'Enfant withheld his physical plan for the city and, in so doing, effectively torpedoed the commissioners' plans to sell lots to raise money for construction. Few wealthy men would buy a lot without seeing the overall plan for completion of the new city.

These and other conflicts made it clear that L'Enfant could not work with the commissioners. The strife between commissioners and L'Enfant came to a head on November 17, 1791, when two of L'Enfant's men began the demolition of Daniel Carroll of Duddington's house, on L'Enfant's order. On March 14, 1792, the commissioners sent him a letter informing him that his services were no longer required and offering him a settlement of 500 guineas [£525, or approximately $1,400] and half a lot for his past services.[2] L'Enfant refused the offer. He shortly left the city.

Ultimately, the commissioners adopted many of L'Enfant's ideas despite this basic conflict between a gentleman class of construction amateurs and a professional architect-engineer. Eventually, by 1800, the commissioners even agreed that they had to borrow money to complete the public buildings, and then President Jefferson organized the second public building campaign under the supervision of a single individual architect-engineer, Benjamin Latrobe.

L'ENFANT'S CONSTRUCTION PLAN FOR THE CITY

RATIONALE FOR THE PLAN

After completing the physical plan for the new city in the winter of 1791–1792, showing the location of streets, avenues, parks, lots, and other urban features, L'Enfant turned his attention to developing a construction plan that would implement his physical plan. He began his twenty-two-page construction plan with a statement of why it was important to show substantial construction progress in the coming year: "The approaching season for renewing the work at the Federal city and the importance of progressing it so as to determine the balance of opinion on the undertaking, to that side to which it already favorably inclines, require, that exertion should be made to engage in it from the beginning in such a degree of vigor and activity, as will disappoint the hopes of those who wish ill to the business, and encourage the confidence of the well disposed, it becomes therefore necessary to call your attention on measures of most immediate moment to determine."[3]

L'Enfant was well aware that Washington desired to have decisions regarding the new federal city made by the three commissioners from the gentleman class that he had appointed, Daniel Carroll, Thomas Johnson, and Dr. David Stuart. In his construction plan, L'Enfant explained why he was directly approaching Washington, rather than submitting the plan to the commissioners:

Knowing you wished never to be applied to on the subject of business intrusted to the management of the Commissioners, I would decline troubling you at this moment, when other affairs must engross your time; were it not that I considere the commencement of the work next season, will be but the beginning of the grand operations of the plan & conceive a permanent organization of sistem for continuing all future operations to be of absolute necessity and wish it may come from you in the first instance to prevent difficulties, which, (without such organization) will arise in the prosecution of the work, & must by interfering with the progress prove constant sources of new importunities to you.[4]

It was a sensitive subject. The commissioners, probably viewed by L'Enfant as rank amateurs, had done little to move the vision of the new city to reality. For L'Enfant, the first year of the new federal city, 1791, had been a disaster. As he wrote Washington: "Everything yet remains to be done for establishing a regular mode of proceeding — no adequate means of supply provided — no materials engaged proportional to the work to be effected . . . the neiborhood of the city offering no kind of resources at least none to be depended on."[5]

L'Enfant was pointing out that the new federal city, unlike Philadelphia, New York, or Boston, had no lumber dealers to supply the needed joists and small-dimension lumber called scantling, no water-powered saw mills to cut that timber, few established quarries from which dimensional building stone could be bought, no brickyards that could be engaged in burning brick for the needed federal buildings. Before construction could begin on the President's House and the Capitol, building materials had to be cut, sawn, quarried, burned, or otherwise provided. The commissioners had taken little action to supply the needed building materials.

In addition to building materials, workers, particularly skilled workers, needed to be recruited from other cities or even perhaps Europe. The commissioners had not taken actions to supply these needed workers: "No measures [have been] taken to procure the necessary number of men to employ. . . . Assistance wanted must therefore come from a distance." Not only did these workers need to be recruited from other cities, but they needed to be recruited in the very near future if any meaningful work was to be done in 1792. L'Enfant's report to Washington was dated January 17, 1792, and the traditional construction season would begin in five and a half weeks, on March 1: "The season [is] already far advanced, the demand for such hands as might be procured, will increase in proportion as the winter passing will afford them employment at home."[6]

Also, prices of materials and provisions would increase as the beginning of the 1792 construction season approached: "Materials will be dearer when an indispensable necessity for them is known, & provisions more difficult to obtain."[7]

For L'Enfant there was "no time then to be lost." He explained, "These are the considerations which lead me to demand your particular attention to the enclosed statement of work and estimates of expences."[8]

DETAILS OF THE CONSTRUCTION PLAN

L'Enfant's construction plan proposed to Washington was for all the buildings and other structures needed in the new federal city. The plan was to begin in 1792 and continue until completed in 1796—four years before 1800, the scheduled date for the relocation of the federal government from Philadelphia to Washington.

L'Enfant's construction plan describes sixteen steps needed for providing the necessary materials and infrastructure to the new city and for completing the principal buildings. This effort would require a huge workforce, estimated by L'Enfant at 1,070 men for the first year and a larger number annually over the next three years (1793–1796).[9]

The first step, according to L'Enfant, was to excavate the foundations of the President's House and the Capitol: "1st To continue clearing the cellars & begin laying the foundation of the two principal buildings and bring these forward to such a stage as they will be safe from injury the next winter. The digers to continue afterwards employed in shaping the adjacent grounds."[10]

The excavated earth and clay would then be used for molding and burning bricks. In this case the commissioners had issued instructions to L'Enfant, on September 24, 1791: "Resolved that Majr. L'Enfant be instructed to employ on the first Monday in October next one hundred and fifty Labourers to throw up Clay at the President's House and the house of Congress, and in doing such other work connected with the post road and the public buildings as he shall think most proper to have immediately executed."[11]

Although digging could commence, the commissioners had no approved design for either the President's House or the Capitol. Digging the foundation for these two buildings without an approved plan risked over- or underdigging and the potential of wasting scarce money.

By the time L'Enfant's physical plan was completed, Washington and Jefferson had decided to use a design competition to determine the size, shape, and appearance of these buildings. However, the design competition was not announced until early in 1792. In Washington's letter to Jefferson of August 29, 1791, he set out a number of issues concerning the planning and construction of the federal city on the Potomac. He asked Jefferson how to best develop the needed building plans and how to supervise the work, and Jefferson responded: "By advertisement of a medal or other reward for the best plan—see a sketch or specimen of advertisement."[12]

It was not until March 6, 1792, that Jefferson wrote to the commissioners with a suggested announcement for the advertisement for the design competition:

A PREMIUM of 500 dollars or a Medal of that value, at the option of the party, will be given by the Commissioners of the federal buildings to the person who before the _____ day of _____ next shall produce to them the most approved plan for the President's house to be erected in the city of Washington & territory of Columbia. The site of the building, if the artist will attend to it, will of course influence the aspect & outline of his plan & its destination ~~of the building~~ will point out to ~~the artist~~ him the number, size & distribution of the apartments. it will be a recommendation of any plan ~~that~~ if the central part of it may be detached & erected for the present, with the appearance of a complete whole, and ~~the other parts added~~ be capable of admitting the additional parts in future if they shall be wanting.[13]

To this Washington added the following penciled note. "I see nothing wanting but to fill the blanks and that I presume the Comrs. will do, unless, after the words 'destination of the building' is added 'and situation of the ground' for I think particular situation wd. require parlr. kind or shaped buildings."[14]

The lack of design plans for either the Capitol or the President's House also hindered the initiation of needed site work. L'Enfant addressed the site work needed:

Number of Men	2nd.
150	planting the wall of the terrace supporting each of these building & forming the gradual assent to the Federal Square, either of these must be rised in the mean time as the foundation of the building with which they are connected.[15]

PROJECT PLANNING AND CONSTRUCTION EFFORTS

The third step for L'Enfant was to build a wharf on the Potomac and to begin work on the Washington City Canal. These were necessary to deliver building materials to the construction sites.

Number of Men	
	3rd
	Wharfing the bank of the Potomac to form the end of the canal and from thence to dig & the canal up to the Federal Square. to
300	effect this in proper season three hundred men will be required four months. the men to be afterwards employed at the other end of that canal on the Eastern branche.[16]

L'Enfant recommended building this canal even though he knew that the commissioners four months earlier, on September 8, 1791, had disapproved it, along with the building of a bridge over the Eastern Branch and construction of the wharves on the Potomac below Rock Creek.[17]

The fourth and fifth tasks set out by L'Enfant dealt with grading the new city—work that was not systematically undertaken until after the Civil War—and improving the city's roads:

Number of Men	Number of Teams	
		4th
		To Reduce the two streets on the side of the president park & garden to a proper graduation, the excavation of which will be wanted to fill up the warfing & bank of the canal, two objects which must be carried in
30	10	concert, for this object 10 teams will be wanted and 30 labourer.
		5th
		To reduce some of the principal streets in such parts as may difuse the advantage thro the various property and bring them to the state of good turnpike Roads,
200	10	two hundred men and 10 teams will be wanted.[18]

L'Enfant's sixth task was to improve internal communications within the new federal city, particularly the construction of three bridges, one over Rock Creek and two over the new canal, and the construction of two wharves for landing building materials:

Number
of Men

6th
To build three good stone bridges one over rock creek and two over the canal that over Rock Creek being immediately necessary to engage in to effect a communication with the post Road & for establishing a necessary intercourse will employ fifty men. Filling up the abutment & adjoining warfs will be effected by reducing the post Road. A warf next to that bridge one near the end of the canal on the Potomac & another on the East branch at the nearest communication with the Federal and president Squares must be established for landing materials & for an equal encouragement of improvements in those parts. streets leading to these must be reduced & will serve to the warfing.[19]

50

On March 29, 1792, a little over two months after L'Enfant sent his construction plan to Washington, the commissioners contracted with Leonard Harbaugh of Baltimore, a builder inexperienced in bridge building, to build a stone arch bridge over the mouth of Rock Creek.[20] The bridge, subsequently modified to three arches, was poorly built and had to be repaired, bringing criticism and ridicule down on the commissioners.

L'Enfant also called for sixty men to continue the "aqueducts already done." This entry was curious, as there is no other reference to aqueducts being built. The building of water-supply aqueducts was not begun in Washington until the 1850s, when Montgomery C. Meigs began the Washington water-supply system from Great Falls into Washington. Probably the aqueducts that L'Enfant was referring to were earthen channels or bored wooden pipes to bring water from existing springs to the two principal work sites for the use of the workmen.

Number
of Men

7th
Aqueducts already begun must be continued in various places to convey the water to such places and in such quantity as will be of general use to the city, an object to be done so early as to be compleated before any material improvement are begun for which sixty men will be required.[21]

60

Transporting the building materials from the wharves to the construction sites would require fifteen teams of horses and ten laborers:

Number of Men	Number of Teams	
		8th
		The transporting of material from the three entry places to various parts where they are to be used will
25	15	employ 15 strong teams and 10 labourers.[22]

The first year twenty-five laborers would be required for brickmaking, to be increased in subsequent years:

Number of Men	Number of Teams	
		9th
		The quantity of brick wanted in the first instance will
		employ twenty five men and as many labourers &
	35	drivers but considering the quantity of bricks that will
		be necessary in prosecuting the building an increas-
		ing number of brickmakers will be wanted after the
52	2	first year.[23]

Two horse-powered mills would also be needed to grind the plaster-of-paris cement:

	10th
	Two mills must be erected to grind and pound plas-
	ter of paris cement and clay, Four horses and six men
	must attend these mills[24]

And he recommended the construction of a water-powered sawmill:

Number of Men	Number of Teams	
		11th
		A water mill for sawing various kind of plank will be
		of great advantage if possible to be obtained in the
16	2	vicinity, but a number of sawyers — ten — must be
		employed for this purpose.[25]

A water-powered sawmill, available in settled cities such as Philadelphia and New York, was never erected by the commissioners, although at one point the commissioners directed James Hoban to investigate wind-powered sawmills.[26] Wood for the new federal city would be cut by labor-intensive pit saws.

Latrobe estimated that four scows would be required for shipping the building stone, to be attended by twenty boatmen and thirty laborers for loading and unloading the stone:

12th
Two large scow of a particular construction for the purpose of transporting stone, of large dimensions and two other for smaller stones must be constantly employed and will require twenty boatmen.[27]

Number
of Men

13th
50   The exploring the stone and assisting to load the boats will require thirty labourers.[28]

The building stone would require twenty stone cutters, attended by ten laborers. By the summer of 1792, this number was increased to forty stone cutters attended by sixty laborers and was subsequently increased over the four-year construction period.

Number
of Men

14th
Twenty stone cutters will be indispensable to work the stone for the building
30   ten labourers must attend them—and the increasing demand for this wrought stone will require additional number of hands the succeeding years.[29]

Number
of Men

15th
As soon as the materials are collected in sufficient quantity round the buildings, which will not be before the 4th Jully, the twenty
80   masons must be increased to 40 with the adition of 60 labourers, _____   that number to be increased in proportion to the progress of the
1043   building.[30]

Here L'Enfant is specifying that twenty masons (including stone cutters, stone carvers, and stone setters) be hired in the next six weeks (i.e., by March 1, 1792) and another twenty masons by July 4, 1792. Masons were the scarcest of all of the building tradesmen. There were virtually no masons within Georgetown or Alexandria; these masons would have to be recruited from established cities such as Philadelphia, New York, and Boston. Eventually, the commissioners had to recruit masons from Europe. Laborers, on the other hand, were readily available in the new federal city, if not free whites, then black slaves.

> 16th
> The various kinds of iron which must be readily supplied requires that two shops be erected with two fires for a master and 4 smiths with proper tools & stock.

5
17 overseers
 2 wagon men
 3 commissarys

———

1070 men[31]

For the coming construction year, beginning March 1, 1792, L'Enfant envisioned a very large workforce:

Carpenters (including four foremen)	40
Masons (including four foremen)	40
Brickmakers (including two foremen)	25
Stone cutters (including two foremen)	20
Blacksmiths (including one foreman)	5
Mechanics	117
Drivers (including two masters)	41
for ten teams of four oxen each and	
for twenty-nine teams of three horses each.	
Pit sawyers	10
Boatmen (including one master)	20
Overseers for the laborers	17
Commissary and two assistants	3
Laborers	849
Total workforce	1,187[32]

It was an enormous workforce for a semipopulated backwater. Many of the skilled workers would have to be recruited in Europe. Housing and work shops for these men would have to be constructed. And the cost for the first year would be high—L'Enfant estimated $300,000. This was a competent estimate, but it far exceeded the money that the commissioners had available. L'Enfant's plan to cover the shortfall was to borrow the money needed, using the building lots as collateral. He proposed to Washington borrowing $1 million for construction operations, a loan over twenty years at 5 percent.[33]

It may have been possible at the time to borrow such a large sum from Europe. The French Revolution had not yet triggered a continental war. This situation changed soon, however. Approximately two years later, James Greenleaf tried, but failed, to borrow approximately $400,000 from the Dutch to pay for the six thousand building lots that he, Robert Morris, and John Nicholson had bought from the commissioners. L'Enfant's financing plan was essentially the same approach that the commissioners attempted later, in 1796, to fund the construction in the new federal city. In that year they received an authorization from Congress to borrow a maximum of $500,000 against their remaining lots. Although Congress authorized the loan, it was not possible to obtain it. By 1796 the French had invaded Holland, and the money the Dutch previously had available for loan overseas went to increased taxes and military expenditures or simply was not loaned because of the uncertainty of the times.

Washington never responded to L'Enfant's construction plan. Two months later L'Enfant was released from his position. But many of the elements included in L'Enfant's construction plan were subsequently adopted by the three initial commissioners (Daniel Carroll, David Stuart, and Thomas Johnson) and their subsequent replacements (Gustavus Scott, William Thornton, and Alexander White). Eventually the commissioners adopted L'Enfant's proposal to borrow money to complete construction. Nonetheless, construction began on the new federal city without L'Enfant.

The conflict between the three commissioners and L'Enfant is usually interpreted as being due to L'Enfant's intransigence or his volatile artistic nature. However, similar conflicts occurred over the next thirty years between future commissioners and future architect-engineers, suggesting that deeper divisions than personality separated these individuals. The people changed, but the issues remained.

With the departure of L'Enfant, work on the new federal city continued, but at a languid pace. Instead of being completed within four years, as envisioned by L'Enfant, work on the new city extended over the full ten years. At the end of the ten years, the President's House and the Senate wing of the Capitol were mostly finished, but work on the House of Representatives wing had been barely begun. Of the two planned administrative buildings, only one was completed. As L'Enfant had expected, financing of the new city continued to be a major problem. Jefferson's three-legged stool of land sales and grants from Maryland and Virginia never was adequate, and the commissioners eventually had to turn to L'Enfant's recommendation to borrow the needed money to support construction.

2

Financing the Federal City
The Difficulty of Collecting Adequate Funds

The completion of the work will depend on a supply of the means. These must consist either of future grants of money by Congress which it would not be prudent to count upon—of State grants—of private grants—or the conversion into money of lands ceded for public use which it is conceived the latitude of the term "use" & the spirit & scope of the Act will justify.

THOMAS JEFFERSON, November 29, 1790

THE RESIDENCE ACT: AUTHORIZATION WITHOUT APPROPRIATION

The Residence Act, locating the nation's new capital on the Potomac River, represented an authorization without an appropriation: Congress provided no money for construction of the public buildings. Nor did Washington or Jefferson think it was likely that Congress would provide any money. Jefferson wrote, "It would not be prudent to count upon."[1] Backers of other competing cities for the new federal capital, such as Philadelphia and New York, would oppose appropriating funds for that purpose, and so Washington and Jefferson needed a plan for funding the new capital.

Washington and Jefferson did manage to secure pledges of money for the new city from the states of Maryland and Virginia. Jefferson suggested that the landowners at the site of the new capital transfer half of their lands to the federal government, which then could sell these lots to finance the buildings and other expenses of the new federal city.

Jefferson's funding strategy might be compared to a three-legged stool. If any leg did not fully support the goal of building the new city, the entire stool would topple. The first two legs consisted of the grants from Maryland and Virginia. The third, and most important, leg was the money from the sale of land transferred from the landowners to the US government. Jefferson explained why the self-interest of the landowners would make them willing to donate their lands: "When the President shall have made up his mind as to the spot for the town, would there be any impropriety in his saying to the neighboring landholders, 'I will fix the town here if you will join & purchase & give the lands.' They may well afford it from the increase of value it will give to their own circumjacent lands."[2]

The problem with this approach was that Washington and Jefferson had only developed a very rudimentary cost estimate of how much money would be required. They had no idea how much land might be required to fund construction of the new federal city. Nonetheless, they went forward with the land-for-money proposal.

At Washington's request, in November of 1790,[3] Thomas Jefferson and James Madison met with three of the largest landholders in the vicinity of the proposed location of the future federal capital to sound them out on this proposal. Jefferson's note of the meeting with William Deakins, Benjamin Stoddert, and Charles Carroll indicated that the land-for-money proposal was well received.[4]

Jefferson's "land-for-money" proposal seemed feasible, since a rudimentary estimate had indicated that only a modest amount of money was necessary to build the new city. He envisioned fewer than fifty buildings in total. He estimated that £200,000 ($533,400) was required to build those public buildings.[5] He also estimated that this amount would be easily raised by land sales.[6] Unfortunately, Jefferson's estimate lowballed the final cost by approximately half of the total $1,047,167.35 spent by the commissioners between 1791 and November 18, 1800.[7] It was certainly much less than the total expenditure, including the moneys required to finish the Capitol after 1800. And Jefferson greatly overestimated the amount realized by land sales.

Once Washington and Jefferson determined the basic funding strategy, they selected three commissioners to supervise and implement the federal construction program. Daniel Carroll, Thomas Johnson, and David Stuart were appointed January 21, 1791. None had experience or training in supervising large construction efforts. The federal construction program developed at a slow pace, because the unpaid landowners held commissioners' meetings infrequently.

LAND SPECULATORS

The initial financing strategy ensured that funding problems would continue throughout the duration of the first federal building program. By 1793 the commission had raised only $16,000 through the sale of lots. That year's audit indicated a serious financial shortfall (table 2.1), which was intensified by the difficulty in obtaining $40,000 of the $120,000 promised by the State of Virginia, because of that state's budgetary problems.[8] With very few lot sales and the problems encountered with Virginia's grant, the commissioners lost two of the three legs of Jefferson's planned financial stool. This financial situation set the background for the

TABLE 2.1. Financial status of the commissioners as of 1793

	JEFFERSON ESTIMATE	AUDIT OF 1793
State of Maryland Grant	120,000	72,000
State of Virginia Grant	120,000	120,000
Proceeds of Land Sales	533,400	16,000
Total	733,400	208,000
Expenditures through 1793		$134,335

Source: Proceedings, Audit by Robert T. Hooe and David Ross, October 31, 1793, in *RDCC*, roll 1, 202–203.

By the end of 1793 the commissioners were running out of money. This table overstated the commissioners' financial solvency, as they had difficulty recovering $40,000 of the funds expected from the State of Virginia. It was the precarious situation of their financial solvency that set the stage for the land sales to Greenleaf and to Greenleaf and Morris at the end of 1793.

contract eventually made by the commissioners with three land speculators: first with James Greenleaf and later with Greenleaf, Robert Morris, and John Nicholson. Though these agreements were intended to resolve the commissioners' financial difficulties, they instead intensified their problems.

On September 17, 1793, James Greenleaf, a New York merchant and land speculator, arrived at what later became Washington. He immediately ingratiated himself with important persons involved in developing the new capital city. He provided Tobias Lear, George Washington's former secretary (1785–1793) with seed money for Lear's firm, and he bought from commissioner Thomas Johnson fifteen thousand acres of his land in Frederick County, Maryland for fourteen thousand dollars. The commissioners saw Greenleaf as a wealthy individual with a ready source of cash. However, Greenleaf's wealth was largely dependent on European investors looking for opportunities in the new republic.

On September 23, 1793, only six days after Greenleaf arrived in Washington, the three commissioners (Johnson, Stuart, and Carroll) sold him three thousand lots in the new federal city at a price of £25 ($67) each, in Maryland currency, for a total of approximately $200,000. Greenleaf benefited from wholesaling the lots at bargain prices—for example, Jefferson had earlier noted that unsold lots in Georgetown sold for £200 ($533) and estimated that lots in the new city should sell for £100 ($267).[9] Only three years later, in 1796, commissioner Alexander White reported that the average price for lots sold by the commissioners to individuals had been $285 per lot.[10] Not only did the commissioners sell these lots to Greenleaf at a very low price; they also provided Greenleaf exceptionally favorable terms of sale: he was to pay the commissioners in seven equal yearly payments, without interest, beginning May 1, 1794.

In return for such favorable prices and terms, the commissioners received Greenleaf's contractual promise to loan the commissioners £1,000 per month, current money of Maryland ($2,667), at 6 percent interest, beginning May 1, 1794, and to continue doing so each and every month until January 1, 1800, or until all the public buildings were complete. In addition, by the contract Greenleaf promised to build ten houses per year, each house being two stories tall and with a ground plan covering twelve hundred feet, and not to sell any lots before January 1, 1796, without a contract stipulation that the buyer would build a similar house on every third lot within four years from the date of the sale.[11]

Three months later, on December 24, 1793, the commissioners entered into a second contract with Greenleaf, who this time was acting as Robert Morris's agent and partner (later to be joined by John Nicholson). This was not a separate contract but a consolidation of the previous contract of September 23. In the new agreement, Greenleaf and Morris agreed to purchase an additional 3,000 lots, but at £35 per lot (approximately $93 each). The price for all 6,000 lots, calculated at £30 per lot, Maryland currency ($80 each), came to a total sales price of $480,000. The terms of payment were identical to those in the September contract, seven equal yearly payments at no interest beginning May 1, 1794. And Morris, through his agent Greenleaf, also agreed to construct ten two-story houses each year with a ground plan covering twelve hundred feet each, with the stipulation that sales contracts would require the buyer to build a similar house on every third lot within four years from the date of sale.[12] The commissioners also agreed that 4,500 of the 6,000 lots could be selected by Greenleaf and Morris from those available south and west of Massachusetts Avenue; these were the more desirable lots, which sold, on an average, at three times the price of those north and east of Massachusetts Avenue.[13] Washington was appalled by these sales. He wrote Daniel Carroll, one of the original commissioners:

> You will recollect no doubt that I yielded my assent to Mr. Greenleaf's first proposition to purchase a number of lots in the Federal City (altho' I thought the price he offered for them was too low) because at that time [lot sales] seemed to be in a stagnant state, and something was necessary to put the wheels in motion again. To the second Sale which was made to him, my repugnance was greater, in as much as the necessity for making it was not so apparent to my view — and because another thing had become quite evident — Viz: that he was speculating deeply — was aiming to monopolize deeply, and was therefore laying the foundation of immense profit to himself and those with whom he was concerned.[14]

Washington's criticism of the commissioners stung deeply, especially since it came on top of criticisms of their other actions, such as the construction of the

PROJECT PLANNING AND CONSTRUCTION EFFORTS

James Greenleaf, by Stuart Gilbert. Arriving in the new federal city September 17, 1793, Greenleaf seemed to be the answer to the commissioners' financial problems. Lot sales, needed to provide money for the construction of public buildings, had badly lagged. Greenleaf offered to buy a large number of lots—3,000—for which he received very generous terms from the commissioners. The agreement was finalized September 24, 1793. Three months later, on December 24, 1793, Greenleaf, in conjunction with his new partner, Robert Morris of Philadelphia, purchased an additional 3,000 city lots from the commissioners, also on very favorable terms. The agreements called for Greenleaf and Morris (later including John Nicholson) to make seven equal annual payments beginning on May 1, 1794. Greenleaf and his partners failed to make the annual payment due on May 1, 1795, and subsequent payments as well. The result was that the commissioners were deprived of badly needed construction funds and that the title of the lots involved became clouded, thereby depressing future lot sales. Courtesy of the Library of Congress.

Robert Morris, by Charles Willson Peale. Sometimes referred to as "the Financier of the American Revolution," Morris engaged in massive land speculation in subsequent years. He was a partner with James Greenleaf for the purchase of six thousand city lots, based on the agreement with the commissioners made on December 24, 1793. Later John Nicholson was added to the partnership. The failure of these partners to meet their required second of seven annual payments, due May 1, 1795 (and subsequent payments), deprived the commissioners of funding for the needed public buildings, clouding the title on city lots so it was difficult to resell them. Courtesy of the Library of Congress.

bridge over Rock Creek that failed and had to be rebuilt. These criticisms led to the early resignation of the first three commissioners.

OUT OF MONEY BY SPRING 1794

By the spring of the following year, 1794, the commissioners had expended all the money originally granted to them except the $40,000 yet to come from Virginia. On April 22, 1794, shortly before commissioners Johnson and Stuart left the board, Colonel Deakins, the treasurer to the board, gave the commissioners a statement of their precarious financial condition. The original grants had all been expended by February 1794, and Deakins had to borrow $8,000 to sustain current operations. There were no hopes that the commissioners would receive the $40,000 owed them by the State of Virginia, and worse yet, construction operations for 1794 were estimated to cost $10,000 per month.[15] Deakins's loan continued construction, but at a slow pace.

John Nicholson, by Charles Willson Peale. Nicholson was the third partner who, with Greenleaf and Morris, purchased thousands of lots in the new federal city. By the end of the 1790s, all three were incarcerated in Philadelphia's Prune Street debtor's prison. Although Greenleaf and Morris were released, Nicholson died in that prison on December 5, 1800. Courtesy of the Art Institute of Chicago.

In the following year, 1795, Thomas Law arrived from England, having amassed a large amount of money in India. He bought 500 lots from Greenleaf at $266 per lot, for a total cost of $133,000, making a profit of $93,000 in one year for Greenleaf. Upon hearing of this, Washington was even more appalled. It was evident to Washington that Greenleaf was profiting greatly from his speculation — receiving money that could, and should, have gone to the commissioners. Washington was not wrong.

In addition to the 6,000 lots bought from the commissioners, Greenleaf had bought up 1,341 of the 10,136 privately available lots from the original proprietors — all at bargain prices.[16] This was not an absolute monopoly, but it constituted a massive proportion of the total, giving Greenleaf and his partners' great influence in the sale of future lots and the direction of the new city.

Washington feared that Greenleaf's portion of lots endangered the financing of the new city's construction program. He wrote to the commissioners: "The sum which will be necessary to compleat the public buildings and other improvements in the City, is very considerable. You have already, if I mistake not disposed of more than a moiety of the Lots which appertain to the Public; and I fear not a fourth part of the Money necessary for that purpose, is yet provided."[17] Washington was also offended that speculators, and not the commissioners, were reaping the profits of these land sales: "The persons to whom you have sold are reselling to others (subjecting them to the conditions to which they are made liable themselves) and this they are doing to an immense profit. Lately, a Gentleman from England, has paid, or is to pay £50,000 for 500 Lots. — Will it not be asked, why are speculators to pocket so much money? Are not the Commissioners as competent to make bargains?"[18]

This lack of money was an important consideration for the commissioners, but it was not their only problem. Construction of the Capitol lagged. By the beginning of 1795, it was apparent that great progress had been made in building the President's House, but little progress was evident in the construction of the Capitol. Washington, in a letter to the commissioners in January 1795, encouraged them to emphasize construction progress on the Capitol over the President's House.[19] However, accelerating progress on the Capitol would take more money, not less, leaving Washington and the commissioners in a deeper predicament.

THE DEFAULT OF THE SPECULATORS

By the end of the 1795 construction season, the commissioners had almost totally exhausted all available funds. They had received or were scheduled to receive $779,185 (table 2.2). The great bulk of these moneys came from a single source, Greenleaf, Morris, and Nicholson, who had promised to pay the commissioners $480,000 in seven installments for the six thousand lots purchased at $80 apiece (table 2.3). In May 1795 Greenleaf, Morris, and Nicholson failed to make their second scheduled payment,[20] and they continued to lapse on subsequent payments. Worse still, the group stopped construction on the houses that their contract required.

When Greenleaf and his partners bought these lots, they were counting on European capital to fund the payments. But the European wars in the wake of the French Revolution had dried up their lending sources, leading to insolvency for Greenleaf, Morris, and Nicholson. Their insolvency not only deprived the commissioners of money for construction operations but also clouded the legal title on the city lots to be sold. Lots that Greenleaf and his partners could not pay for also could not be resold. A deep enmity developed between Greenleaf and Nicholson, and a newspaper battle broke out, each warning potential lot buyers that the titles of city lots up for sale were not clear. The newspaper battle had an additional adverse effect on the already sparse land sales. Because titles to the lots were not clear, buyers avoided making purchases, including purchases of the commissioners' lots that were still available.[21] The Greenleaf-Morris-Nicholson contracts had clearly been a financial disaster for the commissioners.[22]

TABLE 2.2. Moneys received by the commissioners through January 1796

THE FUNDS OF THE CITY ARE:

The Virginia donation	$120,000	
The Maryland donation	72,000	
Amount of sales to Morris & Greenleaf	480,000	
Sales of other lots	96,652	
Harbaugh, Lee, Deakins, and Cassanove's note (arising from the sale of property, the original cost of which is charged in the article of expenditures)	10,533	$779,185
Deduct money expended	343,783*	
Deduct money due the Bank of Columbia	30,000	
Deduct money due the original proprietors, for squares appropriated	12,000	384,783**
		$394,402

Source: District of Columbia Commissioners, "Memorial of the Commissioners," 1:142–144, 143.

*This figure contains a mathematical error; actually $338,565.52 had been expended (not $343,783.00).
**This sum is $1,000 below the actual amount.

After Greenleaf, Morris, and Nicholson failed to make their May 1, 1795, payment and halted construction of the twenty houses they were required to build each year, the commissioners complained to President Washington. Washington referred their letter to his old friend Robert Morris. On September 21, 1795, Morris responded to Washington that he could not make the required payments.[23] He wrote Washington that although he was mortified by being in arrears to the commissioners, he had nothing to do with the £1,000-per-month loan promised to the commissioners by Greenleaf. Morris was unaware of any of the promises made that induced the commissioners to give Gree£nleaf such favorable terms. He also explained to Washington, "[I] was not until the receipt of your letter acquainted with the necessity there is for supplying the Commissioners with money, and imagined that a little delay was not of any *real* importance."[24] He further wrote, "I see the matter now in a very different light & will immediately commence my remittances & continue them until my part of the arrears are discharged, that part is $15000, Mr. Nicholsons $25000."[25] This did not happen. In the same letter, Morris identified that payments to the commissioners could not recommence because of the wars in Europe and especially the recent French invasion of Belgium and Holland.[26]

TABLE 2.3. Future payments expected from Morris, Greenleaf, and Nicholson for 6,000 lots sold to them in 1793

OF THE LAST MENTIONED SUM, $378,191 ARE NOT YET DUE, BUT WILL BECOME DUE IN THE FOLLOWING PROPORTIONS, VIA:

In the year 1796	$84,539 19
In the year 1797	80,719 09
In the year 1798	75,789 93
In the year 1799	68,571 42
In the year 1800	68,571 42
	$378,191 05

Source: District of Columbia Commissioners, "Memorial of the Commissioners," 1:142–144, 143.

Greenleaf, Morris, and Nicholson had intended to pay for the lots with borrowed money, Greenleaf with money from the Netherlands and Morris and Nicholson with money from England and Ireland. The European wars eliminated this possibility. Money that might have gone to the United States for investment went to European armies and taxes. The wars affected Greenleaf, Morris, and Nicholson's financing in other ways as well. France and other countries prohibited emigration, thereby reducing the number of European immigrants who would have purchased land in America. The violence, disruption, and social disorder of the time further discouraged potential European overseas investments.

Morris knew that in the light of nonpayment, the commissioners would attempt to resell the lots contracted to Greenleaf, Morris, and Nicholson. He offered his view that the commissioners would be better off not to sell those lots and that things would improve.[27] They did not.

The immediate effect of Greenleaf, Morris, and Nicholson's nonpayment was that it clouded the title of a large number of building lots, making them more difficult to sell. This situation continued for eight years. In 1801 the Griswold committee of the House of Representatives reported the dilemma and recommended that Congress pass a law to clear the title to these lots so that they could be sold.[28]

By the end of 1795, Jefferson's three-legged funding strategy for the new federal city was defunct and the commissioners were out of money. They had spent $338,565.52 through November 1795. The commissioners had not much to show for their expenditures in the first four years of construction of the new federal city. As table 2.4 indicates, much of the money had been spent for infrastructure or developing sources of building materials, and not on the public buildings. Although progress had been made on the President's House—the walls were within six feet of the eaves and much of the interior work was well advanced[29]—little progress had been made on the construction of the Capitol.[30] All of the work completed had been on the Senate wing (the north wing), and no work had been initiated on the House wing (the south wing), above the foundation.

CONGRESSIONAL LOAN GUARANTEE

The commissioners turned to Congress for help. In their appeal to Congress, formalized in their memorial—their formal request for assistance—and in reports to Congress in 1796, the commissioners estimated that an additional $400,000 would be required to complete the Capitol; an extra $100,000 was needed to complete the President's House, and $200,000 was needed for the government administration buildings that would be required. So they requested an additional $700,000.[31] Their communications with Congress indicated that even at this late date, the commissioners still thought that some, perhaps a substantial amount, of that $700,000 might come from the Morris and Greenleaf contracts.

The commissioners expected little additional revenue from land sales, but they still held a substantial number of lots that could be used as collateral for a loan for the funds needed to complete the federal buildings. In January 1796 the commissioners informed Congress about the value of these unsold lots:

CAUTION.

—

THE Subscriber, having large concerns with Robert Morris & John Nicholson, and having an interest in many valuable Estates, of which the legal Title is in them, or one of them, or subject to their joint or separate order, thinks fit to CAUTION and FOREWARN all persons against purchasing from them, or either or them, any Lands in the states of *New-York, Pennsylvania, Maryland, Virginia,* north & South *Carolinas, Georgia* or *Kentucky,* or Lots of Ground in the *City of Washington,* or making payments for any such Lands purchased within the last twelve months, without first ascertaining whether or not the same is free from claims on the part of
JAMES GREENLEAF.
Philadelphia, March 16, 1797.   77--2w.

Caution against Morris and Nicholson.
Washington Gazette, March 25, 1797, 1.

The real property of the city consists in the 4,694 lots, exclusive of water lots, averaging 5,265 square feet; 1,694 of these are choice lots, from which Morris & Greenleaf were excluded in their selection. The average value of these lots, taking the prices at which they have been sold from the commencement of the city, is $285 per lot, amounting to

	$1,337,790
Water property, 3,500 feet front, at $16 per foot	56,000
	$1,393,790 [32]

The commissioners expected that once the federal government moved to Washington in 1800, increasing demand would greatly increase the value of these unsold lots.

TABLE 2.4. Expenditures of the commissioners through November 1795.

GENERAL ACCOUNT OF EXPENDITURES BY THE COMMISSIONERS OF THE FEDERAL
DISTRICT, FROM THE 12TH OF APRIL, 1791, TO THE 1ST OF NOVEMBER, 1795.

Bridge at James's creek	$342 04
Bridge over the Tiber	788 04
Wharf at Eastern Branch	1,017 51
City of Washington	15,311 42
Capitol	78,035 29
Engraved plans of the city and territory	370 37
Expense of commissioners' office	5,895 76
Freestone quarries	28,749 96
Hospital for sick laborers	1,196 26
President's house	97,329 82
Public ground, paid sundries	1,766 67
Surveying department	18,838 97
Stone quarry in the city	291 56
Temporary buildings	3,018 46
Wharf account	1,858 30
Wharf log account	6,189 12
White oak swamp, Virginia	1,948 78
Buildings on square No. 728, balance	573 33
Causeway account at the bridge over Rock Creek	6,297 14
Provision account	11,809 27
Stone wharf at the south side of the bridge over Rock Creek	315 73
Slate quarries	58 29
Bridge over Rock Creek	12,568 83
Canal from Tiber to James's Creek	4,974 75
Utensils for various works	1,376 91
Expense of expresses	61 67
Surveying on the line	2,986 25
Post road	1,518 27
Account of commissions	1,887 58
Account of discounts	2,722 74
Account of advances made on contracts	25,762 39
Brick making account	3,209 09
	$338,565 52

Source: District of Columbia Commissioners, "Memorial of the Commissioners," 142–144, 143–144.

Therefore, they proposed to Congress that Congress guarantee a loan to the commissioners, using the unsold lots as collateral, so as to carry the construction on the public buildings forward.[33]

The commissioners submitted this proposal to Congress on January 8, 1796. On January 25, the House committee assigned the proposal reported favorably on the commissioners' request to guarantee a $500,000 loan, not exceeding $200,000 in any one year.[34] In arriving at these sums, the committee was estimating that the commissioners would require $140,000 each year through 1800 but that $40,000 per year could be gained from the sale of lots.

Commissioner White, a former congressman, was the commissioners' representative before Congress, but his status did not prevent the loan guarantee from being a hard sell. Many in Congress criticized the commissioners' efforts to date, including their approval of the plans for the President's House and the Capitol without a cost estimate. White explained to his fellow commissioners, Scott and Thornton: "But what shall I say when I am asked for an estimate of the expenses of the Capitol. What a Field for cavil and declamation when it can be said that the plan of such a Building was accepted without any knowledge of the costs?"[35]

Commissioner White was also quite anxious about asking Congress for the loan on the basis of the security of unsold city lots whose titles were cloudy at best. He explained his strategy to his fellow commissioners: "I attended the Committee at 7 o'clock last night the Attorney General was present. After much conversation they agreed to report a Bill, a rough copy of which is enclosed. I gave no information with respect to the title of the lots, but that the Trustees had never conveyed. Had I introduced the maze of Maryland Laws, we never should have got through them."[36]

To the Public.

MR. JAMES GREENLEAF, in his notification of the 15th inst. published in some of the Newspapers of this City, observes that he has "large concerns with Robert Morris and John Nicholson," which unfortunately is true; but he goes farther and asserts "that he has an interest in many valuable estates, of which the legal title is in them or one of them, or subject to their joint or separate order." When this assertion was made, it would have been doing justice to himself, to us, and to the Public (whom he warned) if he had described the estates in which he claims an interest with us. As he has not done that, justice, we must endeavour to do it. Mr. Greenleaf has an interest with us (or rather had, for we have since his notification sold and conveyed our shares) in an estate at the great Fall of Potomac, purchased of General Lee. He is in erested with us in a contract for 300,000 acres of land in the state of New-York, & in another for 100,000 acres of land in the state of Virginia, neither of which have yet been carried into full effect.

He has an interest with us in certain lots in the city of Washington, which were sold previous to our purchase of his remaining share or interest in that city, some of which lots have not yet been conveyed to the purchasers, and in order that these may be known, we insert at the foot hereof such of the numbers of the said lots as we have account of, and of the squares in which they are contained. If Mr. Greenleaf has an interest with us in any other lands or lots, we call on him to name them: in the mean time we deny that he has other interest with us in valuable estates in New-York, Pennsylvania, Maryland, Virginia, North and South Carolina, Georgia, and Kentucky, or Lots in the City of Washington. His notification must therefore have been made for the purpose of creating suspicion and distrust as to our titles to the property we possess. Purchasers, however, will not be deterred by this attempt to injure us, as they no doubt will require perfect title, even if Mr. James Greenleaf had not published his notification.

ROBERT MORRIS,
JOHN NICHOLSON.

Philadelphia, March 21, 1797.

N. B. This address would sooner have appeared had not our time and attention been engaged by more important business.

Complaint against Greenleaf. *Washington Gazette*, March 20, 1797.

Despite these problems, and luckily as a result of White's careful positioning, on March 31, 1796, the House of Representatives approved a bill called the Guarantee Bill, guaranteeing a loan to the commissioners of up to $300,000. Passage through the Senate was going to be more difficult. Before the Senate could address the loan guarantee for the new federal city, it first had to resolve the issue of ratification of a treaty between Great Britain and the United States, in which the United States agreed not to impede the right of free access of British citizens by any further treaties with the American Indians. This treaty was an extension, in the eyes of Britain, of the Treaty of Paris, which had ended the Revolutionary War. Many American opposed ratification of this treaty. But despite general opposition, the treaty was ratified by the Senate on May 9, 1796, and the Senate moved on to consider the Guarantee Bill, passing it by a vote of 16 to 7.[37]

Passage of the Guarantee Bill was only the first step in resolving the commissioners' financial problems. Next they needed to obtain the loan. Almost immediately following the passage of the Guarantee Bill, the commissioners sent a power of attorney to the firm of Wilhelm and Jan Willink of Amsterdam to borrow 512,000 guilders.[38] As Greenleaf, Morris, and Nicholson had earlier learned, borrowing European money was impossible owing to the French Revolution and subsequent European wars. During 1796 the commissioners were only able to keep construction going by the acquisition of two loans of $42,000 and $40,000 from the Bank of Columbia. The bank agreed to provide $10,000 per month to the commissioners beginning August 1, 1796. For the total of $82,000 the commissioners were required to give their personal short-time notes, endorsed by Robert Peter, Notley Young, Thomas Law, Uriah Forrest, and Francis Loundes.[39]

The commissioners tried other sources of money. On October 31, 1796, they applied for a loan of $100,000 from the Bank of the United States in Philadelphia but were informed that "the bank was averse to making long term loans and, moreover, was much taxed to accommodate its local patrons, and declined to make the loan."[40]

Besides the two loans from the Bank of Columbia, the only loan the commissioners could obtain was for $200,000, in United States 6 percent stock, at par, from the State of Maryland. For this the commissioners were required, in addition to the guaranty of Congress, to give bonds in their individual capacities. The cash resulting from the sale of this stock, including the interest accrued, amounted to only $169,873.43, and the interest paid on it was $39,000. So for the $200,000 loan, only $130,873.41 could be used on the construction of the public buildings.[41]

On November 21, 1796, the commissioners again turned to the Maryland legislature for a second loan. They authorized commissioner Scott to pursue that loan, and President Washington wrote to Governor Stone in support. On December 14, 1796, the Maryland legislature passed a resolution for a loan of $100,000 in United States 6 percent stock. The commissioners were required to give bond in their official capacity in the sum of $100,000 and in their individual capacities in twice that amount.[42] Without these loans, which took some time to convert to cash, the commissioners had no funds, having expended virtually all of the money received between 1791 and May 16, 1796.

LABOR PROBLEMS

On top of continued money problems, the commissioners began to have serious labor problems. In October of 1797, they could not meet their payroll. This event shattered the labor peace that the new commissioners had achieved at the work site since 1795. On October 3 they wrote their architects James Hoban and George Hadfield, "The present situation of our funds renders it impossible to pay the time-Role due yesterday, or to discharge the arrears of last month."[43] The workers bitterly protested.

The commissioners thought that the lot sale of October 18, 1797, would resolve their insolvency, but this sale did not provide adequate funds to cover the commissioners' obligations. Some money was found from lot sales and other sources to pay the workers, and the labor problems receded; but they appeared again later.

The commissioners decided to turn to Congress again for financial aid and prepared another memorial. On November 25, 1797, they transmitted the new memorial to President John Adams. Action on that memorial was suspended, however, as landowner Colonel Uriah Forrest reported that he thought the Maryland legislature would approve a second loan of $100,000.[44] Accordingly, the commissioners drew up a memorial, which they presented to the Maryland Assembly on December 13, 1797. On December 22, the Maryland legislature approved the second $100,000 loan to the commissioners on the same terms as the first.[45]

THE SECOND CONGRESSIONAL LOAN GUARANTEE

Although temporarily helpful, the second Maryland loan was still not adequate to meet the commissioners' construction expenses. They decided to again approach Congress. And again they drafted commissioner White to undertake this task. He traveled to Philadelphia in February 1798. White and President Adams decided that the president would not submit the commissioners' request as his but would submit to Congress the commissioners' memorial requesting the $100,000 loan. They did so on February 23, 1798, only two and a half years before the federal government was scheduled to relocate to the new city and with no substantial work having been achieved on the Capitol.[46]

On April 18, 1798, Congress passed the requested act authorizing the loan to the commissioners for $100,000. The act called for this loan to be paid in two yearly installments and to bear 6 percent interest. All the unsold public lots in the city were declared to be subject to sale for the repayment of the loan.[47]

In the following year, 1799, the commissioners sought and received a $50,000 loan from the State of Maryland,[48] to be secured by such real and personal security as the governor and council of the state should approve. The terms of security were met by the execution on February 28, 1800, of a bond signed by commissioners Scott and Thornton and joined in by Colonel Uriah Forrest and General James M. Lingan. In addition, Colonel Forrest gave a mortgage on his farm of 420 acres of land in Montgomery County.[49] Of this $50,000 loan, only $40,488.36 could be applied to building the federal city.[50]

Little time now remained to complete the public buildings. Relocation of the government from Philadelphia was scheduled to take place in a little over a year and a half, in November 1800, and the first meeting of Congress was planned for November 17, 1800.[51]

RELOCATION OF THE FEDERAL GOVERNMENT TO THE NEW CITY

The new capital was still incomplete when Congress and the government moved to the new city from Philadelphia. What the congressmen and senators found when they arrived was a mostly completed President's House and a mostly completed north, or Senate, wing of the Capitol. Work on the House wing had been suspended when it became obvious that adequate funds were lacking. Of the two administration buildings, the Treasury Department, located east of the President's House, had been completed, but the War and State Building, located west of the President's House, had not.[52] They also discovered that the US government owed considerable money for the unfinished capital. On January 1, 1801, the commissioners calculated that the government owed a total of more than $360,000 (table 2.5).

THE GRISWOLD COMMITTEE

It was obvious to all that the commissioners had expended a great deal of money without completing their construction task. Congress authorized a committee, chaired by Representative Roger Griswold, to look into the matter. They directed the committee to "inquire into the expenditure of money made by the commissioners of the city of Washington, the disposition of public property made by them, and generally into all the transactions of the commissioners which relate to the trust confided to them by the President of the United States."[53]

On February 27, 1801, Griswold and his committee reported that the commissioners had expended more than $1 million.[54] With regard to the achievements of the commissioners, Griswold and his committee simply reported: "It may, however, be proper to remark that the principal objects of expense have been the Capitol,

TABLE 2.5. Debts of the commissioners as of January 1, 1801

A LIST OF DEBTS DUE FROM THE COMMISSIONERS OF THE CITY OF WASHINGTON
ON THE 1ST OF JANUARY, 1801, AS CORRECTLY AS CAN BE ASCERTAINED.

To whom due.	*For what due.*	*Amount due.*
The State of Maryland	Loan of United States, six per cent. stock, at par, principal	$250,000 00
The United States,	Loan	100,000 00
The State of Maryland	Interest for a quarter of a year, ending the 31st ultimo, on $250,000	3,750 00
Bank of Columbia	Loan	3,000 00
Lewis Clephane	Painting Capitol and President's house; account unsettled; balance about	1,560 00
Commissioners and their officers	Balance of their salaries for the quarter ending 31st ultimo	636 46
Small accounts allowed, and not yet paid	Sundries	254 37
	Over payment for lots, 18th ultimo	180 22
Joshua Johnson, Claims for which accounts have not yet been presented, estimated at		1,500 00
		$360,881 05

Source: District of Columbia Commissioners, "Report of the Commissioners," January 30, 1801, 228.

the President's house, and the two buildings erected for the accommodation of the Executive Departments; and that the situation of those buildings, being under the eye of every member of the Legislature, cannot want a particular description."[55]

Sensitive to the criticisms being leveled at them, the commissioners had issued a report a month earlier that although the government owed $360,000, the value of the assets (principally unsold lots but also physical improvements such as the city wharves and the quarries on Government Island) and accounts receivable greatly exceeded the $360,000 owed by the federal government (table 2.6).[56] Griswold and his committee simply responded to the commissioners' report with a vote of no confidence in the commissioners' numbers:

The commissioners undoubtedly possess much better means of judging of the value of the buildings lots belonging to the public than the committee; but it could not escape the observation of the committee that the actual sales which have been made for cash since the board of commissioners was established cannot, in their opinion, support the estimate which the commissioners have made of this property; and, whatever may be the product of sales hereafter, the committee believe that the Government cannot rely upon that fund for completing the objects which the Legislature may deem necessary for the accommodation of the Government.[57]

The short, page-and-a-half report ended with a resolution to abolish the board of commissioners and transfer its responsibilities to the secretary of the Treasury. The secretary of the Treasury would prepare a plan of Washington delineating the streets, squares, and public grounds and draft a bill authorizing the sale of the remaining building lots.[58] No action was taken by Congress on these committee recommendations.

MEETING THE TERMS OF THE GUARANTEE ACT

The necessity of repaying the loans made to the commissioners and guaranteed by the US government that were coming due eventually prompted action by Congress on the Griswold committee's recommendations. At the time of the passage of the Guarantee Act, it had been assumed that once the government had moved to the new federal city, demand for building lots would greatly increase and the sale of building lots would repay the loans. This was not the case. On December 19, 1801, commissioners Thornton, White, and Dalton[59] reported that the December 4 lot sales had yielded only $4,234. The commissioners reported only a balance of $5,880 in their hands.[60]

The Guarantee Act required the sale of the unsold lots to repay the loans. In the presence of little demand, these lots could only be sold at fire-sale prices. It was thought that if they were held for a time, their price would increase. President Thomas Jefferson, in his cover letter to the commissioners' memorial of December 4, 1801, explained that if the lots were sold immediately to pay for interest on the loans, the government would stand to lose, as these properties would appreciate in value.[61]

The alternative to the sale of lots to pay the loans due was to advance moneys from the US Treasury and to sell the lots later, after they had appreciated.[62] This was the plan the government adopted. It was a bailout for the commissioners, who were already criticized for mismanagement of the construction for the new federal city. Although Congress voted to provide the money from the Treasury to pay the loan, they were not happy with the situation.

TABLE 2.6. Schedule of property owned or due to the commissioners as of January 1, 1801

SCHEDULE OF PUBLIC PROPERTY ON HAND BELONGING TO THE FEDERAL SEAT		
1,504 building lots, southwest of Massachusetts Avenue, estimated at $343 each	$515,872	00
3,178 ½ building lots northeast of said avenue, estimated at $105 each	338,747	00
2,648 feet front of lots binding on navigable water, estimated at $12–71 per foot front	25,979	00
Four wharves, cost	3,221	88
Island, containing freestone quarries in Aquia creek, cost	6,000	00
	$884,819	88
Debts due to the commissioners, which are deemed good	144,125	80
	$1,028,945	68
Debts contracted on the credit of the above fund	360,881	05
Fund remaining, clear amount	$668,064	63

Source: District of Columbia Commissioners, "Report of the Commissioners," January 30, 1801, 226.

ABOLISHING THE OFFICE OF THE COMMISSIONERS

On February 5, 1802, Congress directed one of its committees to "inquire into the expediency of discontinuing the offices of the said commissioners."[63] The committee, chaired by Congressman Joseph Nicholson, reported on February 12, 1802, that the board of commissioners should be abolished: "*Resolved*, That from and after the 1st day of March next, the offices of two of the commissioners of the city of Washington ought to be discontinued, and thereafter the powers now vested in the board of commissioners ought to be vested in one only, who ought to discharge all the duties now required to be performed by the whole number."[64]

On May 1, 1802, Congress enacted "an act to abolish the board of commissioners," to take effect on June 1, 1802. The act directed the commissioners to deliver all their official records and property relating to the city to an officer designated as the "Superintendent," to be appointed by the president, to succeed to all powers and duties of the three commissioners. Also, the act directed the superintendent to sell sufficient numbers of the public lots, which had been pledged under the Guarantee Act, to satisfy the two $100,000 Maryland loans as well as the one $50,000 loan as fast as the interest and installments should come due.[65]

This marked the end of the first public building campaign. With the commission abolished, construction halted. It did not resume again until 1803, under the direction of President Thomas Jefferson and architect-engineer Benjamin Latrobe. Unlike the first public building campaign, the second one was funded by annual congressional appropriations.

The first public building campaign was widely viewed as a failure. More than a million dollars had been expended and the needed buildings not finished. The failure of the building campaign was largely attributed to the gentleman amateurs who made up the first and second commissions. A general consensus developed that professional architect-engineer direction was desperately needed to complete what the first public building campaign had left uncompleted.

PROJECT PLANNING AND CONSTRUCTION EFFORTS

Everything yet remains to be done for establishing a regular mode of proceeding — [there is] no adequate means of supply provided — no materials engaged proportional to the work to be effected . . . the neiborhood of the city offering no kind of resources at least none to be depended on.

PIERRE L'ENFANT
to President George Washington, January 17, 1792

Because the site for construction of the new federal city was located at a great distance from established population centers, there were no existing sources of building materials nearby. Sawmills and stone quarries, commonly used throughout New England and the mid-Atlantic states, were nonexistent in the Potomac region. So the commissioners had to create sources of supply for building materials and arrange for their delivery to the building sites: they had to develop stone quarries, build wharves, improve roads, construct bridges, and develop adequate supplies of timber, brick, and lime. Undertaking these infrastructure tasks required the commissioners to expend much money and time. The development of the new federal city was the largest construction program in the United States up to that time, and it required unprecedented amounts of building materials. Unfortunately, without a body of contractors to bear some of the administrative burden of coordinating the infrastructure development, the commissioners approached these tasks ineffectively; for example, they issued and supervised separate contracts for wharf logs and for the construction of the commissioners' wharf. Similarly, they executed separate contracts for brickmaking and firewood in building the Presidents' House. In some cases, such as with the quarry operations at Aquia and the lumbering operation in White Oak Swamp, the commissioners hired their own employees to do the work, which entailed additional expense to provide supervisory personnel, any necessary tools, and housing and provisions for the workmen.

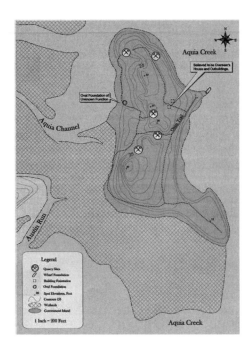

Government Island, Aquia Creek, Stafford County, Virginia. The island was purchased by Pierre L'Enfant for six thousand dollars to provide building stone for the Capitol and the President's House. The principal quarries used by the commissioners are in the middle of the island. Robert Stewart, a stone cutter from Baltimore, owned an acre-sized quarry on the island just south of "Oval Foundation." The island was surrounded by marshland, which made three handlings of the stone necessary: after it was shaped at the quarry, it was loaded onto scows, and finally onto the barges that took the stone up to Washington. Map by Keg Goode, GIS Unit, Stafford County, from field notes by Robert Kapsch, in Kapsch, *Aquia Quarry on Government Island*, 2.

PROVIDING BUILDING STONE

The first building material needed was stone for a firm foundation. In addition, the commissioners decided that the President's House and the Capitol should be built with an exterior stone appearance. The question of whether to use stone bearing walls versus stone-faced brick remained. The former was preferred for its appearance and the latter for its economy. The commissioners finally resolved on September 23, 1793, to construct the public buildings with brick walls and face them with stone.[1]

Luckily, there was no shortage of stone acceptable for foundations. Local landowner William Deakins owned a quarry at Rock Creek, which was under the general supervision of Joshua Greigg.[2] John Mason, probably the wealthiest man in the region, operated one or more quarries up the Potomac River from Georgetown, under the management of Nicholas Hingston. Rock from the quarry was taken to the river's edge with the use of oxen-hauled skids, also called stone boats, and there loaded onto scows and barges by block and tackle and transported downriver. Mason used this quarry in 1795 to supply foundation stone for the Capitol.[3] In 1795 the commissioners operated their own quarry within the federal city, at Hamburg, under the direction of John McNamara.[4] They contracted foundation stone from Phillip Fendall and Lewis Hipkins on October 21, 1791,[5] and again from William Smith on June 6, 1792, for an additional four hundred perch (9,900 cubic feet).[6]

The quarries closest to the new city, however, provided only foundation stone. The decision to face the public buildings with white or light-colored stone, making them in that sense like the great government buildings of the capitals of Europe, required freestone. This decision was made by the commissioners without regard to the cost implications or time commitments—it both greatly increased the cost of the public buildings and lengthened the time required to build them.[7] Freestone, usually sandstone or limestone, was so named because, compared to other types, it could be more easily worked in any direction and thereby cut into the rectilinear blocks necessary for a building's exterior.[8] But there was no white freestone in the immediate vicinity of the new city.

AQUIA QUARRIES AS A SOURCE OF FREESTONE

The commissioners located a source for freestone forty miles south of the new city at the quarries on Aquia Creek (now Stafford County, Virginia), seven miles from the Potomac. George Washington likely would have known of the Aquia quarry, for in 1774 he had used Aquia stone for steps and other elements in constructing his home at Mount Vernon. Aquia stone was also used in the Aquia Church, built in 1751, for tombstones and quoins. The earliest Aquia quarries were on the island referred to as Hissington or Wigginton[9]—later renamed Government Island. The earliest quarry on the island was operated by Robert Stewart, a stone cutter of Baltimore.[10]

The commissioners, through L'Enfant, both leased and purchased several tracts on or near Aquia Creek, enabling them to begin quarrying the freestone. The initial lease was made between L'Enfant, acting on behalf of the commissioners, and John Gibson, a merchant from Dumfries, Virginia. The terms were a yearly rent of £20 ($53) for a period of ten years, commencing on December 1, 1791.[11] Wigginton Island was purchased for six thousand dollars.[12] Other quarries in the vicinity were later rented and/or contracted with to supply the needed stone.

To work the newly acquired quarries on Government Island on Aquia Creek, the commissioners contracted with William Wright of Alexandria on April 10, 1792, to serve as quarry superintendent. His duties were to manage the quarry and supervise the workers.[13]

David Dale Owen, in his report of his 1847 visit, described how the two quarries on Government Island were worked:

FREE STONE QUARRY.

WILL be rented for the year 1800, on the premises, on Tuesday, the 3d day of December next, if fair, if not, the next fair day, a VALUABLE QUARRY, belonging to the estate of Charles Porter, deceased, lying on Aquia Creek, in Stafford county, Virginia.

DANIEL C. BRENT.
21st October, 1799.

Advertisement for quarry rental.
Centinel of Liberty and Georgetown (DC) Advertiser, October 29, 1799.

FREE STONE QUARRY.

TO BE RENTED for the year 1798. A valuable free-stone quarry of excellent quality, lying on Aquia Creek, belonging to the estate of Charles Porter deceased, and well known by the name of MILLARS QUARRY.

I shall attend on the premises on Thursday the second of November for the purpose of renting, in the mean time will shew them to, or receive proposals from any person disposed to rent

A good house, with some land, may be had with the quarry.

Letters by the post directed to me near Dumfries, Virginia, will come safe to hand; Bonds with approved security for the payment of the rent will be required.

Daniel C. Brent.

October 2, 1797.

Millars Quarry. *Washington Gazette*, October 14, 1797.

"Quarrying by Channeling and Wedging." This late-nineteenth-century illustration depicts the traditional method of quarrying sandstone. The overburden was first removed from the rock (*top*). Channels twenty to twenty-four inches wide were cut, usually with stone picks, to isolate the selected block of rock from the surrounding mass (*left*—behind the top-hatted man's right leg). Horizontal drill holes were cut to encourage rock cleavage along stratification (*bottom right*). Iron wedges were pounded into the drill holes along the line where the rock mass was to be cleaved from the surrounding rock (*center*— the two sitting workmen are cutting grooves in the rock for additional wedges). The wedges induced tension and separation. Once a block had been separated from the bed, it was cut into desired sizes and scabbled, or dressed. *Manufacturer and Builder* 23 (1891): 57.

The principal ledge hitherto worked here has a covering of only two or three feet of earth, and lies nearly horizontal, in a vast bed from six to eight or 10 feet in thickness, without the slightest apparent seam. For this reason, though the bed is so near the surface, it has been quarried at great expense, at least by the method hitherto employed, which is to groove it behind and on one side two feet wide, (or sufficiently to receive a man,) in a vertical direction, even to its base, and then "loft" it off in a horizontal direction by the introduction of wedges at the bottom of the bed in the direction of the stratification.[14]

In August of 1806, Latrobe visited Robertson's quarry,[15] one of the private quarries at Aquia Creek. He provided a similar description to Owen's of how Aquia freestone was quarried:

In working these quarries, the workmen having cut the face perpendicularly, first undermine the rock;—an easy operation, the substratum being loose sand.

If the block is intended to be 8 feet thick, they undermine it 5 feet, in a horizontal direction, in order that it may fall over when cut off. They then cut two perpendicular channels on each hand, 1 ft. 6 in. wide, at the distance from each other of the length of the block, having then removed the earth and rubbish from a ditch or channel along the top of the rock, they cut into the rock itself, a groove, and put in wedges along its whole length. These wedges are successively driven, the rock cracks very regularly from top to bottom, and it falls over, brought down partly by its own weight.[16]

Latrobe reported, "The largest blocks however which I have taken out, do not exceed in weight four tons."[17]

TRANSPORTING THE STONE TO
THE NEW FEDERAL CITY

Once quarried, the stone was shipped forty miles up the Potomac River to the new federal city. On May 1, 1792, Captain Elisha Williams, the principal overseer for the commissioners, delivered to the commissioners a letter from a prospective contractor, James Smith, proposing a method to transport the stone to the federal city from Aquia Creek. The commissioners approved the offer and invited him to attend their next meeting.[18] By 1795 he was using three schooners to transport the stone: the *Ark* (27 tons), the *Columbia* (39 tons) and the *Sincerity* (40 tons).[19]

Aquia stone, although suitable for building, was not perfect. Owen, in his 1847 report, summarized the problems with Aquia sandstone: "[Aquia sandstone has been] disfigured by conspicuous holes, pebbles, or stains, such as are almost universal in the blocks to be seen in the public buildings in Washington, where this material has been used. . . . Its color, occasionally approaching that of marble, is in its favor, though time and the weather change it for the worse."[20]

Cutting a new block at the quarries was a major investment in time and money, and it was difficult to recruit and keep adequate numbers of stone cutters. As a result, the quarries could not keep up with the demand for stone at the building sites. Shortage of building stone meant idled workers and work delays. The commissioners pressured the quarries to deliver more stone, and in response the quarrymen frequently shipped inferior or cosmetically repaired stone to Washington.[21]

Shortages of building stone from Aquia led the commissioners to propose a

Drilling Stone, by unidentified National Park Service artist, possibly Donald Demers. To drill the holes required for quarrying sandstone, a masonry drill or "jumper" is driven into the stone by sledgehammers, with the drill being turned one quarter after each blow. The worker holding the drill would have a tube, usually formed of leather, to occasionally blow the dust from the hole. Access to a blacksmith would be required to sharpen the drill. The drawing illustrates quarry operations at Seneca. National Park Service, *Chesapeake and Ohio Canal*, 68.

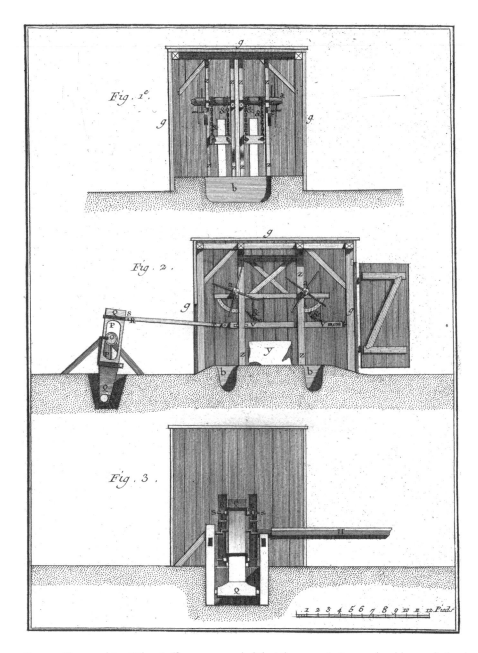

Stonecutting machine. When Jefferson responded that the commissioners should consult the drawings in Diderot's *Encyclopédie Méthodique*, it was this drawing to which he referred. Power is delivered to the stonecutting machine from the water wheel and gearing The rotating motion of the drive shaft is transformed into back-and-forth horizontal motion by what might be called the rocker device. The rocker device changed the circular motion into horizontal motion, which was transmitted to the stonecutting machine moving the cutting blade across the stone. The cutting blade was probably of soft iron and smooth. Actual cutting of the stone would have been by wet angular sand sprinkled under the smooth iron bar moving across the stone's surface. *Moulin à Scier les Pierres en Dalles,* by Benard Durex, from Denis Diderot, *Encyclopédie Méthodique* (Paris, 1751).

stonecutting machine as a technological solution to the problem. Stonecutting machines were used at that time in Europe and elsewhere in the United States, but there were none near the new federal city. Samuel Millikin of Philadelphia sent a proposal for building such a machine to the commissioners.[22]

The commissioners had previously heard of the possibility of such a machine from a Philadelphia newspaper and had asked Jefferson to look into it for them.[23] Jefferson referred them to Denis Diderot's *Encyclopédie Méthodique*,[24] for a drawing of such a machine (see illustration on page 56). Unfortunately, because the new federal city was in a backwater, the commissioners didn't have access to *Encyclopédie Méthodique*. They wrote back to Jefferson on December 5, 1792 asking his opinion on Milliken's proposed stonecutting machine.[25] On December 13 Jefferson replied that the machine in *Encyclopédie Méthodique* was superior to Milliken's.[26] This is the last mention of mechanizing the stonecutting operation. Since neither machine was built, stone shortages continued to plague and delay construction operations in the new federal city.

With the walls of the President's House and the north wing of the Capitol completed in 1796, the commissioner-supervised quarrying operations on Government Island were terminated and supplanted through contracts. Questions over whether to directly supervise the quarrying or to contract it out continued into the second public building campaign. Early in 1804, the architect of the Capitol, Benjamin Latrobe, reopened the quarry at Government Island, but he preferred contracting for the stone.[27] Latrobe felt that even if stone could be quarried at the public quarries and by men directly working for the public at lower cost than what would be needed to pay contractors, the risk involved did not justify the use of these quarries.[28]

There was a second source of freestone in the Potomac River valley that could have provided an alternative to Aquia sandstone—the Seneca quarries along the Potomac River west of the mouth at Seneca Creek, twenty-three miles above Washington. Stone from Seneca was one-half the cost of stone from the Aquia quarries. Further, whereas Aquia sandstone tended to deteriorate when exposed to weather and time, Seneca sandstone had the opposite characteristic: it was indurate, meaning that its durability and resistance to weathering increased with time. Seneca sandstone was soft to cut in the quarry and extremely durable once installed. David Dale Owen, who investigated the sandstones of the Potomac River in 1847 for possible use in constructing the new Smithsonian building, reported on this remarkable characteristic of Seneca sandstone.[29] In an 1872 congressional hearing, architect Adolph Cluss provided an anecdote on the durability of Seneca sandstone:

I used [Seneca stone] first in the reconstruction of the Smithsonian Institution after the fire and there was the remarkable fact connected with that experience, it was during the war, at the place where the Quarries are now located [i.e., Seneca, MD] was so far on the disputed ground that all work stopped and it was difficult to get stones from the Quarry. After a good deal of trouble I heard that a consid- erable quantity of stone from the original building had been left over; I looked for it and found that the Government had built temporary sheds over it. They were up in the first ward. After some negotiations we succeeded in having these sheds removed and we bought these stones at a low rate. After we had bought them I found that by exposure for fifteen years they had become so hard as to make it difficult to work them. There was considerable trouble between the sand stone cutters and the granite cutters; the person employed as Master Stone Cutter tried to have the work done by the Granite Cutters whose tools are Constructed for the purpose of working hard stone; but the sand stone cutters struck against that; they considered that it was their privilege to work this stone as it was sandstone. Finally in order to prevent trouble after considerable delay we preferred to go up and open the Quarries which we worked at our own expense and got the stone.[30]

Owen went on to recommend the use of Seneca sandstone in building the new Smithsonian structure. But it would not have been possible to use Seneca sand- stone in the first public building campaign, because in the 1790s the Potomac by- pass canal around Great Falls in the Potomac River had not yet been completed; it was not completed until 1802. Shipping costs would therefore have been exorbi- tant. After 1802 much construction still remained to be done in the new federal city, and once Seneca sandstone could be economically shipped by water, it came to be extensively used in the city, especially as pavers, for which a strong, durable stone was needed and its dark red color was not considered a detriment.[31]

BRICKS, MORTAR, AND LIME FOR THE NEW FEDERAL CITY

Besides building stone, bricks were essential for construction. On May 3, 1792, the commissioners entered into a contract with Anthony Hoke to produce brick near the site of the President's House.[32] Burning bricks required firewood, and on June 5, 1792, the commissioners made separate arrangements to provide the firewood.[33] Brickmaking continued both at the President's House and at the Capitol.

Mortar was needed for building with bricks, and mortar required lime. Securing an adequate supply of lime at reasonable expense was a problem for the commis- sioners. On October 22, 1791, they directed their secretary-clerk, Thomas Munroe, to advertise for four thousand bushels of lime.[34] Although lime was readily available in Philadelphia,[35] shipping would greatly increase the cost. The commissioners were able to find an inexpensive supply of lime nearby, in Frederick, Maryland. On Octo- ber 22, 1791, they wrote to Thomas Jefferson that "the Eastern lime was exorbitant and expensive and that the Frederick lime was one fifth superior."[36]

Despite this optimistic report, and even though there was a nearby supply, trans- porting it was still a problem. Just as for Seneca sandstone, obtaining adequate supplies at a reasonable cost depended on inexpensive transportation of it, and that depended on the opening of the Great Falls bypass canal of the Potomac Company (not completed until 1802). Without the Potomac Canal, Frederick lime, also called stone lime, was expensive owing to the expense of shipping over land. Alternative supplies were investigated, including the use of shell lime, one-third less expensive than stone lime but not as satisfactory.[37] When the commissioners again advertised for lime, this time stone lime, they had to offer more money for it.[38]

STRUCTURAL TIMBER FOR THE PUBLIC BUILDINGS

Structural timber was required for joists, framing, and trusses. Trees were, of course, available extensively throughout the new federal city and the surrounding countryside, but this was second-growth timber, not suitable for the large structural members needed for the public buildings. White oak, considered the strongest of all wood, was needed for structural work. A large supply of white oak was located in White Oak Swamp in Virginia, below Aquia Creek in what is now King George County, northeast of Richmond. On January 1, 1793, Hoban went to examine it and reported favorably on it to the commissioners. Early that year the commissioners made arrangements to fell, cut, square, score, and transport that timber using their own workers.[39] But soon, as was the case with the quarrying and brickmaking operations, the commissioners decided to utilize a contractor for the necessary felling, squaring, scoring, and transporting instead of directly supervising the work. In December 1793, James Hoban presented to the commissioners a contract for providing timber, to be entered into with Henry Lee.[40]

PROVIDING SKILLED WORKERS FOR CONSTRUCTION

By 1793 the problem of securing adequate building supplies for the first public building campaign had been largely resolved, primarily through the use of contractors. Building began in earnest, particularly on the President's House and the north wing of the Capitol. Whereas the workers who provided the basic building materials were largely unskilled, the workers at the building sites needed to be skilled in carpentry, stonecutting, stone carving, and other types of work. However, the new city was in a semipopulated backwater, and skilled workers were not available. The commissioners would have to recruit, transport, house, and retain workers from elsewhere. In some cases, they could be recruited only from Europe.

The skilled workers that the commissioners had the most difficulty in recruiting were those that received the highest wages. Each construction site had a social hierarchy, determined by salary, among the workers. Because payrolls were open (they had to be signed by each worker, acknowledging receipt of public funds), the salaries of all, and hence also their status, were known by all.

TABLE 3.1. Pay scales for various work groups in the new federal city as of 1798

The Commissioners	Gustavus Scott William Thornton Alexander White	$400/Quarter
Other Commission Officials (including architects and architect-engineers)	James Hoban & George Hadfield, architects Thomas Munroe, clerk William Brent, assistant clerk Elisha Williams, principal overseer	$350/Quarter $312.50/Quarter $200/Quarter $212.50/Quarter
Skilled workers: Foremen	Henry Edwards, foreman of stone carvers (Capitol) James Dixon, foreman of stone carvers (Pres. House) Pierce Purcell, foreman of carpenters (Pres. House) Barnard Crook, foreman of stone cutters (Pres. House) Joseph Middleton, foreman of cabinet makers (Pres. House) Lewis Clepham, foreman of painters (Pres. House) Robert Brown, foreman of bricklayers (Pres. House)	25s/day ($3.33) 17s6p/day ($2.33) 17s6p/day ($2.33) 15s/day ($2.00) 15s/day ($2.00) 15s/day ($2.00) 12s6p/day ($1.67)
Skilled workers: Senior Journeymen	Stone cutters: Alexander Wilson; Andrew Shields; Hugh Sommervelle; Alexander Reid. Carpenters: Joseph Hoban	12s6p/day ($1.67) 12s6p/day ($1.67)
Skilled workers: Journeymen	Stone carvers Stone cutters Carpenters and Joiners Bricklayers Cabinet makers Painters	12s3p/day ($1.63) 10s/day ($1.34) 9s/day ($1.20) 11s/day ($1.47) 8s4p/day ($1.11) 10s/day ($1.34)
Overseer of Laborers	Bennett Mudd, overseer	4s4p/day ($0.57) Excludes provisions
Laborers	Various	2s4p/day ($0.31) Excludes provisions

Source: Miscellaneous Treasury Records, RG 217, NARA, 1798 Payrolls, boxes 4–5.

The list in table 3.1 shows the hierarchy of workers on the building sites. Of the four social groups, the gentleman class, including the commissioners and their officials, held the highest positions in this hierarchy. Next came the architect-engineers, not quite considered gentlemen, and next the skilled workers. The laborers and slaves (the last category in the table) occupied the lowest rung. Commissioners received almost five dollars per day reimbursement, more than fifteen times what a common laborer was paid (excluding cost of provisions). The list also indicates the basic division that characterized the building trades: the clear distinction between the gentleman class, paid by the quarter, and the artisans and laborers, paid by the day. The pay rates reveal the great divide between the skilled and the unskilled laborers.

These pay rates, as of July 1798, marked the high point of wages for the skilled workers. The various trades were usually strongly organized. As the first public building campaign progressed, there were consistent efforts by the commissioners to reduce the wages of these skilled workers, efforts that were strongly resisted.

The wage rates shown also reflect a hierarchy based on skill. The stone carvers occupied the highest rung of that hierarchy, and foremen such as Henry Edwards and James Dixon were clearly at its pinnacle. The bricklayers fell surprisingly low on the ladder. Each building trade had its own hierarchy, with the foreman at the top, followed by senior journeymen, journeymen, indentured servants, apprentices, and, in the case of the carpenters, slaves at the bottom.

The building trades were internally organized by the master-journeyman-apprentice triad. Many of the building trades had formal trade organizations, which excluded those in the laborer category. Whereas the distinction between master and journeyman was not great (a master could and did work as a journeyman and a journeyman could aspire to and did become a master), the distinction between journeyman and apprentice was great. No man could become a journeyman in the trade without going through an appropriate apprenticeship. Because they were not yet fully trained in their trade, apprentices were not considered valuable. And by requiring an apprenticeship as a prerequisite to working in the trade, the trade could restrict the number of men entering the trade and therefore the supply of qualified men. In this way skilled artisans of the various building trades could maintain their high wages.

Although data on the number of workers engaged at the construction sites is not available for the early years of construction (1793–1794), reliable data exists for the years 1795–1800, subdivided by trade. For example, table 3.2 indicates the size of the workforce at the President's House for the years 1795–1800. To these numbers would have to be added workers who delivered building materials, supplied provisions, and performed other services. It was a large and experienced workforce, particularly in the stone-carving, stonecutting, and bricklaying trades. Because the future federal city did not have a native population of skilled workers, the commis-

TABLE 3.2. Estimate of workforce at President's House, 1795–1800

TRADE	1795	1796	1797	1798	1799	1800
Stone carvers	5	3	5	0	0	0
Stone cutters	19	12	13	14	0	2
Bricklayers	4	5	3	1	1	0
Carpenters	15	14	22	15	9	34
Cabinet makers	0	0	0	4	4	0
Painters	0	0	0	3	3	3
Sawyers	8*	8*	8*	2*	2*	2*
Laborers	24	7	2	2	2	2
Plasterers	0	0	0	0	0	6
Total	75	49	53	41	21	49

Source: U.S. General Accounting Office, General Accounting Office Accounts of the Commissioners of the City of Washington, 1794–1802, RG 217, NARA, seven boxes.

The workforce at the President's House peaked in 1795, at approximately 75 workers. It fell to a low of 21 workers in 1799 as workers were reassigned to the Capitol so as to complete the Capitol (north wing) in time for the move of the federal government to Washington in 1800. The workforce at the Capitol was somewhat larger. In addition to these men, there was an unknown number of workers not employed at the construction site but at the Aquia quarries quarrying stone, at White Oak Swamp felling and cutting timber, and in similar occupations.

*Estimate.

sioners had to bring them from elsewhere. The largest numbers of skilled workers were in the largest cities, especially Philadelphia and New York, but both Washington and Jefferson had an inherent distrust of workers from Philadelphia or New York, owing to the competition of those cities with the new federal city to be designated the national capital.[41] For certain building trades, such as stone carving and stonecutting, there were very few skilled workers anywhere in the United States. For these reasons Washington and Jefferson encouraged the commissioners to look overseas for experienced labor.

RECRUITING SKILLED WORKERS FROM EUROPE
On March 6, 1792, Jefferson suggested to the commissioners the idea of importing European workers, particularly Scottish and German workers.[42] Two days later Washington also wrote to the commissioners about the need to import Scottish and German workers.[43] This was a one-two punch. On March 14 the commissioners adopted Washington's and Jefferson's idea and asked for additional suggestions

on how such a plan could be implemented.[44] Jefferson advised that Hermen-HendDamen, an Amsterdam merchant, could produce as many workers as the commissioners desired and deliver them to Richmond at a cost of $28 each.[45] He followed this suggestion with other ideas of how to recruit workers.[46]

On June 2, 1792, the commissioners again wrote to Jefferson, for his assistance in procuring more than one hundred needed workers, especially stone cutters, masons, and bricklayers.[47] At this time the commissioners were still thinking about importing from Europe unskilled laborers as well as skilled workers, an indication that they had not yet made up their collective mind on the use of slaves as laborers on the public buildings. This uncertainty continued into the next month, when the advertisement for immigrant workers authorized by the commissioners on July 5, 1792 included laborers.[48] The terms offered by the commissioners were generous. They proposed a daily rate of $2 per day (15 shillings) for the stone cutters and masons and $1.60 per day (12 shillings) for laborers, rates quite a bit in advance of the rates prevailing in 1798 that are shown in table 3.1. For these wages the workers would have to work twenty-six days a month—they got Sundays off.[49]

The commissioners took other actions to recruit European workers. On August 3, 1792, they wrote to shipmaster William Prout, asking him to inform all Irish immigrants coming in his vessels that there was a demand for laborers in the new federal city.[50] The commissioners also acted to recruit native talent. On September 4, 1792, they placed an advertisement for brick masons in the Georgetown and Baltimore papers.[51] The commissioners also used their own workmen to recruit other workmen. John Hunter, one of their hands, was going to travel to Philadelphia to get his wife and family and then on to New York. Although Washington and Jefferson had concerns about recruiting workers from the other cities attempting to secure the federal city, Hunter was authorized to recruit, while in Philadelphia and New York, up to twenty masons. For each mason that could be brought to the new federal city, Hunter would receive one day's wages.[52] In addition, the commissioners tried to procure fifty stone masons from Britain and France;[53] but their efforts to recruit French workers resulted in failure, owing to the war restrictions on the emigration of French workers.[54]

By January 1793, the commissioners were beginning to feel confident about their ability to recruit an adequate workforce for the coming construction season. In response to Jefferson's offer at the end of 1792 to recruit workmen in Connecticut,[55] the commissioners responded that they had an adequate supply of carpenters and didn't require many masons at this time.[56] They again expressed this optimism to Jefferson in a letter dated April 11, 1792, in which they informed Jefferson, "We might probably have 2000 mechanics and labourers here on very short notice."[57]

One can see this new optimism about their ability to attract workers in the tighter conditions mentioned in the commissioners' advertisement for labor authorized on January 3, 1793. This advertisement differs from the one issued on July 5, 1792, in that the prices listed for stone cutters and masons are much lower than in the earlier one. The commissioners no longer guaranteed a specific wage but indicated that the prevailing wage would be paid, and they made allowances for workmen to bring their wives with them. Finally, the advertisement of January 3, 1793 makes no mention of recruitment of laborers.[58] By this time, the commissioners had resolved that slaves would make up at least a part of their laborer workforce—the workforce would consist of part slaves, part whites, and part free blacks.[59]

The confidence that the commissioners had in securing the necessary workmen precluded the adoption of an innovative plan to recruit German workers, proposed to the commissioners in 1793 by Charles Firrer, of Dumfries, Virginia, himself an emigrant from Germany. Firrer's plan, put before the commissioners on February 26, 1793, had the following four elements: for each worker the commissioners would pay passage, provide a suit of clothes annually, provide rations, and, at the end of two years, provide a lot in the federal city. This plan played to the commissioners' strengths, utilizing the lots in the federal city in lieu of cash payment.[60]

Firrer's proposed daily wage was $1.15 (8 shillings 7 pence in Maryland currency) (table 3.3). As compared with the salaries listed in table 3.1, this daily wage was less, and in some cases considerably less, than what the commissioners paid their skilled workers. Firrer's overall plan would have cost $24,670, as compared to previous proposed salary outlays of $60,000. Firrer estimated that rations would cost £28 per person over two years, £14 ($37) per year or about ten cents a day. His plan illustrates an enlightened view of reimbursement for potential worker-immigrants and indicates something of the living conditions of skilled workers at that time. The commissioners did not accept the plan, but they did, in some cases, adopt Firrer's idea to make payment to workers in land rather than wages, such as was done with James R. Dermot.[61]

FIRST NEED: LABORERS FOR EXCAVATION

Because construction could not begin until sites for the buildings' foundations were excavated, the first need in 1793 was for excavation laborers. For this task the commissioners relied on a mix of slave and free white labor. As they explained to Jefferson, this practice helped keep the white workers in check, as they knew they could be replaced by slaves.[62] The commissioners typically paid slave owners sixty dollars per year for slave labor, and the owners supplied the slaves with clothing and blankets.[63]

Serious excavation began in 1793. Payrolls for 1793 and 1794 are not extant, but the payrolls from 1795 indicate that a large number of laborers continued to work at the site in that year, using wheelbarrows and shovels under the direction of overseer Bennett Mudd, to complete the excavations for the foundations. Virtually all of these laborers were illiterate, as indicated by their marks on the monthly payrolls.[64]

These laborers were at the lowest rung in the social structure of the building

TABLE 3.3. Cost estimate of Charles Firrer's plan to import 100 workers from Holland

	MARYLAND CURRENCY (£)
Transport from Holland @ 6 guineas per head	1050..0..0
Clothing in Two Years @ £10 per head	2000..0..0
Rations @ £28 in Two Years	5600..0..0
	8650..0..0
Extra expences in procuring the hands	600..0..0
Total expence for 100 Mechanics in Two Years	9250..0..0
100 Mechanics was at 1 dollar 15 per day, and calculating 300 days, Their wages	22500..0..0
Deduct expence of plan	9250..0..0
Balance in favor	13250..0..0

Source: Charles Firrer to the commissioners, February 26, 1793, in RDCC, Letters Received, roll 9, 214.

site. They received the lowest wages, lived in the temporary barracks provided for them on the public grounds near their workplace, and when sick, went to the public hospital provided by the commissioners. When they died, they were buried in the public cemetery in simple coffins provided by the commissioners. Without an apprenticeship in one of the building trades and without the ability to read or write, they could never become more than common laborers.

THE STONEMASONS
At the other end of the social hierarchy on the building site were the stone carvers, who received the highest wages. Down slightly from the stone carvers in status were the stone cutters, the men who finished the stone delivered from Aquia quarries. Stonecutters were divided into two groups, the stone setters, who worked on the walls and who received a higher wage, and the stone cutters proper, who worked in the stonecutting shed and on the grounds.[65] It was the stone masons who demonstrated the most independence from the commissioners and who later led the frequent work stoppages, slowdowns, and pay raise demands in the first public building campaign.

THE BRICKLAYERS
The bricklayers were fewer in number. The payrolls between 1795 and 1798 indicate that Robert Brown was their foreman. James Maitland and David Tweedy worked during most of this period. William Whitehead and John Lipscomb worked during

1795–1796.[66] Because there were fewer of them and they were easier to replace, they did not exert as much influence on the commissioners as did the stone cutters and stone carvers.

OTHER BUILDING TRADES

Toward the end of construction, the commissioners employed smaller numbers of workmen from other building trades. They included carpenters, cabinetmakers, plasterers, glaziers, roofers, and others. The President's House was painted by Lewis Clepham, foreman, with assistance from journeymen John D. Lourey and Thomas Monney in 1798. Joseph Middleton, foreman of the cabinetmakers, William Middleton, presumably his brother, and apprentices Thomas Brown and Charles Boon(e) constructed and installed doors and similar wooden furnishings in 1798.[67]

REPLACING THE ORIGINAL COMMISSIONERS

By the beginning of 1793, great progress was especially necessary and, finally, could perhaps be made. But the original three commissioners, Johnson, Carroll, and Stuart, were dissatisfied with the extensive criticism that they had received, and so they wrote to Washington of their unhappiness. Washington asked Jefferson what could be done to appease them.

Jefferson detailed the approach that was eventually followed for the remainder of the first public building campaign. New, younger, commissioners were appointed and were compensated for their services.[68] The commissioners eventually received sixteen hundred dollars per year. Daniel Carroll was the last of the original three gentleman commissioners to resign. He did so in May 1795, owing to advanced age and ill health. He died one year later, at age sixty-three. By 1795 there were three new commissioners: Gustavus Scott, William Thornton, and Alexander White. These men represented a younger generation. They oversaw most of the remainder of the first public building campaign. Although younger, they were no more experienced in or knowledgeable about the construction industry than the original three commissioners. Gustavus Scott was a lawyer from Somerset County, Maryland, who represented that county in the Maryland House of Delegates. He was appointed as a commissioner by Washington in 1794, at age forty-one. William Thornton was appointed as a commissioner also in 1794, at age thirty-five. Thornton was a self-described architect who had won the design competition for the Capitol. Trained as a physician, he possessed no experience or training as an architect or a builder. His design for the Capitol had serious flaws. Yet he continued to maintain that he was an architect of the first order, and this insistence became a factor in the future of the public buildings' construction. The last of the new commissioners was Alexander White, from Frederick County, Virginia. A lawyer and ex-congressman, White was appointed as a commissioner in 1795, at age fifty-seven. Problems experienced under the original three commissioners continued and expanded under the next three.

The new commissioners were determined to place the federal construction program on a more businesslike basis. On February 18, 1795, they adopted a series of rules that became the basis of their work for the next five years. These rules established a paper system that would track the payments made for goods and services.[69] The commissioners also decided that all workmen were to be paid a daily salary, thus avoiding the problems of pay "by the piece."

In one aspect, the new rules were quite commendable. They clearly sought to establish accountability. The difficulty with the rules was that they imposed a paper system on top of the building process rather than dealing with the building process itself. They were the type of rules one would expect from lawyers and accountants and not from men involved with architecture and engineering. One shortcoming of the new commissioners was that they rarely left their comfortable quarters in Georgetown to inspect the work going on at the President's House and the Capitol. Substandard work was not detected promptly and thus had to be dealt with at a later time and at great expense.

It was not until 1795 that the commissioners were able to establish the adequate supply of the needed stone, timber, bricks, and other materials for the new city. And it took that long to recruit and transport the needed skilled workers. Beginning in 1795, real construction progress began. But it was not enough to fully complete the federal buildings by November 1800, when the government was scheduled to move to the new federal city. Problems under the first three commissioners intensified under the second three. Labor unrest, work stoppages, cost overruns, and construction delays became endemic.

4

Developing a Commercial Center
Harbor Navigation and River Improvements

The main channel of the Potomack opposite the city, running near the Virginia shore, that part of the city which lays upon the Potomack has only a small channel, carrying from eight to twelve feet of water, until you come within about three quarters of a mile of George Town, when the channel turning between Mason's-Island and the city, gives a depth of water from twenty to thirty feet close in with the shore of the city. This renders the water-lots within that small space very valuable; for any ships that come up the river may here lay within twenty yards of the city, and the boats which bring the produce of the country down the river, may at all times come here deep loaded as they come down; whereas they could not go, thus loaded, down to the eastern branch, unless in very smooth weather.

TOBIAS LEAR, 1793

Washington, Jefferson, and other supporters of the new federal city on the Potomac foresaw its becoming a great commercial center as well as a governmental one. The principal export of the mid-Atlantic states was flour, a commodity that could be shipped in barrels, generally would not spoil, and was much in demand on the sugar islands of the Caribbean, where the much more lucrative sugar production preempted use of the land to grow wheat. To become a great commercial center, the city would have to become a major exporter of flour, rivaling Baltimore, Philadelphia, and New York. With the Treaty of Paris (1783) ending the Revolutionary War, this became even more important, since the treaty greatly expanded the territory of the original thirteen states through the addition of the Northwest Territory and other lands. The Potomac River was the shortest and most convenient route to these areas. It was thought that the wealth of the new territories would flow down the Potomac and through Georgetown and the new city.

At the time of settlement, Georgetown was a deepwater port capable of receiving oceangoing vessels. In 1755 General Edward Braddock, it was thought, had disembarked his troops at what came to be known as Braddock's Rock, just below Georgetown at the present-day Washington, DC, end of Theodore Roosevelt Bridge.[1] Fifty years later Thomas Moore, a self-taught engineer, predicted a bright future for Georgetown owing to its deepwater port: "George Town has loaded about fifty ships and other

View of the Suburbs of the City of Washington, by George Isham Parkyns, ca. 1795. View of the port of Georgetown looking upriver, with Three Sisters Islands in the distance (*center*). Here, at Georgetown, the Potomac River changed from a piedmont river 900 feet wide to a tidal river 5,000 to 6,000 feet wide. It was here, it was thought, that the agricultural wealth of the twenty-thousand-square-mile drainage area of the Potomac River and beyond could be transshipped to vessels bound for Europe, the Caribbean, and elsewhere. Flour was foremost among these agricultural products in the late eighteenth and early nineteenth centuries. At the time of this drawing, the Potomac River was rapidly silting up, preventing larger vessels from approaching Georgetown. Courtesy of the Library of Congress.

large vessels a year. . . . They have a right to calculate upon being visited in future by a much greater number of vessels than at any former period . . . and a proportional number of coasters, packets, and other small craft."[2]

TWO CONDITIONS TO BECOME A COMMERCIAL CENTER

Fulfillment of the expectation that Georgetown would grow into a major export center depended on two conditions: transport to Georgetown from the farms and mills of the hinterland must be developed, and access to the sea had to be assured.

Since 1785 George Washington had worked to develop the Potomac River as the main passage to the nearby valley and the region beyond, through the development of the Potomac Canal. The second condition, access to the sea, was being threatened at the time when the new federal city was being developed. Large amounts of silt were being carried down the Potomac River and deposited in sand bars below Georgetown, preventing seagoing ships from loading at Georgetown's wharves.

At the time of the passage of the Residence Act (1790), flour had replaced tobacco as the principal export of the mid-Atlantic states. Flour was initially shipped to the Caribbean Islands, but after the end of the Napoleonic Wars, increasingly it was shipped to Europe. On September 17, 1790, Thomas Jefferson met with James Madison and George Mason to discuss the development of a major port adjacent to the new federal city. Georgetown was identified as the most likely candidate to become that port. Jefferson reported to Washington:

> He [George Mason] mentioned shortly in its [Georgetown's] favor these circumstances: 1. Its being at the junction of the upper and lower navigation where the commodities must be transferred into other vessels . . . 2. The depth of water which would admit any vessels that could come to Alexandria. 3. Narrowness of the river and consequent safeness of the harbour. 4. Its being clear of ice as early at least as the canal and river above would be clear. 5. Its neighborhood to the Eastern Branch whither any vessels might conveniently withdraw which should be detained through the winter. 6. Its defensibility, as derived from the high and commanding hills around it. 7. Its actual possession of the commerce and the start it already has.[3]

Mason had described Georgetown as being at "the junction of the upper and lower navigation where the commodities must be transferred into other vessels." By "upper navigation," Mason was referring to the work of the Potomac Canal Company, a company founded in 1785 to open the Potomac River for navigation through bypass canals and in-river navigation. Cargoes from boats on the upper navigation would be transferred to oceangoing boats on the lower navigation at Georgetown.

Both upper and lower navigations were troubled. At the time of Jefferson's meeting with Mason and Madison, the opening of the upper Potomac navigation had been delayed but was expected shortly.[4] At the same time, Georgetown's access to oceangoing ships in the lower navigation was being restricted by rapidly developing siltation.

Besides siltation, other environmental changes took place in the Potomac River at Georgetown. Six years before the Jefferson-Mason-Madison meeting, in March 1784, a major ice freeze had struck the region.[5] This monumental freeze and the subsequent flood tore open a new channel between Mason's Island (present-day Theodore Roosevelt Island) and the Virginia shore. George Washington Parke Custis described the freeze and the subsequent ice flood of 1784:

> The ice, in the memorable year of 1784, moved twice: It first descended in vast quantities from the upper Potomac, till it reached the Three Sisters, where it stopped and accumulated in great masses and froze together again; then came the deep snow, followed by a general thaw and violent rains. The second movement carried all before it. The shoving off of a strongly built stone house or stable from the bank, of where now is your town, is a well remembered story of the past. Both branches of the river around Analostan Island were open in those days; the eastern branch being used as the ship channel. The freshet of 1784 tore open the western branch, and formed in the one freshet, a channel way that would have admitted the passage of an Indiaman to Georgetown, being from twenty-seven to thirty-feet depth up to the wharves of the town.[6]

The 1784 ice flood gouged out a new deepwater shipping channel on the west side of Analostan or Mason's Island opposite Georgetown. This new channel became known as the Virginia Channel.[7] In his 1883 report on navigation on the Potomac River to Georgetown, Major Peter Hains quoted a Mr. Lewis of Virginia on the impact of the ice flood of 1784 on the port of Georgetown:

> That before the year 1784, the channel on the western side was so shallow that vessels only of very ordinary draught could pass, while on the Maryland side vessels of great draught of water could easily pass up to Georgetown.
> The uncommon hard winter of 1783–'84 was followed in the spring by the greatest torrents ever known in the Potomac. The bodies of ice were of immense magnitude, and many of them lodged upon the island and under the rocks of its bed. Pressing with a force beyond all credulity it tore the rocks asunder and pressed them over into the river channel, occasioning a rise of 30 or 40 feet on the Georgetown shore.[8]

The 1784 ice flood exacerbated other changes that were taking place, including the increasing siltation of the Potomac River below Georgetown. As the Potomac valley became increasingly settled, more farmers cleared additional land by

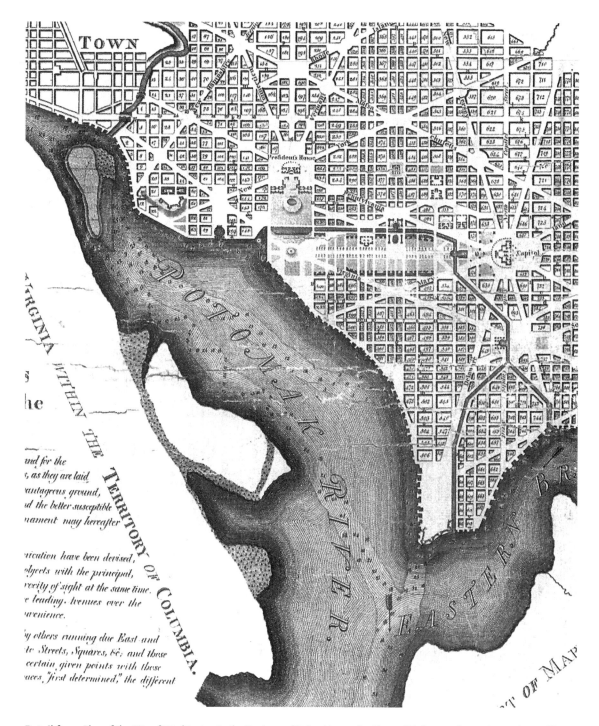

Detail from *Plan of the City of Washington in the Territory of Columbia . . .*, by Pierre L'Enfant, redrawn by Andrew Ellicott. Engraving by Thackara and Valance, Philadelphia, 1792. To address the problem of maintaining a shipping channel up the Potomac River to the port of Georgetown (*top left*), the engineers first needed hydrographic information on the existing navigable channels. Andrew Ellicott was the first to prepare a plan of the three existing channels in the Potomac at Washington using soundings taken by Ignatius Fenwick and Richard Johns: the Virginia Channel (along the Virginia shore), the Georgetown Channel (from Georgetown down the middle of the river), and the Washington Channel (along the Washington shore). There was an additional channel in the Eastern Branch (Anacostia River) adjacent to the north shore and extending up to the Navy Yard. Note the extensive number of hypothetical piers drawn by Ellicott on both the Potomac and the Eastern Branch. Obviously he thought the new federal city had a bright mercantile future. Courtesy of the Library of Congress.

girdling standing timber — killing trees by cutting a circle around their trunks — thereby encouraging soil erosion, runoff, and sedimentation in the Potomac River. The river could carry a great deal of sediment as long as the gradient was high and the velocity of the river fast. But at Georgetown the width of the Potomac increased, the gradient decreased, and the river slowed and lost much of its sediment-carrying capacity; as a result large amounts of sediment were deposited at or just below Georgetown in the form of sand bars. Georgetown and Mason's Island were located at the upper limit of this tidal action and caused further siltation. The depth of navigable waters in the Georgetown Channel below Georgetown decreased,[9] partly because the river width increased from a quarter of a mile at Georgetown to one or one and a half miles wide below the town and partly because of the action of tidewater.[10] The result of the ice flood of 1784 was to reduce Georgetown's ability to function as a deepwater port:

> On the Virginia side the torrent also forced itself and deepened that channel, while it left a vast quantity of mud, rocks and sand in the eastern [Georgetown] channel, which has been constantly accumulating since that period.
>
> The situation of the present bar is at the union of the two arms of the river below the island, and does not permit passage over it of vessels, drawing more than 12 feet water.[11]

IMPROVING NAVIGATION ON THE POTOMAC RIVER

Efforts were called for to improve the Potomac River for navigation, but action was hampered by a lack of knowledge of existing conditions in the river. The first hydrographic map of the Potomac River was not produced until 1792, by Andrew Ellicott.[12]

Ellicott's hydrographic map of the Potomac revealed three navigable channels in front of Georgetown. Hains described these three channels:

> The *first*, commenced at the head of Analostan Island, near the site of the Virginia abutments of the Aqueduct Bridge, passed between Analostan Island and the Virginia shore, followed along the latter as far as the present site of Long Bridge, thence, by a gentle curve, convex towards the Washington side, it pursued its course until it met the waters of the Eastern Branch near Giesborough Point.[13]
>
> The *second*, or middle channel, called the "Swash," was a continuation of the Georgetown channel which swept along the front of the Georgetown wharves, and after passing Easby's Point,[14] curved gently towards the Washington shore and passed through the site of Long Bridge at a place nearly in the middle of the present causeway, uniting with the Washington Channel a short distance below.
>
> The *third* channel, called the Washington or City Channel, was a branch from the Georgetown Channel, which, after reaching Easby's Point, separated into two

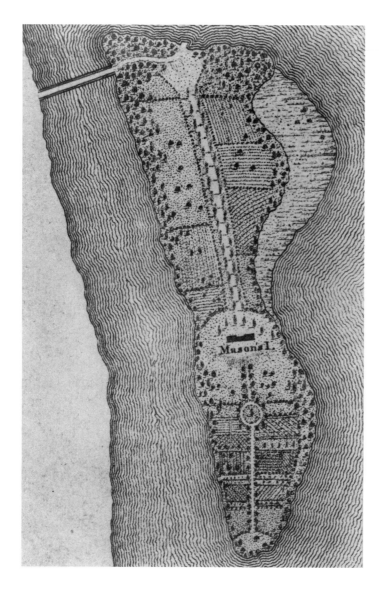

Mason's Island, also called Analostan Island (present-day Theodore Roosevelt Island) in the Potomac River. Detail from "A Map of the City of Washington in the District of Columbia . . .," 1818, Robert King, delineator. Shown on the island is the summer residence of John Mason, the eighth surviving child of George and Ann Eilbeck Mason of Gunston Hall. After the severe ice and flood of March 1784 opened a deepwater channel to the west of this island (*left*, the Virginia Channel), the citizens of Georgetown became aware that their channel to the east of the island (*right*, the Georgetown Channel) was becoming silted up. In 1805 they received permission from Congress to build a dam or mole across the top of the Virginia Channel (*top left*) to divert water to the Georgetown Channel to scour it. This dam was built in 1806 by Thomas Moore. Courtesy of the Library of Congress.

parts—one the Swash, just described: the other, following close along the Washington shore, united with the Swash a short distance below the site of Long Bridge.[15]

The increasing sedimentation limited shipping in these channels. The Corporation of Georgetown tried to dredge the river, but as Latrobe observed, "it did not succeed, the alluvion being again deposited as fast as it was removed. That work has therefore been abandoned."[16] The citizens of Georgetown tried another approach: the construction of a causeway or mole at the upper end of the Virginia channel to close that channel and throw additional water into the Georgetown channel. This solution was thought to have a number of advantages. It would close the Virginia channel to shipping, thereby encouraging ocean-bound ships to use the channel closer to Georgetown, and it would encourage downriver craft to off-load at Georgetown rather than at Georgetown's rival down the river, Alexandria. It would also increase the quantity of water flowing down the Georgetown channel and, it was thought, decrease sedimentation in that channel and perhaps scour away existing sediment. Finally, it was thought that another ice flood flow similar to the 1784 event would further gouge out the Georgetown channel. On January 19, 1805, Congress approved the causeway's construction.[17]

Following the passage of this act, the citizens of Georgetown turned to self-educated engineer Thomas Moore to build the causeway. Moore, who had studied the behavior of rivers and streams, applied his findings with some success to the works of the Potomac Canal Company. In 1806 he constructed the causeway from the Virginia shore to Mason's Island, thus blocking the Virginia shipping channel and throwing more water into the Georgetown channel. Although this increased river flow resulted in more scouring in the Georgetown channel, it also caused more sedimentation immediately below Mason's Island.

DETERIORATION OF THE GEORGETOWN SHIPPING CHANNEL

As time progressed, the Georgetown channel became shallower. In 1841 George Washington Parke Custis discussed what had happened between 1791 and 1841: "The river at Georgetown is but one-half as deep, being thirty feet from shore to shore fifty years ago [i.e., 1791], is now not over fifteen feet."[18]

Sediment was not the only problem. Ships were also throwing their ballast overboard at Georgetown, further reducing the depth of the harbor. The commissioners prohibited this practice in the harbors of Georgetown and the Eastern Branch.[19]

In 1883 Major Hains studied the long-term changes in the shipping channels in the Potomac by comparing Ellicott's river survey of 1792 to Lieutenant Colonel James Kearney's survey of 1834 (there had been no riverbed topological surveys in the intervening forty-two years). He described how the Virginia channel between Mason's Island and the Virginia shore had ceased to exist, primarily because of the construction of the causeway built by Moore. At the downriver end of Mason's Island, where the Potomac became wider and met tidewater, large shoals developed. One shoal extended halfway from the southeastern end of the island to the Long Bridge. This shoal was dry at extreme low tide. Another shoal had developed west of the Long Bridge, more than halfway to Mason's Island. Below the Long Bridge a third shoal had developed, extending eastward one-quarter of a mile. Shoaling had also closed the Washington channel that had extended along the Washington waterfront from Easby's Point downriver, cutting the Washington channel off from the Georgetown channel.[20]

At the beginning of the nineteenth century, the citizens of Georgetown turned to Thomas Moore for suggestions on what should be done. In 1806 Moore published his suggestions for improving navigation in the Potomac channel at Georgetown. He foresaw that construction of the causeway across the top of the Virginia channel would prove to be only a temporary expedient. Unless the commissioners adopted other river improvements, sediment would continue to fill the Georgetown channel.[21]

Specifically, Moore had in mind building wing-wall dams, a device he had successfully used for the Potomac Canal Company. He knew that wing-wall dams would increase the flow and velocity of the main channel, thereby increasing river flow and the river's sediment-carrying capacity. Moore described the wing walls as "artificial planters."[22]

In 1810 Moore contracted with laborers to erect wing-wall dams in the Potomac River opposite Georgetown. This action did not go unchallenged. The Washington Bridge Company, formed in 1808 to build what became known as the Long Bridge (at the approximate location of the present-day Fourteenth Street Bridge), feared that the action of the wing walls might undermine and carry away their bridge. On September 17, 1810, the bridge company sought and received a court injunction against the construction of the dams. Architect-engineer Benjamin Latrobe was summoned to give evidence at the resulting hearing, thus pitting the self-taught engineer Moore against the formally educated and experienced architect-engineer Latrobe. In his deposition, Latrobe stated that the faster current caused by the wing walls might undermine and partially carry away the Washington Bridge.[23] Based on this deposition, a permanent injunction against the construction of the wing walls was issued by the court.[24] The citizens of Georgetown then petitioned Congress to build the wing walls. The Senate passed a bill that provided for three appointed commissioners to sanction "works in the bed of the river." When the bill arrived in the House of Representatives, it was directed to the District of Columbia committee, which called upon Latrobe to again provide expert testimony. Latrobe answered the committee's questions in much the same manner as he had done before the court. The committee asked Latrobe two questions:

1. Will the wing dams proposed to be constructed in the river Potomac by Mr. Moore, as laid down upon the map filed in the office, have the effect of deepening the channel called the middle channel, from the foot of Mason's Island through the bar or shoal, so as to produce an uninterrupted ship channel seaward from Georgetown?
2. If these wing dams produce the proposed effect, will any effect injurious to the Washington bridge, or to the harbors of the city on the Potomac, or in the Eastern Branch, be also produced?[25]

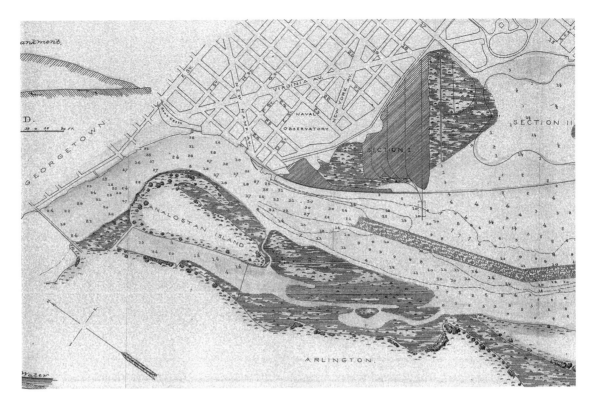

Georgetown Harbor, in 1884. The construction of the dam or mole in 1806 at the head of the Virginia Channel (*left center*) did not eliminate the deposit of sediment (also called shoaling) in the Georgetown Channel immediately downriver of Mason's or Analostan Island (*center*). Instead, sedimentation increased downriver of the island, further obstructing navigation to Georgetown (*upper left*). The Washington Channel (*upper right, from Georgetown downriver*) also became severely obstructed for navigation. Dredging was required throughout the nineteenth century to maintain even minimum channels for navigation to Georgetown. [Major Peter C. Hains], *Potomac River at Washington, D.C. Map Showing Progress of Work as of June 30, 1884*, accompanying *Annual Report of the Chief of Engineers to the Secretary of War for the Year 1888*, U.S. House of Representatives, 49th Cong. 1st sess. (1888).

Latrobe answered that the wing-wall dams would deepen the channel, but wherever they ended, sediment would be deposited. In effect, Latrobe said that the wing-wall dams would just move the problem of sedimentation downriver, still restricting navigation to Georgetown.[26] He stated: "There will be, in twelve or fifteen years, an equal quantity of alluvium deposited in the new channels. . . . But the navigation for ships to Georgetown will be equally obstructed, with only this difference, that ships will not be able to approach as near to the town as they now do."[27]

Latrobe also wrote that the wing-wall dams proposed by Moore would do harm to the Washington Bridge, the Long Bridge: "The injury which may be done to the Washington bridge, will consist in deepening the river, in the new channel, in such

View of the City of Washington . . . Taken from Arlington House, The Residence of George Washington P. Custis, Esq., by Fitz Henry Lane, ca. 1838. Georgetown (*far left*) and the Washington Bridge, also known as the Long Bridge (*far right*), can be seen. Immediately across is the mouth of Tiber Creek. The location of the channels can be judged by the location of the ships on the river. The majority of the ships shown are close to the Virginia shore, in the deeper and wider Virginia Channel. The smaller Maryland channel is on the far shore. Also shown are the President's House (*left center*) and the Capitol (*right center*). Courtesy of the Library of Congress.

a manner, as that the piles on which the bridge is built, may not be sufficiently deep in the mud to be held upright against the weight of any fresh [i.e., freshet or spring flood], or logs, or ice, brought down by the current; the mud itself being washed away by the increased rapidity of the current in a place where no such current ran or had been expected in the erecting of the bridge."[28]

This was enough. Moore's plan to build wing-wall dams was blocked, and it was Latrobe's testimony that had defeated the bill and the construction of the wing walls.[29] In a letter to his son, Latrobe wrote, "The bill fell thro', as it ought to have done. I suppose I shall never be forgiven by Georgetown. I dont care."[30] It was a victory for the professional architect-engineer over the amateur. But it was also the beginning of the end for Georgetown in its quest to become a great commercial center.

Dredging the channels of the Potomac River continued throughout the nineteenth century. It proved to be expensive and not totally effective. Not all the clay, mud, and debris in those channels were from natural causes. In the 1830s the Alexandria Canal Company built a canal that intercepted the Chesapeake and Ohio

PROJECT PLANNING AND CONSTRUCTION EFFORTS

East and West Branch below Washington, by Augustus Köllner, 1839. The Eastern Branch (Anacostia River), looking south toward the Potomac River (*center, in the distance*). The structure across the Eastern Branch is the Navy Yard Bridge, located at the foot of Eleventh Street, SE. Although the Navy Yard Bridge was equipped with a draw span, little traffic proceeded upstream of the bridge. The post-and-beam construction of this bridge (and others across the Eastern Branch) can be seen under the bridge decking. The posts would have been pile-driven into the river bottom either by horse-operated rams or by steam-powered rams mounted on scows. Courtesy of the Library of Congress.

Canal just above Georgetown, crossed the Potomac River above Georgetown on an aqueduct, and continued downriver to Alexandria. Through much of its route, it paralleled the Virginia, or main channel, not far offshore. Workers of the company made it a practice to dispose of their excavation spoil into the Virginia Channel. From their point of view, this had the dual beneficial effects of inexpensively disposing of the spoil while at the same time blocking the main navigation channel to Alexandria's principal competitor, Georgetown. In 1836 Congress instructed the company to stop this practice and open up the channel where it had been blocked.[31]

WHARVES FOR THE NEW FEDERAL CITY

Besides improving navigation on the Potomac River, it was necessary to build wharves in the new federal city to receive the building stone and other materials. On October 22, 1791, the commissioners directed the secretary-clerk to the commissioners, Thomas Munroe, to insert an advertisement in the *George Town*

Steam Boat Wharf in Washington, by Augustus Köllner, 1839. Although the sedimentation of the Potomac River kept large ships from reaching Georgetown, smaller vessels, especially lighters and coastal vessels, made extensive use of the river throughout the nineteenth century. After the War of 1812, steamboats began to be used on the Potomac. Courtesy of the Library of Congress.

Weekly Ledger for two thousand wharf logs.[32] On November 25, 1791, the commissioners purchased an additional one thousand wharf logs.[33] And on June 4, 1792, they contracted with Absalom Ware and Patrick McMahan to build the wharf for a consideration of £250 pounds Maryland currency ($667).[34] This was the first of four timber crib wharves that were built by the commissioners. The four wharves cost $3,221.88.[35]

On September 17, 1790, when Thomas Jefferson and James Madison met with George Mason, few would have been aware of the development of bars in the shipping channels of Georgetown. But these bars greatly restricted the future commercial growth of Georgetown and the new federal city. In the future, smaller vessels and lighters had to be used to transfer outgoing cargoes from Georgetown to oceangoing vessels in the Eastern Branch or in Alexandria. The inability to ship meant that Georgetown and adjacent Washington would never become the commercial city envisioned by Washington and Jefferson.

Early Infrastructure and Transport Improvements

*Potomac Chain Bridge, having suffered the last year [1811]
by an extraordinary flood, has recently undergone a thorough
repair with improvements that do much honor to the skill of
the indefatigable constructor, John Templeman. As it is very
near this City, members of the Congress and other gentlemen
wishing to improve by their travels may be highly gratified with
a very romantic view of the Little Falls, the lock navigation [of
the Potomac Canal], and this important Bridge, by a short ride
of only four miles from Washington.*

Washington National Intelligencer, April 30, 1812

There were few roads, bridges, canals, or other infrastructure improvements in what was to become the new federal district, at the time of the
enactment of the Residence Act (1790). The new federal city was located
on a peninsula separated from the surrounding countryside by the Potomac River on the west and by the Eastern Branch (Anacostia River) on the
east. Though ferries were available to connect the new city to the country
across either waterway, they were slow, sometimes dangerous, and unavailable when the rivers froze. Only the construction of bridges would
seamlessly connect the federal city to the surrounding country.

L'Enfant had planned two major river crossings for the city, one across
the Potomac "at the place of the two Sisters w(h)ere nature would ef-
fectually favor the undertaking"[1] [at Three Sisters Islands approximately
a half mile above Georgetown] and one across the Eastern Branch. On
September 8, 1791, owing to a lack of money, the commissioners de-
ferred plans to build these bridges, as well as the construction of other
improvements recommended by L'Enfant, such as the Tiber Creek Canal
and wharves on the Potomac at the mouth of Tiber Creek.[2] The necessary
bridges and turnpikes, in addition to the city canal, would have to be fi-
nanced by private interests.

Bridges were also needed within the city. The new federal city was
intersected by Rock Creek and Tiber Creek, and the proposed city canal
would further divide the city.

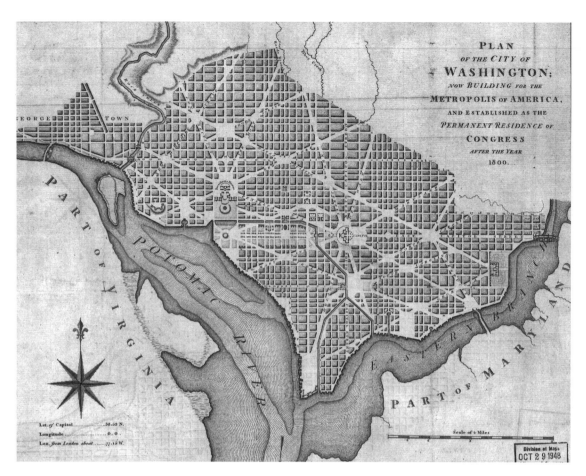

Plan of the Ciy of Washington, by William Bent, 1793. This drawing depicts the bridges and canals origi-nally planned for the new federal city. Architect-planner Pierre L'Enfant proposed a bridge across the Potomac at the Three Sisters (*extreme upper left*) and two bridges across Rock Creek (*left center*), the then existing 1788 bridge at present-day M Street, NW, and the masonry arch bridge begun in 1792 by Leonard Harbaugh at present-day K Street, NW. He also proposed one bridge across the Eastern Branch (*right*). Not shown is the Long Bridge across the Potomac, not built until 1808. Also shown in this map is the proposed Washington City Canal from below the President's House to its division below the Capitol into two branchs: St. James Creek to the west and the Eastern Branch to the east.
Courtesy of the Library of Congress.

BRIDGES IN THE NEW FEDERAL CITY

ROCK CREEK BRIDGE

The most urgent need was for a bridge across Rock Creek to connect the existing town of Georgetown to the building sites in the new city. Supplies and provisions for the workers at the President's House and the Capitol came almost entirely from Georgetown. Wrote the Duc de la Rochefoucauld-Liancourt after his visit to the new

PROJECT PLANNING AND CONSTRUCTION EFFORTS

city in 1797, "Workmen and merchants all live in Georgetown, so that is where every-body goes for supplies. The few stores open in [the] Federal City are very expensive and miserably stocked."[3] Fords of Rock Creek existed but were inconvenient. In 1788 a small bridge had been built across the creek at the approximate location of where M Street now crosses. But a larger bridge, sized to carry carriages and wagons, was needed. On March 15, 1792, the commissioners authorized an advertisement for bids from builders for such a bridge, to be located where the present-day K Street, NW, crosses Rock Creek.[4]

By March 29, 1792, the commissioners had awarded the contract for the new bridge to Leonard Harbaugh for £3250 ($8,668).[5] Harbaugh was not an engineer but a builder, originally from Pennsylvania, who arrived in the new federal city via Baltimore. The contract was for a single masonry arch bridge over Rock Creek. The cornerstone for the eastern abutment for the new bridge was laid on July 4, 1792.[6]

No measure of water flow had been made, and it soon became apparent that the single arched opening of the new bridge would not be adequate to pass the tidal and flood waters of Rock Creek. On September 1, 1792, the commissioners entered into a new contract with Harbaugh for two side arches of smaller spans for an addi-tional £2000 ($5,334).[7]

In addition to this three-arch bridge, a causeway eighty feet wide and one-fourth of a mile long was built to connect Georgetown to the new bridge and the federal city. The bridge and causeway were described by Chancellor James Kent upon his December 1793 visit: "A causeway 80 feet wide and one-fourth mile long and a stone bridge of 3 arches leads from George-Town to the City of Washington. They were built by the Commissioners of the Federal District and cost £13,000."[8]

The causeway crossed a then-existing tidal marsh between Rock Creek and the foot of Wisconsin Avenue (then called High Street). For approximately fifty years after its construction, this extension was known as "Causeway Street."[9] Later it was named K Street, NW.

The bridge was to be 135 feet long and 36 feet wide. The center arch was faced with thirteen ring stones, or voussoirs, of Aquia sandstone, each inscribed with the name of one of the original thirteen states of the Union. The center stone, or key-stone, was inscribed "Pennsylvania," and it is said that this is how Pennsylvania came to be known as the "Keystone State."[10] Just below the bridge was a "water-gate," a semicircular quay approximately 30 feet long of cut stone for the formal reception of dignitaries arriving by boat in Washington. The bridge and quay were intended to be part of an ornate water entrance to the new federal city—a "Watergate."[11]

The commissioners were soon criticized for spending too much on the bridge, not having it completed sooner, and not providing an opening sufficient to pass the waters of Rock Creek. Local land owner Benjamin Stoddert wrote George Washing-ton: "The instance of mismanagement I mean to adduce respects the Bridge over Rock Creek—It might have been completed by Novr [1792]—It will not be com-

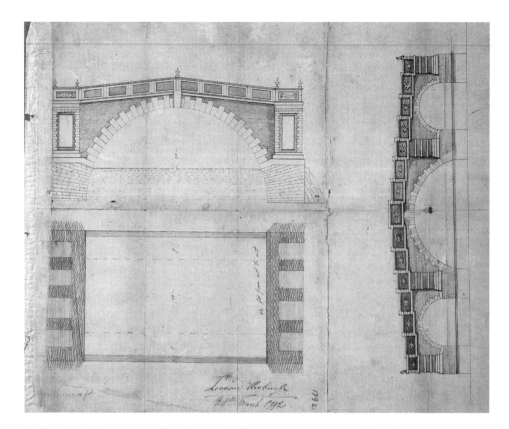

Rock Creek Bridge at present-day K Street, NW (between Georgetown and Washington), by Leonard Harbaugh, builder, March 28, 1792. Shown is the center arch, elevation, and plan, without state monuments in the voussoirs (*left*), and elevation with side arches and state monuments in place (*right*). After construction, the central arch of this bridge sagged as much as three feet, and the bridge had to be rebuilt. Courtesy of the National Archives; also reproduced in the Historic American Engineering Record collection, Library of Congress.

pleted till the Spring [1793]. The Comrs preferred the plan of a single Arch because it appeared cheapest—after some progress had been made in the work—too much to admit of correcting the Error without delaying the completion of the Bridge, & incurring some addition expence, it was apprehended a single Arch would not vent the water of the Creek, & two more were in consequence agreed for."[12]

Almost immediately other problems became apparent. For example, the central arch developed a three-foot sag. Wrote Tobias Lear to George Washington:

The Stage, in which I was, came through part of the city & over the bridge this afternoon, and I was very sorry to find that the report of the upper part of the Middle Arch having settled considerably, was too true. It appears, so far as I could

PROJECT PLANNING AND CONSTRUCTION EFFORTS

judge by the eye in passing it in the stage, to have settled at least 3 feet. The beauty of the Arch is totally ruined & how far it is injurious to the utility & durability of the work I am not able to say; but I think it must be much.[13]

By October 16, 1794, the commissioners wrote to Harbaugh directing him to fix the bridge.[14] On December 15, 1794, James Hoban complained to the commissioners that the bridge had not yet been fixed.[15] Some historians have written, incorrectly, that this bridge was then removed within four years and not rebuilt; that instead, the old bridge at M Street, two blocks above, was rebuilt about 1800.[16] The available records indicate that the bridge was repaired by Harbaugh and continued in use for many years. For example, in 1816 D. B. Warden described this bridge as one of the two across Rock Creek: "There are two bridges. That nearest the mouth has three arches, is a hundred and thirty-five feet in length, and thirty-six feet in breadth. The other [i.e., the earlier wooden bridge at M Street] at a distance of six hundred and fifty yards, is supported by piles, two hundred and eighty feet long, and eighteen feet wide."[17] The bridge was also described as late as 1864 by "Viator" in *The Washington Sketch Book*.[18] In addition, there are several early-nineteenth-century illustrations of this bridge (see plate 5).

In 1796 the commissioners modified the bridge by adding a drawbridge. George Walker, a proprietor and critic of the commissioners who maintained that the commissioners did not have title to the lands they were selling, published a criticism of their decision to build a drawbridge: "The expensive and unnecessary Draw, now erecting upon the Bridge across Rock-Creek, must also be for some private purpose; for it is evident to every one, that it cannot answer any public purpose."[19] Walker was suggesting that the commissioners were using public funds to build a draw in the bridge to enhance the value of the lots they owned above the bridge, that is, to convert inland lots into more valuable waterfront lots. Three days later the commissioners responded to Walker's criticisms in the same newspaper: "Whether the draw in the Bridge over Rock-Creek is necessary or unnecessary, the Commissioners leave to every passenger to form his own opinion."[20]

The bridge contractor, Harbaugh, had come well recommended to the commissioners,[21] although he was a builder and not an engineer. But selecting him as bridge contractor was a bad choice. For example, he made no attempt to measure the flow of Rock Creek under the proposed bridge. Had he done so, he would have discovered the need for adding the two side arches. Although the three-foot sag in the bridge was repaired, this was an inauspicious beginning for the commissioners and the new federal city. In any case, by 1794 the funding pattern for roads and bridges had been established: public funds for bridges and roads within the city, private funds for bridges across the two major rivers and for turnpikes beyond.

Besides Harbaugh's K Street Bridge across Rock Creek and the two small bridges built by the commissioners, one over James Creek (St. James Creek) for $342.04 and one over Tiber Creek for $788.04, the new city needed additional public-funded bridges.[22] One of those city bridges was the F Street bridge, for which an advertisement appeared in 1807:

> Proposals will be received by the subscriber until Saturday next the 14th inst. for building a BRIDGE of Stone and Brick, on North F street, between 9th and 10th streets west, agreeably to a plan left with captain James Hoban. The persons offering to contract are to furnish the materials, & the work to be completed by the 1st day of December next.
>
> THOMAS H. GILLISS,
> Commissioner for the 2d Ward.
> November 11 — 2t[23]

BRIDGING THE POTOMAC AT LITTLE FALLS

While the commissioners were building a bridge to connect Georgetown with the new federal city, the merchants of Georgetown were developing plans to connect Georgetown with the agricultural hinterland of Virginia. Access to the farms of Virginia would provide wheat and flour to Georgetown for export and greatly assist in making Georgetown a major export center.

It was initially thought that access to the Virginia countryside would be provided by the Potomac Canal. To open the Potomac River to shipping, the Potomac Canal Company needed to overcome two principal obstacles to navigation in the Potomac River: the 37-foot fall in the river at Little Falls three miles above Georgetown and the 76-foot drop in the river at Great Falls, eighteen miles above Georgetown. By 1795 the canal company had completed the Little Falls bypass canal and locks, and a ramp had been constructed at Great Falls for conveying flour from above the falls to waiting boats below the falls. This ramp was a bottleneck and limited the shipment of flour and other goods downriver to Georgetown. The Great Falls bypass canal, which was expected to greatly increase the quantity of flour shipped to Georgetown, was not completed until 1802. Yet the Potomac Canal never lived up to its expectations. Five years later, in 1807, it was estimated that the Potomac River valley, including the Shenandoah, would produce 600,000 barrels of flour a year shipped on the Potomac River and canal. But that spring only 45,000 barrels had been shipped by the end of April, with perhaps another 45,000 barrels to be shipped during the remainder of the year.[24] The problem with the Potomac Canal and using the Potomac River for shipping was the great variation of water levels; the river and the canal could be used reliably for shipping only during the abundant river-flow months of spring.[25]

Various solutions to improve navigation on the Potomac were proposed. One observer recommended "swelling" the water over the river shoals by means of low dams built in the river.[26] Another suggestion was to decrease the draft of the boats

The City of Washington in 1800, by George Isham Parkyns. View from Arlington. The bridge over Rock Creek built by Leonard Harbaugh is shown as a single-arch structure on the far shore (*left*). Courtesy of the Library of Congress.

using the Potomac. Another was to build a still-water canal adjacent to the Potomac River (this canal was eventually built as the Chesapeake and Ohio Canal). There was also another alternative, to bridge the Potomac River and to bring flour and other products from the Virginia countryside into Georgetown through new roads and across a new bridge.

L'Enfant had suggested bridging the Potomac River, just above Georgetown, at Three Sisters.[27] Numerous attempts were made to build a bridge at Three Sisters; one was in 1826,[28] and another was in 1852.[29] The problem with the Three Sisters location was that the river was too wide to be crossed with one- or two-span bridges using the bridge technology of the day. The largest single span built in the United States at the time did not exceed 340 feet, and the crossing at Three Sisters was many times that distance.[30] A bridge with smaller multiple spans would be possible, but only if cofferdams (temporary enclosures to hold back the water while workmen excavated for a pier foundation in the river bed) could be constructed in the Potomac River. Building the piers would require construction of the cofferdams in deep water, defined as ten feet or more, so as to muck out the mud on the river bottom and provide a solid foundation for the required masonry piers. Investigations revealed that the Potomac River at Three Sisters was deep, as deep as seventy-five feet, making cofferdams impractical. In 1828, in investigating Three Sisters as a suitable crossing for the aqueduct of the proposed Alexandria canal, engineer James Geddes reported: "From measurements and soundings of the Potomac river, made in pursuance of directions of the Corporation of Georgetown, by Mr. George D. Avery, the place called the 'Three Sisters' is shown to be a very unsuitable one, for a bridge of any kind. There is a distance of two hundred and sixty-two feet, which is over twenty feet deep—deepest part, seventy-six feet. In this deepwater, the bottom is reported 'soft,' and the depth, to a solid foundation, is unknown."[31]

Downriver of Three Sisters, the river widened and its depth decreased. A bridge crossing was potentially possible immediately below Three Sisters, between Georgetown and Mason's Island (present-day Theodore Roosevelt Island), where the Georgetown Ferry (also called Mason's Ferry) crossed the Potomac. But this site would have also required intermediate piers built in deep water. The most feasible bridge site for a single-span bridge was just above Little Falls at Fendall's Mills, three miles upriver from Georgetown. There it would be possible to span the Potomac with a single-span bridge of approximately 120 feet.

The First Bridge at Little Falls The Georgetown Bridge Company was authorized by the Maryland Assembly on December 29, 1791, to build such a bridge.[32] Four years later, the Maryland Assembly granted the same company a charter to build a road, sixty feet wide, from Georgetown to the new bridge at Little Falls (now called Chain Bridge Road).[33]

For several years the company did little. Then in June 1795 it advertised the sale of 400 shares of stock at $200 each, a total capitalization of $80,000.[34] Only 245 shares of the company's stock were sold. Assuming that the bridge would produce high tolls, the company went ahead and began construction, using a $30,000 loan to make up the shortfall.[35] The company contracted with a well-known New England bridge builder, Timothy Palmer: "An Artist, eminantly distinguished by the Bridges he has lately built over the rivers, the Merimick [sic—Merrimack], in the State of Massachusetts, and Piscataqua, in New Hampshire, has undertaken the erection of the Bridge, and engages its completion before the end of next year [i.e., 1796]."[36]

The bridge built by Palmer was a wooden king post truss structure with a king post approximately 15 feet tall and with a bridge span of approximately 120 feet. Architect-engineer Benjamin Latrobe, who visited and painted a picture of the bridge in 1798 (see plate 8), reported that it had been prefabricated in New England of white oak and shipped to the Potomac River site in pieces.[37] The bridge components were then assembled at Fendall's Mill,[38] located on Pimmit Run in Virginia at the western end of the new Little Falls bridge.

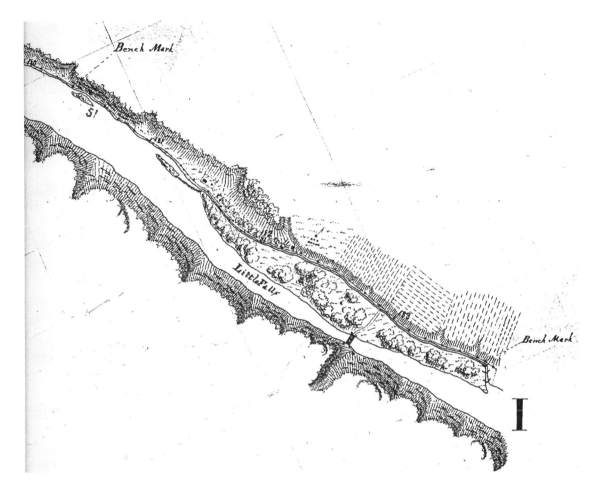

Detail from Survey of the Potomac Canal, 1825 (Williamsport, Maryland, to Harper's Ferry), Lieutenant N. B. Bennett et al., delineators. Shown in this drawing is the topography in the vicinity of the Little Falls bridge (Chain Bridge). The high bluffs of Virginia can be seen (*bottom*) as well as the somewhat lower bluffs of Maryland (*above the river*). In between was flat land. Initially a land bridge crossed the Little Falls Canal of the Potomac Canal Company (*shown*) and the flat land. Despite this level land, the high bluffs constricted floods, which led to high water levels, forty or more feet above the mean water level. Any bridge built here would have to be above that mark. This plan, drawn for the Chesapeake and Ohio Canal in 1825, shows the fourth bridge built at this location and the road up to the bridge. The three previous bridges had been destroyed by floods. The fourth bridge, like the third bridge built here, was also a chain bridge, erected and opened for traffic in April 1812. Courtesy of the National Archives.

Advertisement for laborers.

Centinel of Liberty and George-town (DC) *Advertiser,* July 8, 1796.

The timbers for this bridge had to be large, from first-growth forests in New England. Bridge historian-engineer Frank Griggs explains that Palmer had used timbers as large as 16 by 18 inches in section and more than 50 feet long in the construction of his Piscataqua Bridge.[39] The wood used for the Little Falls bridge would not have been as long, but the timbers were probably almost as large in section. Total cost for the bridge and abutments was $84,000.[40]

Selecting the head of Little Falls as the site for this bridge solved one problem, how to bridge the Potomac River in a single span. But it introduced a second problem: how to protect the bridge from floods. During flood events the river at Little Falls could rise 40 or more feet in twenty-four hours. The officials of the bridge company, not having the benefit of river-gauging stations, would have known this through interviewing longtime residents. From these interviews they determined that bridge abutments 40 feet high would place Palmer's bridge above future floodwaters. The bridge would therefore require two very tall, 40-foot-high, masonry abutments.[41] These abutments were massive, rubble-filled, rough cut, ashlar-faced structures bolted together. At the time they were probably the largest manmade structures in the region. Their construction was very expensive, fully one half or more of the total cost of construction of the bridge.

Stone for these two abutments was locally abundant. Philip Richard Fendall, the owner of the local mill, provided the needed stone from his four local quarries, using his three stone scows.[42] In addition to the wooden main span, two smaller structures were built as land bridges to connect the main structure with the Maryland side of the river. Begun in 1795, the bridge was completed and opened for toll traffic on July 3, 1797.[43]

This first bridge at Little Falls was not technologically innovative; wood truss bridges of great length had been built by the Romans and probably even earlier.[44] It was nonetheless impressive. John Shippen, visiting the bridge in 1801, described the bridge, the abutments, and the surrounding terrain:

Three miles from Georgetown, just at the head of tide water, and at what are called the little falls, a bridge of single arch crosses the Potomac. It is composed of wood; erected by one Palmer, from Connecticut. I am told it was formed by him in Connecticut and shipped in pieces. The abutments are a huge pile of massy square stones bolted together with great iron pins, and melted lead, a novel sight to me, and I take it, capable of resisting the most swollen floods of water. The Virginia side a high bank; Maryland side, low base of immovable, large and deep rocks. The fact is, here the Potomac is narrow and deep.[45]

On Monday, April 23, 1804, a large flood almost reached the top of the 40-foot-high abutments. The main span of Palmer's bridge was spared, but the two smaller wood-frame spans leading to the bridge from the Maryland side were swept away. The bridge became unusable, and the company was deprived of its revenue.[46] Financially, this left the company with its initial capitalization expended, with a debt of $15,000 to $16,000 on the original $30,000 loan, and with little opportunity to sell unsold shares. Repairs would cost $5,000 to $6,000. In addition, the main span, which had never been protected from the elements by sheathing or with chemical preservatives, had deteriorated and required repair and replacement. The wood repair and replacement of the main span would cost another $3,000 to $4,000.[47] The masonry abutments were unharmed by the flood.

The flood of April 23, 1804, compounded the company's other problems. Two years earlier, in 1802, Edward Gantt and other stockholders had brought suit against the company to sell the bridge, to "raise money to pay a large sum of money due from the said company to sundry persons."[48] After the flood, the company's directors summarized the situation for their stockholders: the bridge was a valuable property, but it could yield a high rate of return only if the stockholders came up with more money to repair it.[49]

Advertisement.

BY an Act of the Legislature of Maryland, entitled an Act, for erecting a BRIDGE over Potomac River, the subscribers are authorised and appointed to open books for receiving and entering Subscriptions for the said undertaking—Notice is hereby given, that books will be opened at the house of Mrs. SUTER, in George Town, on the first Monday in July next to receive subscriptions to the number of Four Hundred Shares, at two Hundred dollars each Share.

Previous to any call for money, there will be at least six weeks notice in all the papers of this state, and the Alexandria and Philadelphia papers.

Fifth enacting clause of the law.

"*And be it enacted*, that for and in consideration of their great risk, and the expenses to be incurred be the said proprietors, not only for the building of the said bridge, but for keeping the same in continual repair, the said bridge and all its profits, shall be, and the same is hereby vested in the said proprietors, their heirs and assigns forever, as tenants in common, in proportion to their respective shares, & it shall and may be lawful for the said Directors, at all times hereafter, for the term of fifty years, to demand and to receive such reasonable tax or toll, as they may from time to time agree on, and require, provided they shall not at any time demand more than two thirds of the present rates of ferriage to and from George-Town, which rates or tolls shall be at all times made public, and shall not be altered or changed more than once in each year; and at the expiration of the said term of fifty years the said directors shall receive such tolls as shall be regulated by the legislature of this state, or of the United States, should the said bridge be erected within the jurisdiction of the United States.

WILLIAM DEAKINS,
JAMES M. LINGAN,
URIAH FORREST.

George-Town, May 8 1795.

Timothy Palmer, an Artist, eminently distinguished by the Bridges he has lately built over the rivers, the Merimick, in the state of Massachusetts, and Piscataqua, in New-Hampshire, has undertaken the erection of the Bridge, and engages its completion before the end of the next year.

May 15.

A call for subscriptions for the new bridge company.
Georgetown (DC) Columbian Chronicle, June 12, 1795.

RATES OF TOLL

TO BE RECEIVED AT THE BRIDGE
OVER POTOMAK RIVER FOR
THE PRESENT YEAR.

	S.	D
FOR A FOOT PASSENGER.		4
A led horse or horse } in droves—each—		6
A horse and Rider		11½
A Cart with one horse,	1	6
A ditto with two horses,	1	10¾
A Waggon with two horses,		8
A ditto with 4 or more horses,	3	9

No additional charge will be made for the drivers of the waggons and carts, and if they return the same day they will re-pass the bridge free of tolls.

	S	D
For a two wheel pleasure car-riage with a single horse, }	2	6
For every additional horse,		6
A four wheeled pleasure carriage and 2 horses, }	4	6
A Ditto with four horses		6

No additional toll will be taken for persons in the carriages or for drivers.

Cattle in droves—per head,	3
Sheep or hogs in droves per do.	2

These rates are fixed in Maryland currency.

JOHN TEMPLEMAN.
ACTING DIRECTOR.
Georgetown, Jan. 1798. 72—

Toll rates for the first bridge at Little Falls. *Centinel of Liberty and George-Town (DC) Advertiser*, February 2, 1798.

The Second Bridge at Little Falls The matter was not resolved until a meeting of the stockholders on May 18, 1805, more than a year after the flood. At this meeting the stockholders authorized the directors to issue no more than fifty additional shares of stock at two hundred dollars per share to develop funds for completely rebuilding the main span and for building a stone ramp up to the new bridge from the Maryland side.[50]

The repairs could be made for only ten thousand dollars because the existing abutments had not been damaged and were to be reused. The old bridge would serve as scaffolding for the erection of the new. The newly constructed masonry ramp was less susceptible to flood damage, but it introduced a steep slope on the Maryland side that had to be traversed to get to the top of the forty-foot-high abutments. To build the new bridge structure, the Georgetown Bridge Company turned to one of the most prominent bridge builders in the country, Theodore Burr. He completed the bridge by the end of 1806.[51] Unfortunately, this new bridge was destroyed a little more than six months after completion, in the flood of June 21, 1807.[52] But, like the flood of 1804, the flood of June 21, 1807, did not damage the bridge abutments; it also left the stone ramp intact.

The Third Bridge at Little Falls To repair the bridge, the company turned to a new bridge technology, the chain-link suspension bridge developed and patented by suspension-bridge inventor James Finley. In Finley's patented bridge, the bridge deck was supported by two iron chains erected over towers. The ends of the chains were secured in the ground with stones and the bridge decking linked to the iron chains by iron pendants.[53] The Chain Bridge at Little Falls would require two towers, each one no less than one-seventh of the span, in this case fourteen or fifteen

PROJECT PLANNING AND CONSTRUCTION EFFORTS

feet tall. The suspension chains, one on each side of the roadway, were to be twice as long as the bridge's span and were composed of iron chain links, fabricated from bar iron, each link ten feet long.[54] Each end of the two chains would be secured in the ground with rocks, what would be called today "dead men." From each end of the ten-foot-long chain links, Finley would install vertical "pendants" (now usually called "suspenders"), from which he would hang the deck of the bridge. The deck structure could be one or two tiers; two tiers would add rigidity to the structure. The chain bridge constructed at Little Falls only had a single level.

One of the major advantages of Finley's chain bridges was cost: chain suspension bridges avoided the cost and necessity of building intermediate piers; they were therefore much less expensive, as Finley pointed out: "An estimate on these principles for a bridge of 500 feet between the abutments, with only one pier, will not amount to seven thousand dollars, exclusive of abutments and pier. Compare this with the Philadelphia Schuylkill bridge[55] of the same extent, which cost sixty-five thousand dollars after abutments and the *two* piers were completed; total expense, three hundred thousand dollars."[56]

Whereas the builders of the first two bridges at Little Falls, Timothy Palmer and Theodore Burr, were talented and innovative craftsmen, they were not bridge engineers. Finley, however, was. He was aware of advances made in Europe in the science of strength of materials, and he used that knowledge in developing calculations used for designing and detailing his bridges.

The new bridge at Little Falls was completed in February 1808. The *Washington Federalist* gave the new bridge a rave review:

Flood of April 23, 1804, damages first Little Falls bridge. *Washington National Intelligencer*, April 27, 1804.

WASHINGTON CITY.

FRIDAY, APRIL 27.

There was a confiderable frefh in the Potomac on Monday laft—The water rofe to a great height, and moft of the wharves in George town were covered. It is with pleafure we learn very little injury was fuftained—And though report was bufy in detailing the wonderful fights in the Potomac: and men floating on logs through the river crying for fuccour in vain; horfes, cattl, fheep and hogs fwept down the ftream without the poffibility of efcape; veffels finking by the force of the whirlpools. Fifhermen carried away entangled in the very nets they had prepared for the finny tribe and a thoufand ftories equally alarming, yet we find on enquiry there was no foundation for i.h. It is true the fcene at the Little falls was fufficiently terrific to have excited a thoufand fears and apprehenfions.—It was at the bridge the greateft damage was fuftained—there the water rofe to a height nearly with the top of the abutment which is almoft 40 feet above the river in common tides—and by its great force carried away what is called the land bridge on the north fide of the Potomac—but the main arch with ftood its moft fierce affaults.—Large trees and great quantities of timber were borne down the river and the fhores were lined with adventurers endeavouring to fave fuel for themfelves and their neighbours—and it is with fatisfaction we ftate much was faved and to-day wood has been purchafed at half the ufual price.

The frefh fubfided yefterday—the damage fuftained by the bridge company which we underftand is not great, is the only lofs of any confequence we have heard of—The neceffary repairs will no doubt foon be made and the communication reftored—the interruption however cannot laft long.   *Wafh. Fed.*

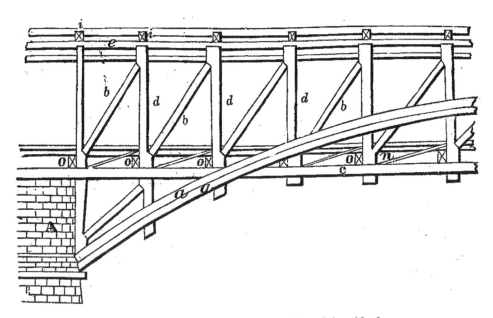

Fig. 134—Represents a side view of a portion of a rib of Burr's bridge.

a, a, arch timbers.
d, d, queen-posts.
b, b, braces.
c, c, chords.

e, e, plate of the side frame.
o, o, floor girders on which the flooring joists and flooring boards rest.
n, n, check braces.
i, i, tie beams of roof.
A, portion of pier.

Detail, Burr arch. The second bridge constructed at Little Falls was a Burr arch, a combination wood truss and wood arch structure invented by Theodore Burr and widely used in highway bridges at the time. This bridge, constructed on the abutments of the first, opened in November 1806. It collapsed on June 21, 1807, as a result of a flood. D. H. Mahan, *An Elementary Course of Civil Engineering*, 1860.

We feel much regret in being obliged to state, that the New Bridge over the Potomac, at the Little Falls, has fallen in. It was some time last week noticed to be giving way, and Sunday evening, it fell into the water with a tremendous crash. All the materials of which it was composed will be saved.

Wash. Fed.

Flood of June 21, 1807, destroys second Little Falls bridge.
Washington National Intelligencer, June 26, 1807.

We are happy to announce that a Bridge is again completed across the Potomac near the Little Falls, three or four miles from the City and Georgetown. . . . The bridge just finished is constructed on the principles of that built a few years ago by Judge Findley [*sic*], near Union-Town, in Pennsylvania, near Jacob's Creek. . . . There are four chains to support this bridge, two on each side of it — the pendents are hung on the chains alternately; about 5 feet apart, on each side, so that each chain receives a pendent in every ten feet. . . . We hear that Judge Findley is about obtaining a patent for its invention, which is perhaps the most important that the world has yet discovered in the art of bridge building.[57]

But like its two predecessors, this bridge too was soon lost to a flood, in November 1810. Wrote a local newspaper account:

Thirty-six feet above the usual level of the water, was erected a few years since a chain bridge, the third that has been erected there. The other two, however, were not chain bridges. During the freshet the water rose quite to the bridge: the wood and logs lodged against it, and for a long time it obstinately refused to yield. The wood increasing, at length it tilted one side upward and was swept away — the two principal chains were so well secured at the abutments that they remain.

When the water rose to this height the river became about a fifth of a mile wide, and the torrent poured over the trees at the height of between 10 and 15 feet. This is the third bridge which that company has lost within 9 years.[58]

This was the third bridge of the Georgetown Bridge Company destroyed by floods in six years (1804–1810), not nine years as stated in the news account. It was too much. The company had only a narrow financial base in the merchants of Georgetown, then a small hamlet. The population of Georgetown was only 3,000 when the first bridge was opened, and only 5,000 when the third bridge failed.[59] With the loss of three bridges between 1804 and 1810, the company lost the confidence of its few supporters. When the directors failed to receive a vote of two-thirds of the stockholders to rebuild, the company was no longer a viable entity.

The Fourth Bridge at Little Falls The backers of a bridge to Virginia regrouped. The stockholders of the Georgetown Bridge Company formed a new company, the Georgetown Potomac Bridge Company. On January 23, 1811, these organizers petitioned Congress for permission to incorporate so that they could erect yet another bridge at Little Falls. A month later, on February 22, 1811, President James Madison signed the act of incorporation. The stockholders of the new company met May 6, 1811, and unanimously voted to rebuild the bridge at Little Falls.

A pressing issue for the new company was whether to rebuild the Little Falls

Chain Bridge at Little Falls above the Potomac River, looking downriver with Virginia to the right. The first chain bridge (the third bridge built at this site) followed the design of James Finley and was built in 1808. The height of the bridge over the river, forty feet, indicated the expected flood level at this constrained spot of the Potomac River. The great height above the river greatly increased the grade approaching the bridge from the Maryland side (*shown, left*). *Family Magazine or, Monthly Abstract of General Knowledge* 4 (1839): 197.

Bridge using Finley's suspension bridge patent. By 1810 Finley had erected eight chain-link suspension bridges, including the one at Little Falls.[60] He advertised his new bridge technology in an article in the magazine the *Port Folio*,[61] which did not mention that two of the eight bridges had failed. On June 18, 1811, Finley published a letter in the Washington paper explaining these bridge failures. He attributed the failure of the chain bridge at the Falls of the Schuylkill, at Manayunk, Pennsylvania, to heavy loading and a failure of the coupling piece that joined two parts of the chain together. Finley attributed the other bridge failure to the bridge builders' deviating from his design for anchoring the chains.[62]

Finley had not supervised the construction of these bridges. Rather, he had licensed the use of his invention, the chain-link suspension bridge. He was therefore not responsible for failures that deviated from his design. The chain bridge at Little Falls was erected under the supervision of dentist, entrepreneur, and bridge builder John Templeman, an acting director of the Georgetown Bridge Company.[63] Templeman was both an agent for Finley and the supervisor of construction, per-

Chain Bridge at Little Falls, by Augustus Köllner, 1839. This view looks toward Virginia. It is difficult to tell whether this image shows the first chain bridge (the third Little Falls bridge, built 1808 and destroyed 1810) or the second chain bridge (the fourth Little Falls bridge, built 1812). They were very similar, except that the first bridge had a level deck and the second had an arched deck. Compare with the illustration on page 96 . Courtesy of the Library of Congress.

haps functioning much as a modern general contractor does. Finley described his relationship with Templeman: "In March 1808 I entered into an agreement with Mr. John Templeman, of Georgetown, Maryland, by which he was to receive one half of all the monies arising from what permits or patent rights he could dispose of for and during the terms of five years." The relationship between Finley and Templeman was not entirely harmonious. In his article in the *Port Folio*, Finley wrote of patent infringement on the part of Templeman.[64]

The arrangement with Finley was lucrative for Templeman, as he built other chain bridges. On June 18, 1811, for example, a letter from the directors of the Essex Merrimack Bridge Company in Massachusetts was published in the Washington paper, probably at the instigation of Templeman, stating how satisfied they were with Templeman's performance in building a 244-foot-span chain bridge over the Merrimack River.[65]

The new bridge company decided that the fourth bridge at Little Falls was also to be a chain suspension bridge based on Finley's design, with erection supervised by

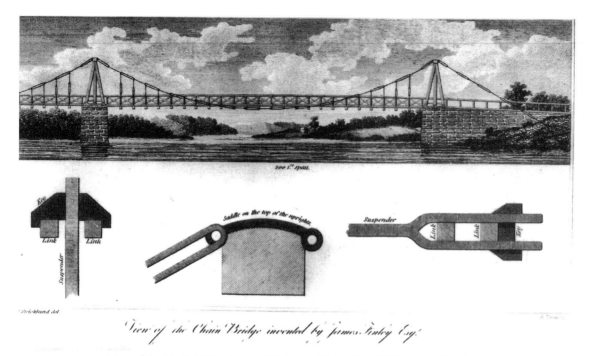

View of the Chain Bridge Invented by James Finley, Esq., William Strickland, delineator. A promotional elevation of Finley's chain suspended bridge. The greater span that Finley's design made possible reduced or eliminated the need for expensive and difficult-to-build intermediate piers. The truss structure supporting the double deck shown would have added rigidity to the structure. In James Finley, "A Description of the Chain Bridge; Invented by James Finley; of Fayette County, Pennsylvania," *Portfolio* 3, no. 6 (June 1810): 440–453, 440.

John Templeman. The bridge had a span of 128½ feet and a width of 16 feet.[66] It was opened in 1812. This chain bridge survived subsequent floods, such as the June 2, 1836, flood, described by Jacob Payne of Georgetown as being "three feet higher at the Chain bridge than he had ever seen it, and [he] had observed every freshet of any consequence for twenty to twenty-five years."[67] Four years later, on February 13, 1840, it was destroyed by yet another flood.[68] Four more bridges were subsequently built at this location, which remains an important Potomac River crossing.[69]

The crossing of the Potomac at Little Falls was only one link in a route from the central Virginia valley to bring flour and produce to Georgetown. A turnpike needed to be constructed from Drane's Tavern (Dranesville) to the bridge. The Falls Bridge Turnpike Company was formed to build and operate this route.[70]

WASHINGTON BRIDGE (LONG BRIDGE)
The Chain Bridge at Little Falls was intended to serve the merchants of Georgetown. The merchants of Alexandria also wanted a Potomac River crossing, and there

was a need for a bridge to carry the post road to the southern states. The second crossing of the Potomac River in the new federal city was the mile-long Long Bridge between the foot of Maryland Avenue and Alexander's Island (also called the Potomac Bridge and, later, the Fourteenth Street Bridge). On February 5, 1808, Congress authorized the Washington Bridge Company to build this bridge.[71] Here the Potomac was almost a mile wide but not deep. The Long Bridge, constructed with the post-and-beam method, was described as being built on "a number of timber piles, driven into the bed of the river, [with] beams . . . laid thereon to form the carriage road, which is planked from end to end." The contractor was a New Englander, William Mills, who had built several similar bridges in New England.[72] Construction began in June 1808, and the bridge was opened to traffic on May 20, 1809, at a cost of nearly $100,000.[73]

Georgetown Potomac Bridge Company calls on its stockholders for share payments to rebuild the Little Falls bridge. *Washington National Intelligencer*, May 9, 1811, 3.

Whereas the chain bridges of Finley's design at Little Falls were considered the latest technology and the beginning of bridge engineering in the United States, the Long Bridge was of traditional post-and-beam construction, a bridge type with a two-thousand-year history.[74] The difference between the Chain Bridge and the Long Bridge was the difference between an engineer-designed structure and a builder-built structure. Architect-engineer Latrobe explained this difference: "I have said constantly to Dr. May [Dr. Frederick May, a director of the Washington Bridge Company] and others, that such a mode of application was the only proper one, that the erection of bridges was a principal object of the profession of a civil Engineer, altho I have also said and still think that such a bridge as was to be erected over the Potowmac [i.e., the Long Bridge] required only a New-England bridge builder."[75]

Bridge builders had existed for hundreds of years, but engineering was a newly emerging profession. In describing the new profession, Latrobe explained "that an Engineer is not merely an overseer of Laborers working under the direction of Managers, but a Man whose means of subsistence and fame depend on his reputation to whose skill and integrity the most important undertakings of Mankind must nec-

"The Long Bridge Leading Across from Washington to Virginia." Washington or Long Bridge during the Civil War, after its reconstruction. *Harper's Weekly*, May 18, 1861, 317.

essarily be confided, and who therefore must be allowed a certain power, without which no Man in his senses will incur responsibility."[76]

Latrobe estimated that the piles of the new Long Bridge were driven through mud 16 to 27 feet deep.[77] They were driven into the mud river bottom by man-, steam-, and horse-powered pile drivers. In some areas, where the mud flats were usually exposed by the river, the bridge was built on thirty-nine earth and stone piers 25 feet square.[78] The total length of the bridge was a little less than a mile— 4,984 feet, with 2,659 feet of the bridge carried over the flats on the earth and stone piers.[79]

The dimensions of the bridge decking were defined by Congress's act of February 5, 1808: it had to be at least 36 feet wide with a 6-foot pathway for pedestrians and had to have two drawbridges, one at least 30 feet wide over the main or Virginia channel and one 15 feet wide over the Maryland channel. Each leaf of the two drawbridges was to be 20 feet wide.[80] When the bridge opened in 1809, it was the longest in the country. A contemporary newspaper provided a detailed description of the bridge:

The Washington Bridge is the longest in the United States; and, we believe, in the world. Its length is upwards of 5000 feet, and, with the abutments, is a mile. It is 36 feet wide, 29 feet of which is appropriated for a carriage way, and a foot way on each side, of 4 1-2 [feet wide], for foot passengers, separated from the

Draw at Long Bridge, Washington City, by Augustus Köllner, 1839. Courtesy of the Library of Congress.

carriage way by a light railing. The bridge is supported by 201 piers, 25 feet open-ing. Each pier is composed of five piles on the flats, 6 in the channel, and 7 at the draw. These piles are driven from 20 to 30 feet into the mud, and strongly braced and bolted. — The floor is supported by 16 stout stringers and it is covered with four inch plank. The railing is strong and durable, and the bridge is lighted by 20 lamps. There are two draws; one of 25 feet, opening in the little channel; the other, of 35 feet, in the large channel. The construction of these draws, the man-ner in which they are raised, are extremely ingenious, and combine strength with great facility and ease in raising. A boy of ten years old will be able to raise one leaf. At the widest and principal draw, there is a wharf on each side, projecting out at right angles, with the bridge, 100 feet, of great strength, and calculated for vessels to lay at and warp through.[81]

Traveler Ann Royall, in her account of her visit to Washington, described the opera-tion of the two draws in the bridge: "The Potomac bridge, at Washington, is a mile in length, and wholly constructed of wood. It contains draw-bridges for the passage of vessels; these bridges are raised by a pulley, and by a single individual."[82]

TABLE 5.1. Tolls for Crossing the Washington Bridge, 1816

A four-horse carriage	1 ½ dollar
A two-horse ditto	1
A four-horse wagon	0 62 ½ cents
A two-horse ditto	0 37 ½
A gig	0 36 ½
A horse	0 18 ½
A man	0 6 ½

Source: Warden, *Chorographical and Statistical Description*, 57–58.

The completion of this bridge was seen as highly significant for the growth of the city as well as for connecting the northern and southern portions of the country, and particularly for carrying the post road to connect North to South.[83] Although seen as beneficial to the new city, it was not very profitable. The bridge cost nearly $100,000 to construct and needed to charge high tolls, which discouraged extensive use of it (see table 5.1). Wrote visitor Royall, "The profit of this bridge is nothing, except when the river is frozen, as the toll is abominably high. The numerous boats which ply the river, save the people a vast expense in traveling."[84]

The Long Bridge was damaged during the British capture of Washington on August 24, 1814—the British burned one end, and the Americans burned the other. But it was repaired in 1816.[85] It survived until a serious ice jam carried away several spans in 1831.[86] At the time there was a great deal of interest in replacing the bridge with a masonry structure. Orange Dibble, a bridge contractor from Pennsylvania who built the Wisconsin Avenue Bridge over the Chesapeake and Ohio Canal, developed plans and a cost estimate of $1,350,000 for a masonry replacement.[87] In 1834, engineer Loammi Baldwin criticized Dibble's bridge proposal as "impracticable" and "vicious"—vicious in the sense that the piers he proposed would not withstand the operation of worms.[88] Baldwin's report to Congress stated, "Mr. Dibble is altogether mistaken in his views and estimates, and . . . no member of Congress will act prudently in trusting to his calculations."[89]

Baldwin's report to Congress killed Dibble's proposal to rebuild the bridge of masonry. It was rebuilt in the same manner as the original bridge (timber beams mounted on timber piles), under the direction of engineer A. B. McLean. The flood of June 2, 1836, damaged the bridge, but it was repaired.[90] It subsequently served as an important river crossing during the Civil War. During that war, the Washington, Alexandria and Georgetown Railroad Company built a second bridge at this location, seventy-five feet downstream and parallel to the earlier bridge, to carry its tracks into

PROJECT PLANNING AND CONSTRUCTION EFFORTS

Washington.[91] The crossing continues to be important for road, railroad, and Metro use.

BRIDGES ACROSS THE
EASTERN BRANCH

Bridges were also needed across the Eastern Branch (Anacostia River). The bridges built there were constructed by companies organized along the same lines as those that built the Potomac bridges, and the Eastern Branch bridges used post-and-beam construction, as the Long Bridge did.

Eastern Branch Bridge Company On December 24, 1795, the Maryland legislature authorized the Eastern Branch Bridge Company to build a bridge from the foot of Kentucky Avenue across the Eastern Branch (see plate 9).[92] The bridge was completed and opened on January 12, 1804.[93] Until the opening of the Navy Yard Bridge (1819), it was known as the Lower Bridge.

Anacostia Bridge Company In November 1797 the Maryland legislature authorized a second company, the Anacostia Bridge Company, to build a bridge across the Eastern Branch in the new federal city. The bridge was to be built at the foot of East Capitol Street. Benjamin Stoddert, Thomas Law, and John Templeman were authorized to receive subscriptions.[94] Subscriptions were lacking, though, and it was not until 1805 that construction began, primarily at the expense of Benjamin Stoddert. The bridge became known as Stoddert's Bridge, later Ewell's Bridge, and still later Benning's Bridge, after it was purchased by William Benning.[95] It was also called the Upper Bridge.

Both bridges were substantially damaged by American forces in August 1814 in a futile attempt to prevent British troops from entering Washington. The following year both companies petitioned Congress for relief for the war damage to their respective bridges.[96]

NOTICE.

TOLLS will be collected for crossing the BRIDGE over the Eastern Branch, from Thursday the 12th inst. Any person wishing to engage by the year will please to apply to the Toll gatherer at the bridge.

THE DIRECTORS.

January 9—2t

Toll announcement for the Eastern Branch (Anacostia River) bridge. *Washington National Intelligencer*, January 9, 1804.

Congress.

HOUSE OF REPRESENTATIVES.

MONDAY, SEPTEMBER 26.

Mr. *Hurlbert*, a new member from the state of Massachusetts elected in the room of Mr. Dewey, resigned, appeared and took his seat.

A petition from the President and Directors of the Washington Bridge Company, and another from the Eastern Branch Bridge Company, praying compensation for the damage done to their bridges by the army under General Winder, were presented and referred.

Memorials to rebuild the Washington Bridge, or Long Bridge, and the Eastern Branch Bridge. In August 1814, in a failed attempt to prevent the British from entering Washington, both the upper and lower bridges across the Eastern Branch were destroyed or severely damaged. They were rebuilt after the War of 1812. *New Bern (NC) Carolina Federal Republican*, October 8, 1814.

Navy Yard Bridge, photographed by Matthew Brady, 1850–1865. The Eastern Branch (Anacostia River) bridges, like the Washington or Long Bridge, were built by constructing a wooden deck on piles rammed into the riverbed. Courtesy of the Library of Congress.

Navy Yard Bridge Company On February 24, 1819, Congress authorized a third Anacostia River crossing, to be built by the Navy Yard Bridge Company.[97] The Navy Yard Bridge crossed the river between Eleventh and Twelfth Streets immediately east of the Navy Yard.[98] Construction on the bridge began July 4, 1819, and it was completed and in use by October 1819. The bridge was 1,600 feet long and 25 feet wide and was built of wooden beams on pilings. Like the Long Bridge, it contained a draw over the navigable channel, which in this case was 24 feet deep.[99] This bridge is usually remembered as the bridge that John Wilkes Booth crossed after assassinating President Abraham Lincoln on April 14, 1865.

WASHINGTON CITY CANAL

Besides bridges and turnpikes, the early inhabitants of Washington were very interested in building canals. The idea of a Washington City Canal, to link the Potomac River and the Eastern Branch, dated to L'Enfant's original plan of the city presented to President Washington in 1792.[100]

Washington City Canal lottery ticket issued to raise money for canal construction. Author's collection.

The canal proposed by L'Enfant would have linked the eastern and western portions of the new city. Access from the Potomac River would have been via the lower portion of Tiber Creek, immediately south of the President's House. When Benjamin Latrobe subsequently developed detailed plans for the canal in 1804, he located the entrance at the mouth of Tiber Creek, at present-day Constitution Avenue and Thirteenth Street, NW.[101] Latrobe's canal would have proceeded due east along what is today Constitution Avenue, NW. At Pennsylvania Avenue immediately west of the Capitol, the canal would have turned 45 degrees southeast for one block and then south to Maryland Avenue, SW. Between K and I (Eye) Streets, SE, the canal would have turned southeast and proceeded to a point between Second and Third Streets, SE, where it would have turned south to meet the Eastern Branch. A second branch of the canal was planned from the Eastern Branch north along St. James Creek immediately east of Greenleaf Point (present-day Fort McNair) and then northeast to where it intersected the first branch at South Capitol Street, immediately south of Virginia Avenue (see plate 10).[102] Advocates for this canal argued that cargoes being shipped down the Potomac River via the Potomac Canal could inexpensively and quickly be forwarded to all parts of the new city.

On September 8, 1791, the commissioners considered building this canal, but they postponed its construction owing to a lack of funds.[103] A year later the commissioners reversed their decision, and work was begun in 1793 on the branch of the canal along St. James Creek, between Tiber Creek and the Eastern Branch.[104] On August 3, 1792, the commissioners ordered that bids for work on this canal be advertised in the local newspapers.[105] On September 1 a contract for the canal was issued to Patrick Whelan. The canal was to extend one and one-eighth mile, from

In purluance of an Act of Congre pafied the 1ft day of M.y 1802, entitled, "An act to abolifh the board of Commiffioners of the city of Wafhington, and for other purpofes."

WE the underfigned have opened a book at Rhodes's tavern Pennfylvania avenue, for receiving fubfcriptions for openii g the CANAL, to communicate from the Potomac river to the Eaftern Branch thereof.

The Report and Plans of Mr. Latrobe, on the fubject, are depofited at Mr. Rhodes's for the infpection of thofe inclined to promote this ufeful and beneficial work.

THOMAS TINGEY,
DANIEL CARROLL, of D'n.
THOMAS LAW,
DANIEL C. BRENT.
Wafhington 11th June, 1804.

Advertisement for Washington City Canal subscriptions. *Washington National Intelligencer*, June 27, 1804.

PUBLIC MEETING.

A meeting of thofe who are difpofed to promote the OPENING of the CANAL through the City is requested at STELLE's Hotel, on Wednesday evening next, at Six o'clock P. M.
May 11—3t

Washington City Canal promoters' meeting.
Washington National Intelligencer, May 11, 1807.

the tidewater of Tiber Creek to the tidewater of St. James Creek. It was to be 12 feet wide at the bottom and 15 feet wide at the top, and the bottom of the canal prism or section was to be two feet below the common water level.[106]

Money, or rather the lack of it, continued to be a problem for the commissioners. To continue the work begun, in November 1795 the Maryland legislature authorized Notley Young, Daniel Carroll of Duddington, Lewis Deblois, George Walker, William Mayne Duncanson, Thomas Law, and James Barry to hold two annual lotteries, each with a goal of $52,500, so as to raise $100,000 to be paid to the commissioners to make the Tiber Creek Canal navigable.[107] An agreement for this lottery between these men and the commissioners was entered into on February 6, 1796.

The first lottery raised $28,000,[108] although no cash awards were paid and what became of the proceeds remains a mystery.[109] Without funds, work on the canal stalled until 1802, when a group of prominent citizens drafted a memorial to Congress asking that Congress incorporate a company to build the canal. As with the bridges across the Potomac and the Eastern Branch, private money eventually had to finance the Washington City Canal.

WASHINGTON CANAL COMPANY
On May 1, 1802, Congress enacted legislation establishing the Washington Canal Company. In the following year, architect-engineer Benjamin Latrobe was engaged to develop drawings and a report describing the proposed canal, and by July 1803 Latrobe and his two assistants, Robert Mills and William Strickland, were working on these plans. This was a busy time for Latrobe, as he had recently been appointed surveyor of the public buildings by President Jefferson and was also engaged as chief engineer for the design and construction of the Chesapeake and Delaware Canal, one hundred miles north of the city. Latrobe's workload may have slowed the development of the plans for the Washington City Canal, which were not completed until February 1804.[110]

Shortly after Latrobe prepared his report and drawings for the proposed canal, Thomas Law published, anonymously, his promotional pamphlet for the Washington City Canal, *Observations on the Intended Canal in Washington*.[111] By mid-June 1804, prospective investors were able to inspect Latrobe's report and drawings together with Law's pamphlet.[112] Despite Latrobe's highly favorable report and plans and Law's pamphlet, there was not sufficient investor interest to organize the company. The canal project remained dormant for the next five years.[113]

The act of 1802 required that the new company be organized within five years of enactment, and since the deadline was not met, the canal company authorization expired in 1807. But interest in the city canal continued. A meeting to promote the canal was held in May 1807.

THE NEW WASHINGTON CANAL COMPANY

At this meeting, Latrobe's "very able report and accurate surveys with various calculations"[114] were presented and referred to a committee composed of Robert Brent, Daniel Carroll of Duddington, John P. Van Ness, Thomas Law, John Davidson, James D. Barry, and George Blagden.[115] Nothing came of the meeting until February 16, 1809, when Congress passed a new act of incorporation for a new canal company, once again called the Washington Canal Company. This act authorized Daniel Carroll of Duddington, George Blagden, Griffith Coombe, Frederick May, James D. Barry, John Law, and Elias B. Caldwell to raise $100,000 in subscriptions for the new company, for the purpose of building the city canal in seven years. Subscription books were opened at Long's Hotel on May 25, 1809. Half of the specified amount was raised, allowing the company to be organized.[116]

On January 10, 1810, stockholders of the new Washington Canal Company elected their first board of directors: Elias B. Caldwell, James D. Barry, Peter Miller, Griffith Coombe, Edmund Law, Daniel Carroll of Duddington, and Frederick May.[117] Caldwell was elected president of the company. On January 19, 1810, Latrobe accepted the position of chief engineer, to be compensated by $2,000 in canal shares if the construction cost exceeded $40,000 or 5 percent of the total if it fell below $40,000 (along with other considerations).[118]

Latrobe met with Caldwell and the directors of the canal company. In a letter to them summarizing that meeting, Latrobe specified the directors' four primary concerns: (1) that the work be done on the most economical scale, (2) that the locks be constructed of wood, less expensive than masonry, (3) that each end of the canal be open or below the locks as far as possible, and (4) that before any expense was authorized, the engineer present an accurate estimate.[119]

WOODEN LOCK WALLS VERSUS MASONRY

The directors wanted to limit the initial cost of the canal, and building locks out of wood instead of masonry was one means to achieve this goal. Although wooden locks would be less expensive initially, both Latrobe and the company directors knew that they were known to have a much shorter service life than masonry locks. The wooden locks constructed for the Potomac Canal at Little Falls in 1795, for example, were already by then (1810) in an advanced state of decay.[120] Nonetheless, Latrobe acquiesced in the directors' choice of wood for the locks;[121] it was a decision he later regretted (see plate 11).

Limiting the required amount of excavation by limiting the size of the canal prism (the cross section) was another way to limit cost. Latrobe recommended a canal prism 20 feet wide at the water line, 11 feet wide at the bottom, and 3 feet deep, with side slopes of two feet vertical to three feet horizontal — a relatively flat slope.[122] The prism's side slopes could not be made steeper (and thus lessen the amount of excavation required) because of the danger that the earthen slopes would slide or slump into the canal. The canal prism decided upon and contracted in April of 1810 was somewhat larger than these dimensions: 27 feet wide at the top and 12 feet wide at the bottom.[123]

Another excavation cost-saver that Latrobe recommended was "to lay the level of the bottom of the cut as high above low Water as possible." The potential problem here was that setting the level of the bottom of the canal too high might disrupt or delay navigation at low tide. But if it was set too low, the excessive excavation would waste money.[124]

Latrobe asked the directors to set the bottom elevation of the canal one foot above low tide. However, the draft contract called for the bottom of the canal to "be level with the low water mark as set out."[125] Tidal action therefore would not flush out the sediment, garbage, and sewage deposited in the canal, and these deposits would greatly restrict the canal's use during low tide.

Latrobe also asked the Board of Directors for authorization to initiate discussions with "two respectable and experienced Canal diggers to induce them to come to Washington in order to view and make offers for contracts for the work." The two contractors Latrobe had in mind were James Cochran and Charles Randle (sometimes spelled Randal); both of them had worked for him on the Chesapeake and Delaware Canal.[126] On April 20, 1810, the company awarded Cochran a $40,000 contract to construct the Washington City canal.[127]

GROUNDBREAKING FOR THE WASHINGTON CITY CANAL
President James Madison and Mayor Robert Brent officiated at the May 2, 1810,[128] groundbreaking for the new canal, near the corner of New Jersey Avenue and E Street, SE.[129] Contractor Cochran concluded the ceremonies by ploughing a deep furrow with a team of six horses.[130]

On June 11, 1810, Latrobe was able to write his father-in-law, Isaac Hazlehurst, "[I am also working on] carrying a Canal of two miles in length thro' the heart of the

city, on which I have 100 Irishmen at work."[131]
By July 17, Latrobe gave Thomas Law a report
of good construction progress:

> We have now got thro about one third of
> the whole work. All the old cut,[132] the cut
> from Walter Jones's along Pennsylvania
> Avenue and thence to the creek and a
> part of the deep cutting[133] is down to the
> level of the towing path, and the deep
> cutting itself across the Jersey Avenue is
> in many places 14 feet deep. On the East-
> ern branch we have as yet done nothing.

The piling in the Tiber extends from the Bridge Eastward to the Lock and many
of the piles of the Lock are driven. One set of Lock gates is finished and the rest of
the framing [is] in forwardness.[134]

Installment payments due. *Washington National Intelligencer,*
July 30, 1811.

But there were money problems. Wrote Latrobe: "Our difficulty is not now to get
the work done but to find money to pay for it. There has been some apprehension
that several Subscribers would be delinquent. Brent and Tingey and Moses Young
have shown reluctance and created alarm, but I hope and believe that they will
still pay."[135]

Subscribers to stock offerings like the Washington Canal Company did not pay all
of the promised money at one time but paid in installments. One share of stock in
the Washington Canal Company required ten installment payments of ten dollars
each.[136] Subscribers might enter into their subscription in good faith, only to later
find they were unable or unwilling to make the required subsequent payments;
they would then default on their payments. A few installment defaults by some
stock subscribers might encourage other subscribers to withhold their payments. If
enough subscribers withheld payment, construction would be halted. This is what
occurred on the Washington City Canal.

Despite these financial problems, Latrobe remained optimistic. On September 1,
1810, he wrote Isaac Hazlehurst that the canal might be opened by Christmas 1811.[137]
This optimism was shared by many. But by the end of the year the canal had not
been completed, and lack of money had halted the work. On December 31, 1810,
Latrobe wrote to canal company president Elias B. Caldwell that he had tried to
persuade the contractor, Cochran, not to discharge his workers for the winter, but
had failed.[138] Cochran soon after left Washington to work on the National Road, a
project that began at Cumberland, Maryland, and extended west. In June of 1811
Latrobe wrote Thomas Law that he had written to Cochran urging him to return
to Washington and finish the canal.[139] This had no effect. On July 8, 1811, Latrobe

WASHINGTON CITY.

TUESDAY, APRIL 23.

DISTRICT OF COLUMBIA.

The following statement presents a view of the population of this district at the times of taking the census in 1800 and 1810 respectively.

	In 1800	In 1810
Washington city	3,210	8,208
Georgetown	2,993	4,948
Washington county, exclusive of towns	1941	2,315
Alexandria		7,227
County exclusive of town	5949	1,325
	14,093	24,023

From this statement it appears that the population of the district is nearly doubled within ten years; whilst that of the city is almost trebled.

The city is in truth in a prosperous situation, and the rapidity with which it is now improving augurs a still greater increase than we have experienced for three years past. The Canal which passes through the centre of the city, uniting the Potomac and Eastern branch, is now nearly finished, and will embellish the city, and benefit the holders of property on its banks more than we fear it will those gentlemen whose public spirit induced them to undertake it. Many new houses have been already commenced this season, and more are in contemplation; so that we may safely say the increase of Washington is as rapid as that of any city whose population commenced under similar circumstances.

Washington City Canal "nearly finished." *Washington National Intelligencer*, April 23, 1811

again wrote to Cochran to urge him to return to Washington.[140]

Latrobe needed Cochran back in Washington not only to complete the excavation but also to finish the two locks, one at the west end of the canal, near the mouth of Tiber Creek, and one at the eastern end of the canal at the Eastern Branch. The eastern lock proved the most difficult, because of a high water table and high tides and because, as Latrobe put it, the lock was "in fact only a box floating in mud." By June 27, 1811, Latrobe was predicting to Thomas Law that the end was in sight in building the lower lock pit,[141] thoughts he later expanded in his July 8, 1811 letter to James Cochran: "The Company have had a terrible piece of works with the Lock at the Eastern branch. I have never been engaged in anything half so difficult and disagreeable. But the worst is now over. We have got our Lockselles laid and are now raising the Walls, and by the 1st of Augt. the Lock will be passable. At the other end all is ready, and if your 490 Yards were cut, which might be done in three weeks if you were here with a gang of 60 Men, the Canal would begin to produce money."[142]

OPENING OF THE WASHINGTON CITY CANAL

Cochran did return to Washington, but not immediately. And the Washington City Canal was not completed in August 1811 but four years later; it was formally opened on November 21, 1815.[143]

Latrobe thought the greatest mistake he made in building the Washington City Canal was his acquiescence to the Board of Directors' desire to build wooden locks instead of masonry locks and their desire to place the two locks of the canal as far inland as possible. In 1815 Latrobe cautioned Charles Randle, who was then considering constructing a canal for the Clubfoot Canal and Harlow's Creek Company of North Carolina, that he should be aware of the "greatest and most expensive of all follies, the building of wooden tide locks." He advised Randle to build masonry locks.[144]

The company's Board of Directors had directed that the two locks of the canal be placed as far inland as possible so that ships would not have to pay lockage fees to use the docks along the lower reaches of the canal, several of which were owned by members of the Board of Directors. Latrobe found that this location of the locks allowed the incoming tide to deposit large amounts of sediment below the locks, which proved difficult to dredge. Even small amounts of sediment blocked navigation, owing to the diminutive depth of the canal and the approaching waterways.[145] Had the locks been installed at the extreme ends of the canal, as Latrobe intended, closing the lock gates during high tide would have kept tide-borne sediment from entering the canal. Opening the lock gates at low tide would have had a scouring effect on sediment brought into the canal from Tiber Creek and other sources. It was not enough to hire an experienced engineer; it was also necessary to follow his advice.

There were other problems with building the Washington City Canal, including slips and quicksand. Slips were embankment slumps of the canal's slopes, where the caved-in earth would have to be reexcavated by the contractor and the embankment reinforced or excavated to a shallower angle to prevent future slips. Latrobe wrote Randle, "You know how Mr. Cochran fared by slips and quicksands in the Washington Canal, and that he would have been ruined had he not been rich before."[146]

Also, in 1812 Elias Caldwell resigned as president of the canal company, to be replaced by Dr. Frederick May. In the autumn of 1813, Latrobe left Washington for Pittsburgh, to return in 1815.

The canal's first full year of operation was 1816, and it showed a profit of only $458 for this first year.[147] The profits of the Washington City Canal did not improve. After the first report, the company began reporting the status of their treasury rather than the amount of profit they generated. By comparing the financial status of the company in these years, the small amount of profit generated by the company can be estimated. In the three years 1824–1826, for example, the canal company reported the following amounts in their treasury: 1824, $182.10¾; 1825, $38.66½; 1826, $48.11¾.[148]

One of the continued efforts that were made to improve the canal's revenue was to link the Washington City Canal with the new Chesapeake and Ohio Canal. This new canal was intended to replace the older Potomac Canal and link Washington with Pittsburgh. On July 4, 1828, ground was broken for the new canal at Little Falls. Later the canal was extended into Georgetown, and by 1831 a Washington extension had been constructed between the lower end of the new canal, at the mouth of Rock

Washington. D.C., Canal, September 21, 1850, by Seth Eastman, American, 1808–1875. The western entrance to the Washington City Canal, looking east along what became Constitution Avenue. Graphite pencil on paper, 12.2 × 19.9 cm (4 ¹³⁄₁₆ × 7 ¹³⁄₁₆ in.). A gift of Maxim Karolik for the M. and M. Karolik Collection of American Watercolors and Drawings, 1800–1875, 54.1711. Photograph © Museum of Fine Arts, Boston.

Creek, and the western terminus of the Washington City Canal. The president and the Board of Directors defined the route of the Washington extension: "From 27th street to 26th street through square south of 12; from the West side of 26th street to the east side of 23d street, in the Potomac River—with little exception—from the east side of 23d street to the west side of 21st street through squares Nos. 63 and 89; from the west side of 21st street [Northwest], on the river, and along the bed of B street [now Constitution Avenue]."[149]

This one-and-a-quarter-mile linkage was connected to the Washington City Canal by a tide lock.[150] It was at that time that the still-existing lock-house at Seventeenth Street and Constitution Avenue, NW—the last remaining above-ground remnant of the Washington City Canal—was built.[151]

DEMISE OF THE WASHINGTON CITY CANAL: 1831–1870
On January 3, 1831, the City Council of Washington passed an act for the city to purchase the Washington City Canal. The mayor was authorized to issue $50,000 of shares at $100 each to provide for the purchase. A deed of July 23, 1831, was signed, and the purchase of the canal was ratified and confirmed by an act of Congress of May 31, 1832. This ratification was passed under the provision "that the said canal shall be finished and completed, of the breadth and depth, and in the manner and within the time hereafter prescribed, and not otherwise." Another provision

Elevation, Lock-House at Seventeenth Street, NW, and Constitution Avenue. D. McGrew, delineator, 1939. This lock-house was built for the connecting canal between the Washington City Canal and the Chesapeake and Ohio Canal. Historic American Buildings Survey collection, Library of Congress.

required that "the said canal, through its whole length and breadth aforesaid, shall have a depth of at least four feet water at all times."[152] These provisions were not met. On March 2, 1833, Congress passed an appropriation of $150,000, "to aid them [the mayor, alderman, and common council of Washington] in fulfilling the objects and requirements of (the act of May 3, 1832, aforesaid)."[153] Problems with the canal continued. By 1846 it was reported that the canal could be used only at high water, owing to the sewage obstructing it.[154] Subsequent appropriations, for $20,000 each, were passed in 1849 and 1851 for "clearing out and deepening that portion of the Washington city canal which passes through and along the public grounds." Despite these acts, the canal was never really finished. In 1868 a Senate committee reported, "[The canal] is still unfinished . . . as to the greatest part of it useless."[155] The report continued, "The canal remains unfinished. It has often, by the accumulation of stagnant sewerage and filth, been a loathsome spectacle if not an absolute nuisance."[156]

Despite these recitations of the long-standing failure of the Washington City Canal, as late as 1868 the senators were still enticed by the old economic arguments in its favor.[157] But finally, almost seventy years after first being authorized by Congress, even the old economic arguments for the Washington City Canal lost their influence, and in 1870 the canal was filled in.

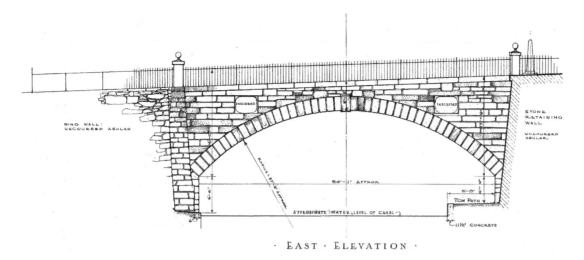

· EAST · ELEVATION ·

East Elevation, Wisconsin Avenue Bridge, Georgetown. D. McGrew, delineator, 1939. One of the ten bridges across the extension of the Chesapeake and Ohio Canal from Little Falls down into George-town. The Wisconsin Avenue bridge is the only one of the ten that survives. Historic American Buildings Survey collection, Library of Congress.

BRIDGES OF THE CHESAPEAKE AND OHIO CANAL

Meanwhile, the extension of the Chesapeake and Ohio Canal from Little Falls down to Georgetown cut off the lower portion of Georgetown from the rest of the community, thus necessitating additional bridges. Although the canal company initially had a policy of building as few bridges across the canal as possible, they eventually yielded to Georgetown and built ten bridges, five masonry and five wooden, across the canal in Georgetown, as well as a wooden bridge across the tumbling dam at the mouth of Rock Creek and a road culvert west of the Foxall Foundry, also west of Georgetown.[158]

Despite the eventual failure of the Washington City Canal, in the first decades of the new federal city, private and public bodies had developed a transportation infrastructure. Initially, the commissioners relied on time-honored approaches for these structures, such as in the use of a builder like Leonard Harbaugh and in the construction plan for the K Street Bridge, with their reliance on rules of thumb. But the cost of these projects was too large to risk failure. Gradually it became clear that experienced engineers, benefiting from newly emerging bodies of knowledge like strength of materials were needed to design them and bring them to completion. Engineers' advice and involvement became an important aspect to ensure that these transportation improvements would be adequate to their use and within budget.

Birds-eye view of Washington, 1861. The transportation improvements to the new federal city included the Potomac Aqueduct of the Alexandria canal (top right); the Long Bridge (*center right*); the Navy Yard Bridge (*bottom left*); and the Washington City Canal (*diagonal, middle*). Courtesy of the Library of Congress.

6 Building Military Defenses for the Capital

Digg's-Point, about ten miles below the city of Washington, is remarkably well calculated for a battery; as all vessels coming up the river must present their bows, for the distance of three miles; and after passing, their sterns are equally exposed for about the same distance; the middle of the channel there is not more than two hundred yards from the point.

TOBIAS LEAR, 1793

A little before you come to Mount Vernon we have the fort on the Maryland side; it appeared to be large, but no one present could give an opinion as to its strength. Governor Barbour, of Virginia, and several other intelligent looking men were on board; they could give no information respecting it, in fact, they seemed to speak of it in terms of contempt.

ANN ROYALL, 1826

With construction of the new federal city proceeding, and with the uncertainty of the European wars weighing on their minds, leaders of the new republic focused on the construction of seacoast fortifications to protect their coastal cities. To plan, design, and supervise the construction of these new fortifications, the United States turned mainly to military engineers educated and trained in Europe. These engineers were adequate for designing and constructing individual fortifications, but they lacked a broader, strategic view of defense of the nation. The defenses envisioned were small, inexpensive, and designed to protect a single town or city, not a region. And they were inadequate. In August 1814, the inadequacy of the defenses was brutally brought home to the residents of the new federal city by the British invasion.

FIRST DEFENSES FOR THE NEW CITY: JONES POINT

On May 12, 1794, Secretary of War Henry Knox sent instructions to French engineer Jean Arthur de Vermonnet to build a coastal battery to protect Alexandria and Washington.[1] John Jacob Ulrich Rivardi,[2] a friend of William Thornton who was subsequently engaged in military construction for the US Army elsewhere,[3] had been assigned the responsibility of building this coastal fortification. Rivardi was a Swiss military engineer who had served in the artillery of the Russian army. But Rivardi was given assignments elsewhere, so Knox turned to Vermonnet. Knox was from Boston and was writing Vermonnet from New York. He would have had little direct knowledge of the geography of the new federal city and, as a result, confusingly described Digges Point, located ten miles below Washington, as the point at the confluence of the Eastern Branch and the Potomac. Wrote Knox, "The President of the United States, who is well acquainted with the river Potomac, conceives that a certain bluff of land, on the Maryland side, near Mr., Digge's, (the point formed by the Eastern Branch and the Potomac,) would be a proper situation for the fortification to be erected."[4] Knox's placement of Digges Point above Alexandria must have appalled the citizens of Alexandria, because the coastal battery to be built would have placed Alexandria downriver of the defenses that might provide protection against attacking ships coming up the Potomac. This confusion was sorted out, and the new gun emplacement, to be built by Vermonnet, was located not at Digges Point but at Jones Point, immediately south of Alexandria; it was the southernmost location within the newly defined District of Columbia.

Knox instructed Vermonnet that the new battery would be inexpensive, not to exceed three thousand dollars, and was to be constructed of earthen parapets. Vermonnet was to determine whether the battery would be *en barbette* (with gun emplacements on top of the parapet rather than in casements in the defensive walls) or not. From his instructions it is clear that Knox preferred to have the guns mounted so as to fire *en barbette*. But if not, or if some were not, then Knox advised Vermonnet that the guns should be enclosed within stout wooden embrasures.[5]

Knox wrote that the new coastal battery was to be protected from land attack by a blockhouse built in the rear, similar to what the Europeans referred to as a Martello tower, that is, a small tower, usually of masonry, with a gun platform on top and accommodations for the troops below. Knox specified that the blockhouse should hold fifty men and should be constructed of timber. The redoubts to defend the coastal battery were to be of sufficient size to hold five hundred men. However, timber casements were not to be erected at this time, except for the very substantial magazine.[6]

Knox provided Vermonnet a great deal of latitude in the design and siting of this fortification. Knox wrote, "Your judgment will also direct what parts of your works shall be protected by frieze, and what by palisadoes; or, whether your redoubts shall have embrasures, or fire *en barbette*, with small cannon." Knox was clear that although Vermonnet had overall responsibility, Vermonnet was to function as the military engineer for this fortification and that someone else would be responsible for supervising construction. Vermonnet was to receive four dollars per day plus expenses for this work.[7]

On June 17, 1794, Vermonnet gave Knox a status report on his work at Jones Point. He had developed a plan for the new fortification; had ordered logs, planks, and other timbers from the Chesapeake region; and had begun building a haul road to bring those materials from the port through the marsh below Alexandria to Jones Point.[8]

On July 5, 1794, Vermonnet sent Knox his plans for this fortification and his plans for developing the haul road through the marsh above Jones Point. Wrote Vermonnet, "The Battery of Jones' point will be a barbette, and calculated for receiving 12 pieces of heavy cannon."[9] On November 5, 1794, Vermonnet wrote Knox that he was going to stop work for the winter on November 15, 1794, and resume work on April 15 of the following year, even though this battery was supposed to have been completed by the end of 1794. Vermonnet explained to Knox that the reasons for this delay were "the absence of militia" and that his appointment was made much later than others.[10]

Construction on the coastal battery continued in 1795, but the fortification was never completed and manned. On January 18, 1796, Secretary of War Timothy Pickering reported to the Senate that the fortifications at Annapolis and Alexandria that had been under Vermonnet's charge were found "unfavorable" and had been "relinquished."[11]

ABANDONMENT OF THE COASTAL FORTIFICATION AT JONES POINT

There the matter remained for a decade. The hostilities of the Napoleonic Wars threatened to involve the United States, and a review of coastal defenses was undertaken in 1806. On February 18, 1806, and again in January 6, 1809, reports to the House of Representatives indicated that the work at Jones Point was in ruin.[12]

The following February, Colonel Jonathan Williams visited the site with a Colonel Burbank. Colonel Williams reported that he had been impressed with the site.

> Having been favored by the company of Col. Burbank, they proceeded to Alexandria by water where they arrived the 2nd instant. They walked to a point about a mile below the city, where they saw the vestige of an old Fort which presented a circular battery in the front, and 2 small bastions in the rear; the whole ditched round in the usual way. The fort did not occupy the whole ground but appeared to the subscriber to be tolerably well designed. The small size of the Bastion in the rear, evidently discovered that only a picket defense by a quarter was there contemplated.
>
> This spot appears constantly well calculated for a water battery with very little expense. It is so low that the guns would be on a level with those of the enemies ship; and it looks down the channel in such a manner that no ship could avoid an attack on her bows and from the first moment that she came within cannon shot, could she keep out of it, while the whole width of the Channel would not permit her to pass at more than 300 or 400 yards distance. On each side of the

fort is a fine harbor where any number of gunboats could ride in perfect waters and complimentary aid the batteries. The command of the high ground in the rear is distant and being in our possession is a circumstance rather in favour than against the position. A fort or rather breastwork battery could be built with little expense, since the earth might be taken from the ditch would be ample to make the rampart and parapet and by proper sluices, the ditch might be full or empty at will.[13]

Despite this favorable report, the War Department had already begun plans to replace the ruins of the coastal battery at Jones Point with a more substantial battery to be constructed at Digges Point on the Maryland shore of the Potomac River. The army's 1807 report to the Senate read: "Potomac—Digges' Point, below Alexandria, is a commanding position; and with a strong battery, covered by a redoubt and two block houses on the highest parts of the adjacent eminence, would, with _____ gunboats, render sufficient protection to Alexandria, the City of Washington, and Georgetown, against the approach of any such naval force as could be reasonably contemplated. Probable expense, _____."[14]

NEW FORTIFICATION AT DIGGES POINT:
FORT WARBURTON

Captain George Bomford was assigned to oversee the construction of the new fortification, which was sometimes called Fort Warburton, after the name of the plantation at that point, and sometimes called Fort Washington. Work began on April 14, 1808, and was completed December 1, 1809. Bomford described the fort, whose plan was copied from that of Fort Madison at Annapolis, Maryland, as "an enclosed work of masonry comprehending a semi-elliptical face with a circular flank on the side next to the Potomac." It was larger than the fortification at Jones Point but still small. The planned wartime garrison was 120 men. An octagonal masonry tower, similar to Martello towers in Europe, was built on the bluff above the fort.[15]

By January 1809, the new coastal defense battery was almost finished. In a report to Congress, the fort was described:

MARYLAND

Fort Washington, on the Potomac, between Alexandria and Mount Vernon, is a new enclosed work, of stone and brick masonry, to which is attached a strong battery of like material. The whole is so nearly completed as to be ready for the reception of the cannon and garrison, which have been ordered, and have arrived at the fort. A stone tower has also been commenced on an eminence that overlooks the fort, and is in considerable forwardness.[16]

By December 21, 1809, it had been completed:

MARYLAND

Potomac River. — *Fort Washington*, an enclosed work of masonry, comprehending a semi-elliptical face, with circular flanks on the side next to the Potomac, mounting thirteen guns, commanded by a tower of masonry, calculated to mount six guns; with a brick magazine, and barracks for one hundred and twenty men, including officers.[17]

On December 11, 1811 it was again described: "On Potomac — Fort Washington, situated at Warburton, on the east side of the river; an enclosed work of masonry, comprehending a semi-elliptical face, with circular flanks, mounting thirteen heavy guns; it is defended in the rear by an octagon tower of masonry, mounting six cannon; a brick magazine, and brick barracks for one company of men and officers."[18]

Fort Warburton was immediately criticized for its shortcomings. Its builder, Captain Bomford, wrote, "Fort Washington [i.e., Fort Warburton] was really an attempt to adopt a standardized plan to an unsuitable site. It violated a fundamental rule in the art of fortification — the fort must suit its site."[19] General James Wilkinson described the fort as "a mere water battery," "useless once a vessel had passed," with an octagonal blockhouse that "could be knocked down by a twelve-pounder."[20] Major Pierre L'Enfant was sent to evaluate the fort in 1813 and reported, "The whole original design was bad and it is impossible to make a perfect work of it by any alteration."[21] Like the fortification at Jones Point, the new fort at Digges Point was a water battery susceptible to attack from the land side. Engineers had been used to supervise its construction, but it was not "engineered." That is, it was built by a standard plan and not designed to meet the vagaries of either site or mission.

FORT WARBURTON IN THE WAR OF 1812

By the summer of 1814, the British fleet was menacing the Chesapeake region and threatening to move north to Washington, Alexandria, and Baltimore. Lieutenant James L. Edwards, then commanding Fort Warburton, on July 25, 1814, reported the deplorable situation of the coastal fortification. He said the Columbiads, capable of firing eighteen-pound cannonballs, had been delivered, but the fort had no way to mount them because it lacked a gin and tackle and the gun platforms built for them were too narrow for the recoil of the cannons; when the cannons were fired, they would run off the platform. He also reported that there was no ammunition for the eighteen-pound water battery to be used against ships. In addition, some of the existing gun carriages were in bad order, although serviceable. The assigned strength of the Fort Warburton garrison at the time was less than half of the 120 men originally intended.[22]

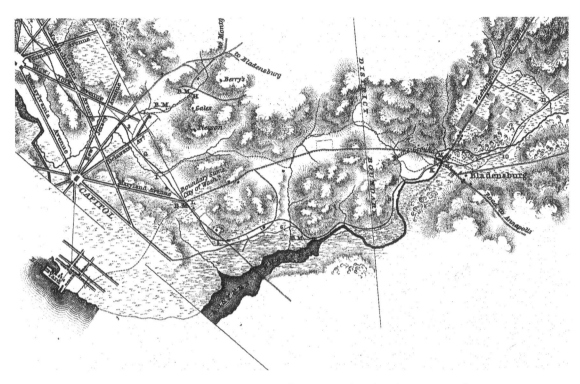

Map of Bladensburg and the approach to Washington. William Howard, civil engineer, 1827. Before the Battle of Bladensburg, there were three bridges across the Eastern Branch (Anacostia River). Both the Lower Bridge (owned by the Eastern Branch Bridge Company) (not shown), at the foot of Kentucky Avenue, and the Upper Bridge (owned by the Anacostia Bridge Company), a short distance above the foot of East Capitol Street (*middle, above the word "Eastern"*), were destroyed by American forces to prevent the British from using them to cross the Eastern Branch. The third bridge was at Bladensburg (*center right, to the west of Bladensburg*); it was not destroyed before the battle. The British arrived from the east, crossed the bridge, and defeated the arrayed American forces. Brigadier General Winder, commanding the Tenth Military District, later wrote, "There was not a bridge on the road . . . the destruction of which would have retarded [the enemy's] advance ten minutes." Once defeated, the American forces fell back along the Bladensburg road. As shown on this map, there was no natural position that would have assisted a regroupment of American forces for the nine miles between Bladensburg and the Potomac River or in the heights above Washington. Detail from William Howard, civil engineer, map of a proposed canal from Baltimore to the Potomac, accompanying his report. [William Howard], *Report on the Survey of a Canal from the Potomac to Baltimore* (Baltimore: B. Edes, 1828).

THE BRIDGE AT BLADENSBURG IN 1861.[1]

"The bridge at Bladensburg in 1861." Of the three bridges across the Eastern Branch, the bridge at Bladensburg was the smallest. It was the only one not destroyed by American forces attempting to block a British advance on Washington. The British troops rushed the bridge and attacked the American forces. From Benson J. Lossing, *The Pictorial Field Book of the War of 1812* (New York: Harper, 1868), 927.

Brigadier General William Winder, commanding the Tenth Military District, was aware of the weaknesses of Fort Warburton. On August 19, 1814, five days before the British attack on Washington, he wrote Secretary of War John Armstrong that ships should be prepared to be sunk at Fort Washington to prevent enemy ships from passing and that boats should be made available to ferry troops across the Potomac as the need arose, so as to reinforce Fort Washington with ground troops.[23] These recommendations were not acted upon.

On the eve of the Battle of Bladensburg, it was widely understood that Fort Warburton was susceptible to land attack. On August 20, 1814, four days before the battle, General Winder ordered Brigadier General Robert Young to take his brigade and establish a defensive position to protect Fort Warburton from land attack.[24] With

PROJECT PLANNING AND CONSTRUCTION EFFORTS

their defeat at the Battle of Bladensburg on August 24, American forces rapidly vacated Maryland east and south of Washington. General Young and his brigade were ordered to cross the Potomac and proceed fifteen miles west of Alexandria. General Winder himself wound up at Montgomery County Court House (present-day Rockville), fifteen miles north of Washington. There were few, if any, American forces between the invading British and the land side of Fort Warburton. The small garrison at Fort Warburton was very much on their own.

Just before the Battle of Bladensburg began Vice Admiral Alexander Cochrane, commander in chief of the British expedition, sent a squadron up the Potomac River under Captain James Gordon as a diversionary move. The squadron consisted of two frigates, three bomb vessels, a rocket ship, and a small tender and dispatch boat.[25] They had difficulty passing the shoals of Kettle Bottom in the lower Potomac, but by August 27, 1814, the squadron had reached the vicinity of Fort Warburton. Without firing a gun, Captain Samuel Dyson, commanding officer of Fort Warburton, ordered the fort to be blown up and abandoned. Dyson had thought British troops were advancing on Fort Warburton from the land side. Later, Dyson was dismissed from the army by court-martial for this act.[26]

BUILDING A REPLACEMENT FORT: FORT WASHINGTON

Five days before the successful defense of Baltimore and Fort McHenry, on September 8, 1814, Secretary of State James Monroe, then acting as secretary of war, ordered Major Pierre L'Enfant to repair or rebuild Fort Warburton. Rebuilding did not begin until the following spring, on March 1, 1815. L'Enfant did not make satisfactory progress and failed to submit his plan of work to the War Department.[27] He was relieved of his duties on July 8, 1815, and was replaced by Lieutenant Colonel Walker K. Armistead. Armistead was instructed to report on work that had been undertaken on the new fort. On July 27, 1815, he reported on the incomplete ravelin. Two weeks later Armistead was instructed to take charge of the fort and prepare a plan for a new fortification, including a permanent barracks for 150 men.[28]

On October 10, 1815, Armistead submitted his proposed plans for the new fort to the War Department. Much larger than the destroyed fort, it included two demi-bastions on the west, facing the Potomac River, separated by a curtain. These demi-bastions would be able to fire *en barbette* or from casemates below. On the east or land side, two bastions were to be constructed. Midway between the two bastions, a caponiere was to be constructed. This was a small fortified structure to permit raking fire down onto attackers attempting to cross the ditch outside the east wall. Between the east and west walls was the parade, with enlisted men's barracks on one side and officers' barracks on the other. Fort Washington was to be a brick and masonry structure. Between the inner and outer defensive walls, and be-

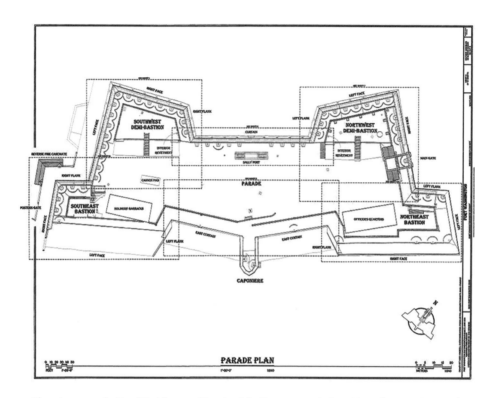

PARADE PLAN

Plan view, parade, Fort Washington, Maryland (built 1815–1824). Greg Harrell, Naomi Herandez, Frank Huto, Anthony Padgett, and Raul Vazquez, delineators, 1996. Fort Warburton, after being destroyed on August 27, 1814, was replaced by a much larger fortification, Fort Washington. The main fort entrance is on the north (*right*). The west curtain between the southwest demi-bastion (upper left) and the northwest demi-bastion (*upper right*) faces the Potomac River. Both demi-bastions are *en barbette* and casemate firing. Midway on the curtain is a sallyport (*center*) permitting access to the ravelin (not shown) between the fort and the Potomac River. The east curtain (*lower*) faces a defensive ditch on the land side. The caponiere in the middle of the east curtain (*lower center*) was a small fortified position designed to place fire on attackers attempting to cross the defensive ditch. Historic American Buildings Survey collection, Library of Congress.

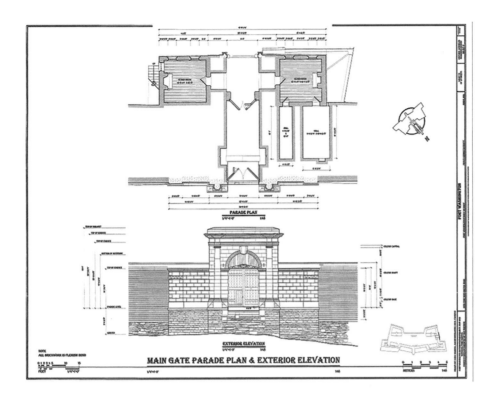

Main Gate Plan and Exterior Elevation, Fort Washington, Maryland. Greg Harrell, Naomi Herandez, Frank Huto, Anthony Padgett, and Raul Vazquez, delineators, 1996. Historic American Buildings Survey collection, Library of Congress.

low the coping stones, was sand. On the inside of the outer walls, brick counterforts were set at right angles to the outer wall so as to reinforce them. On the outside of the inner wall were brick buttresses.

Site work began in October 1815, under the direction of Lieutenant Theodore Maurice. General Joseph Swift recommended on February 12, 1816, that Armistead's plan for the new Fort Washington be approved, and it was approved by Secretary of War William Crawford. Work was begun with the beginning of the new construction season, in April 1817. By 1824 the new fort was essentially complete.[29]

The new Fort Washington never saw enemy action. During the Civil War it served as one of the fortifications surrounding Washington, DC. The garrison was removed in 1872. From 1896 to 1921, the area served as the headquarters for the Defenses of the Potomac. During this period, eight concrete batteries equipped with ten-inch disappearing guns were constructed outside the walls of Fort Washington. In 1921 the military reservation became headquarters of the Twelfth US Infantry,[30] and eventually it became part of the National Park System.

"Fort Washington," Maryland, as seen from upriver on the Potomac River. Fort Washington was the third artillery fortification built along the Potomac River. It was built 1815–1824 to replace Fort War-burton (also called Fort Washington), which was abandoned and destroyed by American forces on August 27, 1814, when approached by a British squadron. Picturesque America (1889).

OTHER MILITARY FACILITIES FOR THE NEW FEDERAL CITY

Two other major military facilities were built in Washington during the early years: the arsenal on Greenleaf Point and the Navy Yard along the Eastern Branch. Neither offered defensive positions against the invading British force.

The arsenal at Greenleaf Point, now Fort McNair, did have some armament as well as its military stores. In 1794 there was a report of only a single gun mounted at the point,[31] but by the time of the British invasion, the battery at Greenleaf's Point included eight 24-pound cannons, one 50-pound columbiad, and one 18-pound columbiad. In addition, smaller cannons on traveling carriages were distributed throughout the city.[32] The large guns were at the point, facing the rivers, and did not offer protection against the British land forces. The British soldiers invaded the arsenal and began their work of destroying the American ordnance found. Here the accounts diverge. Either retreating Americans or invading British forces dumped a large amount of black powder down a well. And either an American set a trail of black powder to detonate the powder in the well, or British soldiers carelessly threw the barrels of black powder in the well, causing a spark to ignite the powder. Either

PROJECT PLANNING AND CONSTRUCTION EFFORTS

way, the resulting explosion was enormous, causing 30 (some sources say 40) British deaths and the wounding of another 47.[33] The blast wrecked adjacent buildings. When the arsenal was rebuilt, the powder magazine was located outside the city at Little Falls, to prevent a reoccurrence.

The Navy Yard on the Eastern Branch (see plate 12) was similar to the arsenal, in that it had no defenses for a land attack. Before the arrival of the British troops, the commandant, Commodore Thomas Tingey, on orders of Secretary of the Navy William Jones, set the Navy Yard ablaze to prevent its ships and stores from falling into the hands of the British.[34]

JEFFERSON'S AND LATROBE'S DRY DOCK AND CANAL

The Navy Yard was to have been the site of ambitious structures envisioned by President Jefferson and architect-engineer Latrobe. Twelve years before Commodore Tingey burned the Navy Yard, President Jefferson had proposed a dry dock at the Washington Navy Yard to store twelve American frigates, each with forty-four guns, under an enormous single-span roof covering the dry dock basin, which was to measure 175 by 800 feet. The single-span roof was to be built after the manner of French architect Philibert de l'Orme, that is, it would be a single wooden arch made up of short timbers laminated to one another. Two very large locks, each of twelve-foot lift, would hoist the frigates into the dry dock basin, where their hulls would be shored up and the water drained from the basin. When a threat appeared, the frigates would be manned, water would fill the basin, and the warships would be locked down into the Eastern Branch to meet the enemy. (See plate 13.) The water for this procedure would have to be drawn from the Potomac River and brought across Washington by a canal, also to be built (see plate 14).[35] This proposal was never approved by Congress.

Although Latrobe's plans for a canal and a dry dock were never executed, Latrobe became a consultant to the Department of the Navy on the construction of the Navy Yard. In 1805, for example, Latrobe advised Captain Thomas Tingey on the construction of the Navy Yard's wharves: "I therefore beg leave strongly to recommend that the Wharf front be erected on piles, and built of solid Masonry, and that the floor of the Wharf be vaulted."[36]

DEFENSIVE MEASURES.

The Secretary at War has sent to this town four double fortified 18 pound cannon, mounted on travelling carriages, with all the apparatus necessary for using them ; five hundred cannon balls and two hundred grape shot, for the above pieces, to be under the care of Major Weeks, of the artillery. He has likewise sent orders to have the gun carriages in the fifteen gun battery in this town repaired, and ready for actual service. A brick building is erecting near the battery, for the reception of the four 18 pounders, &c.

The Secretary at War will be in town on Monday or Tuesday next, on his way to Kennebec.

Reinforcing Greenleaf's Point, Washington.
Washington National Intelligencer, September 9, 1807.

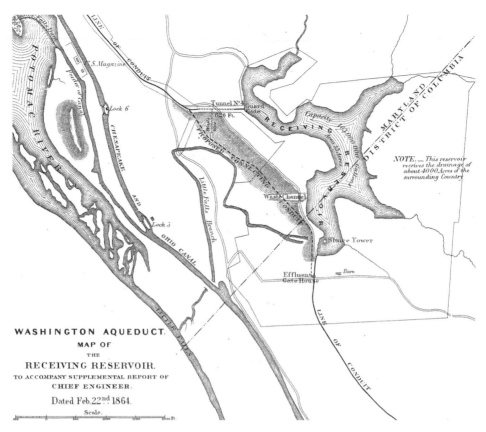

WASHINGTON AQUEDUCT.
MAP OF
THE
RECEIVING RESERVOIR.
TO ACCOMPANY SUPPLEMENTAL REPORT OF
CHIEF ENGINEER.
Dated Feb. 22nd 1864.
Scale.

Arsenal powder magazine location, February 22, 1864. During the British burning of public buildings in Washington on August 24, 1814, powder dumped down a well at the arsenal on Greenleaf's Point (present-day Fort McNair) exploded, killing at least 30 British soldiers and wounding 47 others. Several nearby buildings were severely damaged. When the arsenal was rebuilt, the decision was made to relocate the arsenal's powder magazine outside Washington, DC, near Little Falls. This map, made for the Washington water supply project, shows the relocated magazine (*top left*), situated adjacent to the old Potomac Canal approximated by the feeder canal (*top left*). Courtesy of the Library of Congress.

The military facilities to protect the new federal city were shown to be inadequate by the British invasion of August 24, 1815. Following the war, the defenses were rebuilt and intensified. The first lesson of the British invasion was that a much greater amount of money needed to be spent on military defenses. Another lesson was that the new defenses should be more strategically located. It was at this time that Fort Monroe was begun at Hampton Roads, providing protection for the entire Chesapeake Bay region, not just individual towns and cities. The main base for the US Navy was moved from the Navy Yard in the new city to Norfolk, eliminating the tortuous navigation down the Potomac and providing a more strategic location for the defense of the region.

PLATE 1. Reconstructed topographic view of Washington, DC, as it existed in 1791. Don Hawkins, delineator. The site selected for the national capital was a triangular piece of land between the Potomac River to the west and the Eastern Branch (later named the Anacostia River) to the southeast. Tiber Creek can be seen in the center. To the right is the ridge upon which architect-planner Pierre L'Enfant recommended siting the US Capitol. In the upper left is the port town of Georgetown. The navigable channels of the Potomac are clearly shown (*left*), as is the navigable channel of the Eastern Branch (*bottom right*). The terrain of the new city was well drained, belying the later story that Washington was built on a swamp. Courtesy of Don Hawkins.

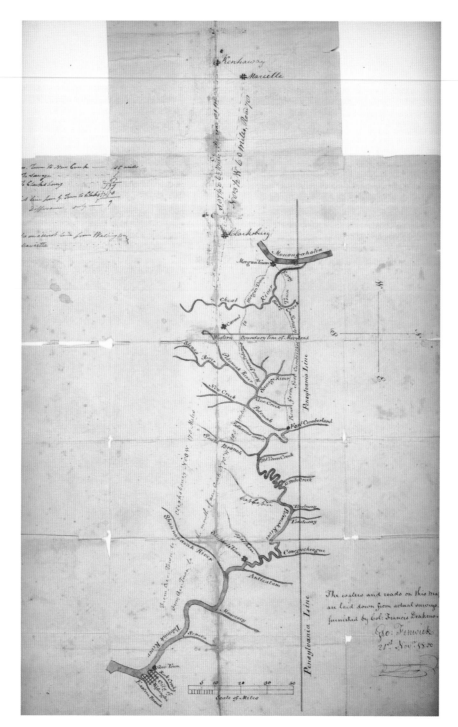

PLATE 2. Map of the Potomac River and wagon road to the Monongahela River. George Fenwick, delineator, from surveys undertaken by Colonel Francis Deakins, November 21, 1800. Washington's vision was that the new capital city would be not only a political center but also an economic center, communicating with the hinterland, especially the new territories along the Ohio River. The shortest route between tidewater and the Ohio River was by means of a canal and river navigation along the Potomac River—the Potomac Canal. This rudimentary map illustrates the practical nature of construction of the Potomac Canal. Only with the beginning of the successor to the Potomac Canal in 1828, the Chesapeake and Ohio Canal, were detailed engineered drawings prepared. Courtesy of the National Park Service, National Capital Region Museum Resource Center, Lanham, MD., Abner Cloud Collection.

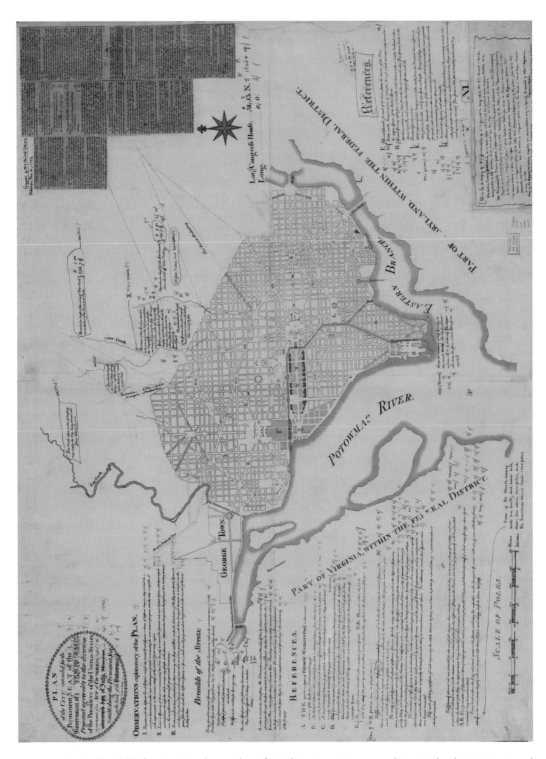

PLATE 3. Facsimile of L'Enfant's original 1791 plan of Washington, DC, prepared in 1887 by the US Coast and Geodetic Survey. Courtesy of the Library of Congress.

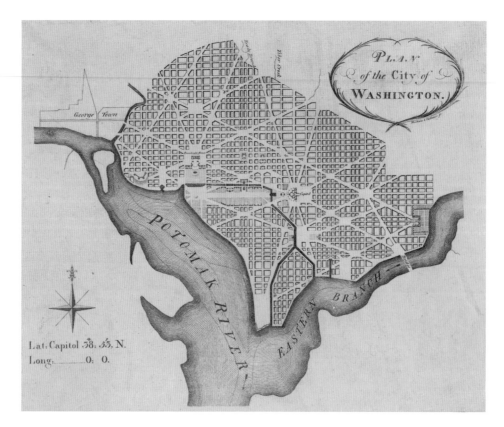

PLATE 4. The first printed version of L'Enfant's plan of the new federal city. Frontispiece of the March 1792 issue of the *Universal Asylum and Columbia Magazine*. Courtesy of the Library of Congress.

PLATE 5. *Georgetown, 1800*, by T. Cartwright, 1801. View down what became Wisconsin Avenue toward Georgetown, at the bottom of the hill, and the new city of Washington, beyond. At the right center is Mason's Island (present-day Theodore Roosevelt Island) and to its right the new channel between the island and the Virginia shore carved out by the ice flood of March 1784. By 1806 self-taught engineer Thomas Moore built a causeway or mole between the island and the Virginia shore (*right*) so as to divert the Potomac River back into the old Georgetown Channel and maintain the town's shipping ability. At the middle left is the bridge built in 1792 by Leonard Harbaugh for the commissioners to connect Georgetown with the new city. Courtesy of the Library of Congress.

East Branch of
Potomac R. Washington.

PLATE 6. *Eastern Branch* (Anacostia River), by Augustus Köllner, 1839. Although sedimentation affected the upper reaches of the Eastern Branch, the channel from the mouth to the Navy Yard remained relatively free from navigation-obstructing shoals. Courtesy of the Library of Congress.

PLATE 7. *View of the City of Washington*, by W. M. Craig, 1816. The view is from what became Wisconsin Avenue, NW, above Georgetown. In the center of this drawing can be seen the three-arch masonry bridge over Rock Creek, connecting Georgetown to Washington, built by Leonard Harbaugh in 1792. Compare with plate 5. Courtesy of the Library of Congress.

PLATE 8. *Potomac River Bridge at Little Falls*, by Benjamin H. Latrobe, 1798. Built in 1797 by Timothy Palmer under contract with the Georgetown Bridge Company, this was the first bridge built at Little Falls. The bridge was 120 feet long, with king post trusses 15 feet high. It was prefabricated in New England of white oak and shipped to the site for erection. The land-bridge approach to the bridge (*right*) collapsed in the flood of June 23, 1804. Because the main wooden bridge had never been clad, it had deterioated after seven years of service. The directors of the Georgetown Bridge Company decided to replace it with a Burr arch when the land bridge was replaced. The old bridge was used as a scaffold to aid in building the new bridge, which was opened by the end of 1807. Courtesy of the Maryland Historical Society, Item ID #1960.108.1.3.23.

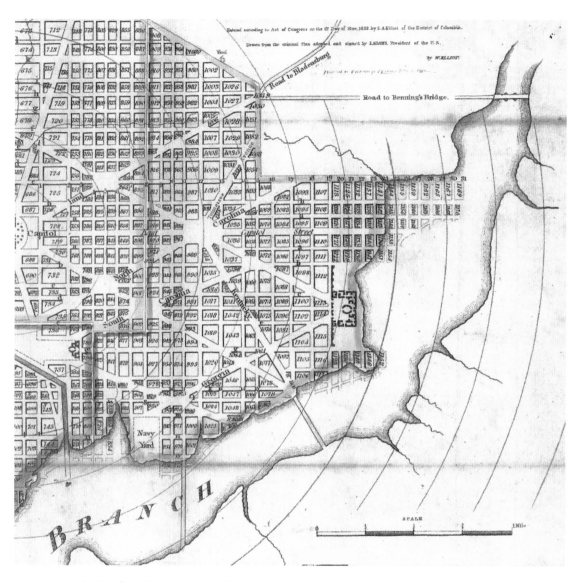

PLATE 9. Detail, plan of Washington, T. G. Bradford, Boston, 1835. Shown are the three bridges across the Eastern Branch (Anacostia River). The lowest (*bottom left*) was also the latest, the Navy Yard Bridge, chartered on February 14, 1819, and opened in October 1819. It became known as the lower bridge. Next upriver is the oldest bridge across the Eastern Branch, built by the Eastern Branch Bridge Company, chartered December 24, 1794 (*bottom center*). Located at the lower end of Kentucky Avenue and opened January 11, 1804, it was originally called the lower bridge. Further upstream (*upper right*) is the bridge of the Anacostia Bridge Company, chartered November 1797. This bridge was built in 1805. Its west end was located a short distance north of the foot of East Capitol Street. It came to be known as the upper bridge and was later called Stoddert's Bridge, Ewell's Bridge, and Benning's Bridge. Not shown is the bridge at Bladensburg, located further upstream. Courtesy of the Library of Congress.

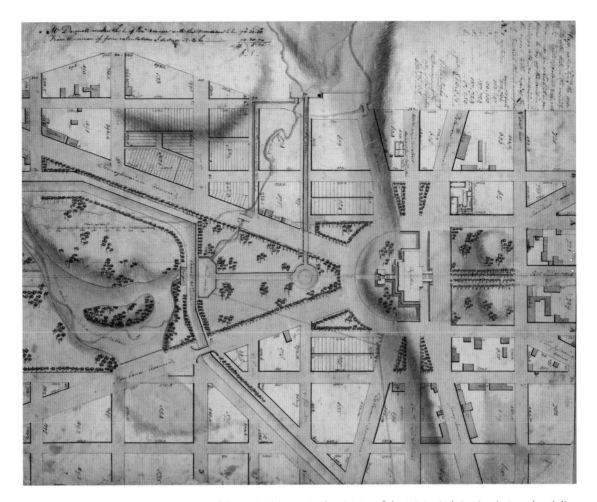

PLATE 10. Map exhibiting the property of the United States in the vicinity of the US Capitol, Benjamin Latrobe, delineator, December 1815. Shown is the Washington City Canal as it turns south-southeast from what is now Constitution Avenue, NW (*top left*), and south in front of the Capitol (*center*) and then again southeast (*bottom center*), where one branch continues east to join the Eastern Branch (Anacostia River) and the other branch swings south along St. James Creek. Courtesy of the Library of Congress.

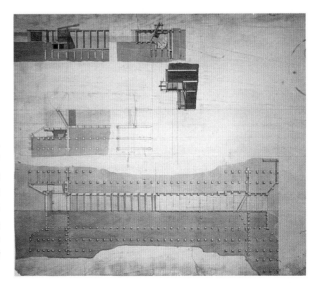

PLATE 11. Elevations and plan view, Wooden Lock, Washington City Canal, Benjamin Latrobe, designer and delineator, 1810. The top left elevation shows the wall of the lock with balance beam and lock gate in place. The bottom plan view shows the full length of the lock with half (*right*) of the lock floored. The drawings in between are details showing the flow mechanism for filling the lock. Courtesy of the Library of Congress.

PLATE 12. *"City of Washington,"* ca. 1833, by William Bennett. The US Navy Yard, the oldest in the nation, is seen on the opposite bank of the Eastern Branch (Anacostia River), right center. The Navy Yard Bridge, built in 1819, can be seen (*far right*). Washington's arsenal was located on Greenleaf Point (now Fort McNair) (*far left*). Courtesy of the Library of Congress.

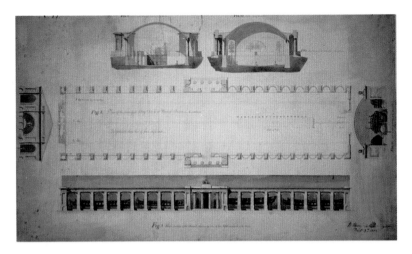

PLATE 13. End sections, plan, and elevations of the covered dry dock at the US Navy Yard, Washington, Benjamin Latrobe, designer, 1802. The covered dry dock was conceived by President Thomas Jefferson as a means of effectively storing unused naval warships. The dry dock was to have been a basin of 175 by 800 feet, covered with a single-span structure and built after the manner of French architect Philibert de l'Orme, that is, through the use of a single wooden arch made up of short timbers laminated together. If it had been built, the dry dock would have held twelve frigates of 44 guns each. Courtesy of the Library of Congress.

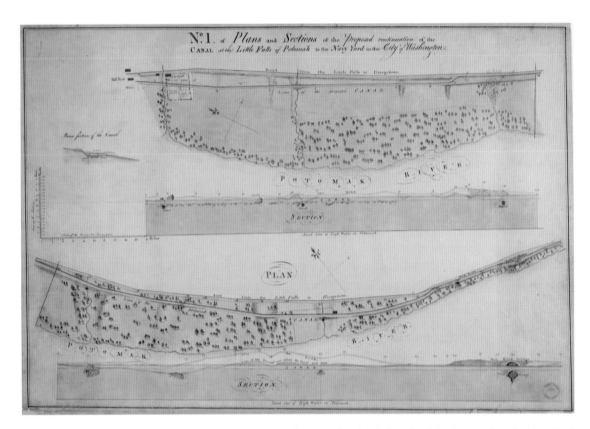

PLATE 14. Plan and section of the proposed continuation of the canal at the Little Falls of the Potomack to the Navy Yard in the city of Washington, Benjamin Latrobe, designer, 1802. The head of the proposed canal to the Navy Yard was at Little Falls, the Potomac River being the only adequate source to fill the basin of the dry dock in a timely manner. Such a canal would also have the advantage of connecting the terminus of the Potomac Canal at Little Falls (*left*) with the Navy Yard as well as with points between. Courtesy of the Library of Congress.

BUILDING THE FIRST WHITE HOUSE

WASHINGTON D.C. 1798

MODERN
WASHINGTON

PLATE 15. Pennsylvania Railroad Advertising Poster, ca. 1920s. In 1800, when the government moved from Philadelphia to Washington, the President's House was mostly but not entirely complete. However, multiple design and construction problems were found in the building. Wrote Benjamin Latrobe in 1804, "It is well known that the Presidents house was inhabited before it was finished; and that it still remains in a state so far from completion, as to want many of those accommodations which are thought indispensable in the dwelling of a private citizen. Of the inconveniences attending the house, the greatest was the leakiness of the roof, which had indeed never been tight." Some of the other problems were that the heavy slate roof was forcing the walls out of plumb and damaging the wooden structural members, that the interior drainage system was poorly detailed and constructed and was dumping rainwater into the building, and that unseasoned wood had been used. In this poster it is unclear what the copywriter meant by "The First White House." Author's collection.

PLATE 16. *The President's House*, by James Hoban, 1793.
Courtesy of the Library of Congress.

PLATE 17. East elevation of the submission for the US Capitol design competition, by Stephen Hallet, 1791. Before they received William Thornton's design, the commissioners preferred the design submission of Stephen Hallet. Courtesy of the Library of Congress.

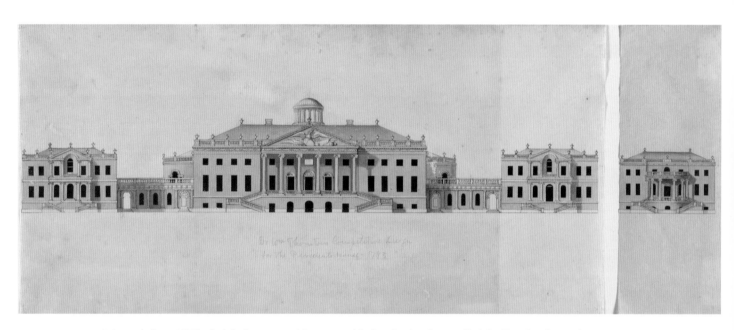

PLATE 18. Intended as a US Capitol design competition entry, this drawing has been called the Tortola scheme, because it was drawn by William Thornton in 1793, when he was in Tortola, British West Indies. It was never submitted for consideration. Courtesy of the Library of Congress.

PLATE 19. US Capitol, east elevation, by William Thornton, 1793 to 1800, following Hallet's modifications. The drawing that won the competition for the Capitol was similar but has not been found and is presumed lost. The Pantheon-like dome links the north and south wings of the legislative assembly building. Courtesy of the Library of Congress.

PLATE 20. Section of the US Capitol, by Stephen Hallet, 1793. Courtesy of the Library of Congress.

PLATE 21. War Department and State Department building, immediately west of the President's House, 1837. It was constructed beginning in 1799 by Leonard Harbaugh, contractor. Author's collection.

PLATE 22. *A View of the Capitol before It Was Burned Down* by the British, by William Russell Birch, ca. 1800. The federal construction program had numerous funding and labor problems. When the federal government moved from Philadelphia to the new city, only the north wing (Senate) of the US Capitol (shown) had been mostly completed. Although foundations for the south wing (for the House of Representatives) had begun, a temporary structure had to be constructed for the use of the House. Even the north wing had not been fully completed by 1800, and stone cutters can be seen working in the foreground (*left*). View to the west. Courtesy Library of Congress.

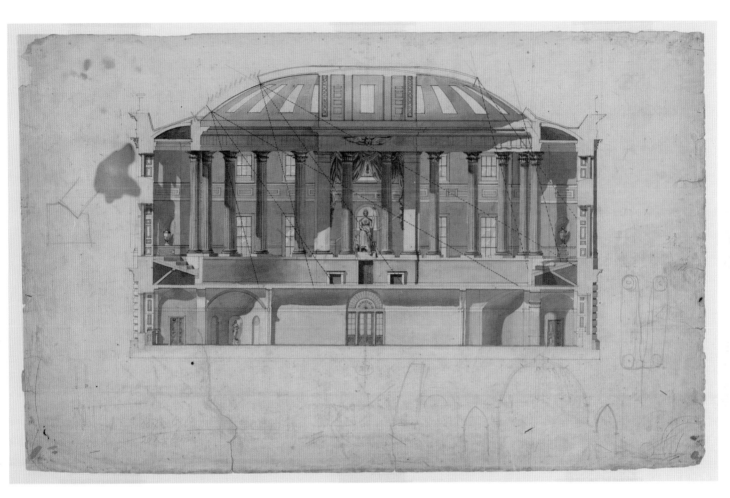

PLATE 23. Section of the south wing, US Capitol (hall of the House of Representatives), by Benjamin Latrobe, 1804. Section looking south, toward the Speaker's chair (*center*). In 1806 Latrobe described the hall: "The House is surrounded by a plain wall seven feet high. The 24 Corinthian columns which rise upon this wall and support the dome are 26 feet 8 inches in height, the entablature is 6 feet high, the blocking course 1 foot 6 inches, and the dome rises 12 feet 6 inches, in all 53 feet 8 inches. The area within the wall is 85 feet 6 inches long and 60 feet 6 inches wide. The space within the external walls is 110 feet by 86 feet." Courtesy of the Library of Congress.

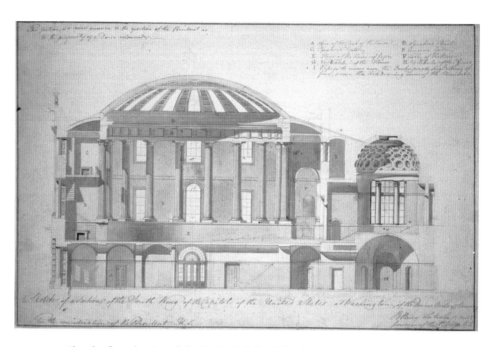

PLATE 24. Sketch of south wing of the Capitol of the United States at Washington, by Benjamin Latrobe. Section looking west. To the right is the small connecting structure referred to by Latrobe as "the recess," to connect the south wing to the central domed section and to provide lateral support for the dome on the south wing. Courtesy of the Library of Congress.

PLATE 25. "Vue de la Halle au Blé," Paris (municipal grain market). It was President Thomas Jefferson's insistence that led to the installation of a glass domed roof over the hall of the Representatives, in the south wing of the US Capitol, based on the de l'Orme–like dome constructed over the Halles au Blé that both he and Benjamin Latrobe were familiar with. Latrobe raised numerous objections to such a roof but eventually installed a modified version. Author's collection.

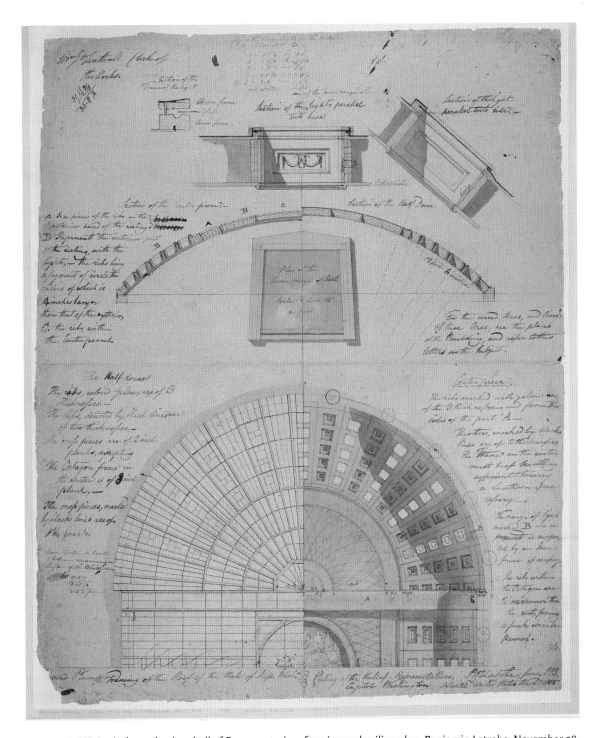

PLATE 26. US Capitol, south wing, hall of Representatives framing and ceiling plan, Benjamin Latrobe, November 28, 1805. Although Latrobe disagreed with Jefferson on the use of a skylight roof similar to that used on Paris's Halles au Blé, he yielded to the president's wishes and designed a domed roof for the hall of the House of Representatives. The roof contained one hundred skylights or "panel lights"—framed skylights mounted between the structural ribs of the dome. Five sizes were fabricated. From below they appeared to be ceiling coffers, although their upper surface was glazed. Venetian blinds controlled the sunlight admitted. Courtesy of the Library of Congress.

PLATE 27. Benjamin Latrobe's perspective of the completed Capitol, including cupolas, from the northeast. Presented to President Thomas Jefferson in 1806. Latrobe preferred cupolas or monitors to illuminate the interior, but Jefferson did not. Wrote Jefferson to Latrobe about the cupolas, "I confess they are most offensive to my eye, and a particular observation has strengthened my disgust at them." Courtesy of the Library of Congress.

PLATE 28. US Capitol from the west, Pennsylvania Avenue, 1814, by an unidentified artist; incorrectly attributed to Benjamin Latrobe. Early in the first building campaign (1791–1802), it was decided to build the US Capitol in three parts: First the north wing (the Senate, *to the left*); second, the south wing (the House of Representatives, *to the right*); and third, the center section, to be domed. By 1814 the center section had not yet been built. Courtesy of the Library of Congress.

PLATE 29. "Capitol of the U.S. at Washington—From the Original Design of the Architect B.H. Latrobe, Esqre./T. Sutherland sculpt.," ca. 1785–1825, Thomas Sutherland. Latrobe never did see the Capitol finished. Courtesy of the Library of Congress.

PLATE 30. "View of Washington," ca. 1852. E. Sachse & Co. This building was enlarged and the cast-iron dome erected, under the design of Thomas U. Walter. Note the Washington City Canal. Courtesy of the Library of Congress.

PLATE 31. *A View of the Capitol of the United States after the Conflagration of the 24th August 1814*, by George Munger, 1814. Courtesy of the Library of Congress.

PLATE 32. A View of the President's House in the City of Washington after the Conflagration of 24th August 1814, by George Munger, artist, and William Strickland, engraver, 1814. Courtesy of the Library of Congress.

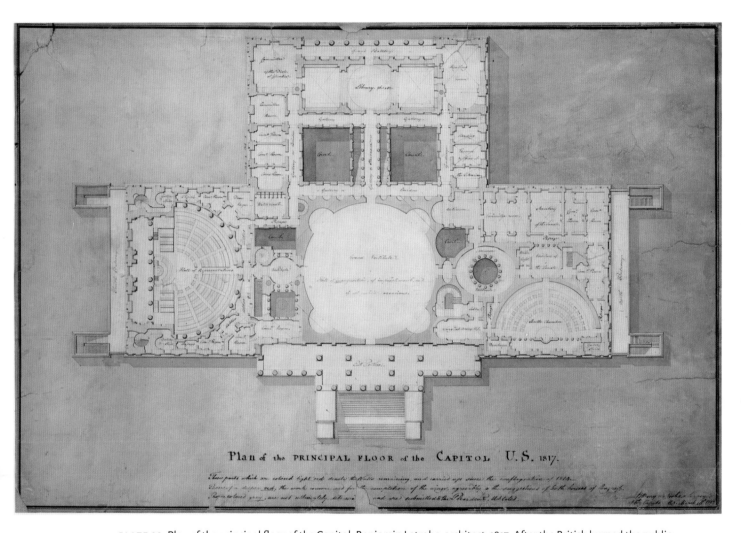

PLATE 33. Plan of the principal floor of the Capitol, Benjamin Latrobe, architect, 1817. After the British burned the public buildings on August 24, 1814, James Hoban rebuilt the President's House according to the original plan. Benjamin Latrobe intended to redesign the Capitol to correct the design and construction errors that had been made in its original construction. Courtesy of the Library of Congress.

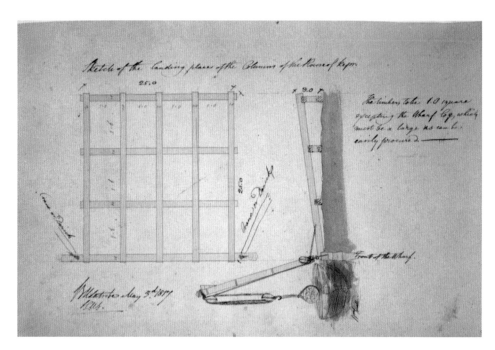

PLATE 34. Sketch of the landing place of the columns of the House of Representatives, Benjamin Latrobe, delineator, 1817. Although the opening of the Great Falls bypass canal in 1802 made it possible to bring monolithic columns from the Potomac Marble quarry upriver from Conrad's Ferry (present-day White's Ferry), it was still difficult to off-load the columns that were 2 feet 8 inches in diameter and 22 feet long and transfer them to wagons for delivery to the Capitol. The columns were tranported in segments, two or three segments for each column. Courtesy of the Library of Congress.

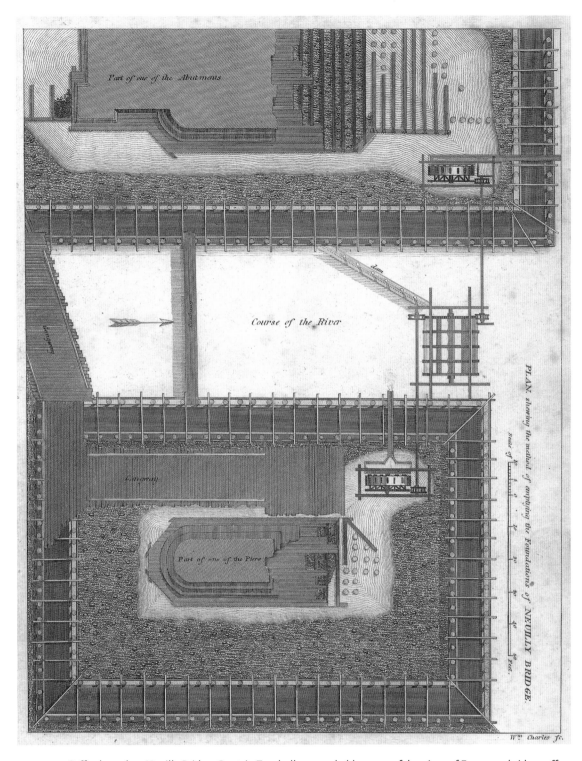

Part of one of the Abutments

Course of the River

Part of one of the Piers

PLAN showing the method of emptying the Foundations of NEUILLY BRIDGE.

Scale of

Wm Charles fc.

PLATE 35. Cofferdam plan, Neuilly Bridge. Captain Turnbull was probably aware of drawings of European bridge coffer-dams that had been published in works such as the American edition of the *Edinburgh Encyclopedia*. Compare this drawing with the illustration on page 246. Built for the Pont de Neuilly, across the Seine at Paris, between Courbevoie on the right bank and Puteaux on the left, it was constructed in 1774 by Jean-Rodolphe Perronet, the founder of the Ecole Nationale des Ponts et Chaussées. A multiple-arch segmental masonry arch structure, it was demolished in the mid-twentieth century. "Plan Showing the Method of Emptying the Foundations of Neuilly Bridge," in *Edinburgh Encyclopedia*, ed. David Brewster, American ed. (Philadelphia: J. and E. Parker, 1832), plate 95.

Plate 36. *Georgetown College*, by Caroline Rebecca Nourse, 1838. View from the heights of Georgetown of the Potomac Aqueduct under construction. Courtesy of the Dumbarton House / The National Society of the Colonial Dames of America, Washington, DC.

Plate 37. *Aqueduct of Potomac, Georgetown, D.C.*, by F. Dielman, 1865. View of the Potomac Aqueduct from Georgetown, looking south along the Potomac River: the Chesapeake and Ohio Canal (*bottom center*); the partially completed Washington Monument (*top left*); Analostan Island, now called Theodore Roosevelt Island (*middle*). At the time of this drawing, the Potomac Aqueduct had been dewatered and was being used as a military bridge across the Potomac River. Courtesy of the Library of Congress.

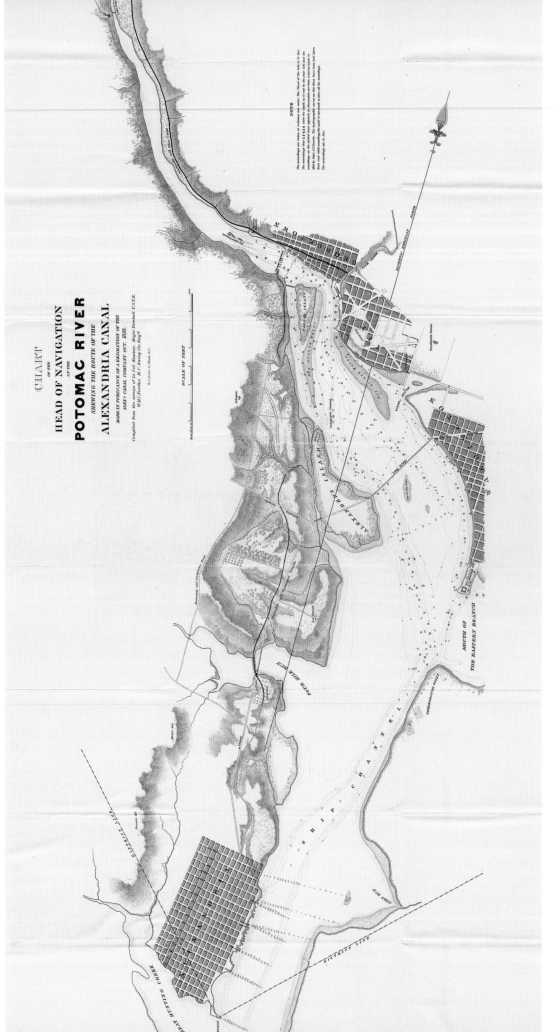

PLATE 38. The route of the Alexandria Canal, William James Stone, delineator. On the right (*center*) is the Potomac Aqueduct, connecting the Alexandria Canal with the Chesapeake and Ohio Canal. The Alexandria Canal proceeded seven miles down the Potomac from the vicinity of Mason's Island (Analostan Island) in what is now Rosslyn, Virginia (*right*) to Alexandria (*left*). Map published by the US Senate, 1841. Courtesy of the Library of Congress.

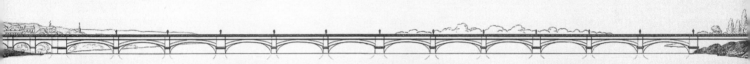

PART TWO   BUILDING CAMPAIGNS:
          THE CITY'S FIRST FIFTY YEARS

The First Public Building Campaign (1791–1802)
The Commissioners Clash with Their Architects and Workers

7

Mrs. Adams expressed the opinion that if the twelve year's work had been going on in New England it would have been better managed and completion much nearer its end.

Quoted by WILLIAM TINDALL, 1914

In building the new federal city, the commissioners' most critical job was to construct the buildings needed by the federal government. The President's House and the Capitol were the most important ones. In addition, several office buildings would be required for the government clerks.

1791–1793: PREPARING TO BUILD

One of the first activities undertaken by the commissioners was to hire Francis Cabot, paying him one thousand dollars for a nine-month contract, to travel to Philadelphia and advise the commissioners on the going rate for workers and materials.[1] A contract at that rate for a nine-month term was extraordinary for the time, when the daily rate for a qualified workman was one dollar per day. For this money, Cabot provided the commissioners only three letters. The real reason he had been hired was revealed in his first letter to the commissioners, from Philadelphia on December 11, 1791. Cabot had actually been hired to lobby for the new federal city and to gain intelligence on what was happening in Congress. At the time there was a very real threat that Congress would change its mind and relocate to New York or remain in Philadelphia. He reported to the commissioners on the very thin congressional support for the new federal city on the Potomac, warning them that the enemies in Congress of the new city were "not only numerous but potent."[2]

{ **WASHINGTON,** }
In the Territory of Columbia.

A PREMIUM,

OF a lot in this city, to be defignated by im-
partial judges, and 500 dollars; or a medal of
that value, at the option of the party, will be
given by the Commiffioners of the Federal Buildings, to the perfon, who, before the 15th day
of July, 1792, fhall produce to them, the moft
approved plan, if adopted by them, for a Capitol to be erected in this City, and 250 dollars,
or a medal, for the plan deemed next in merit
to the one they fhall adopt. The building to
be of brick, and to contain the following apartments, to wit :

A conference room; ⎫ Sufficient to ac-
A room for the Re- ⎬ commodate 300
  prefentatives; ⎭ perfons each.
A lobby, or antichamber to the latter;
A Senate room of 1200 fquare feet area;
An antichamber, or lobby to the laft; ⎬ *Thefe rooms to be of full elevation*

12 rooms of 600 fquare feet area, each, for
committee rooms and clerk's offices, to be of
half the elevation of the former. Drawings
will be expected of the ground plats, elevations
of each front, and fections through the building
in fuch directions as may be neceffary to explain
the internal ftructure; and an eftimate of the
cubic feet of brick-work compofing the whole
mafs of the walls.

The Commiffioners.
March 14, 1792.

Advertisement for the Capitol design competition.
Dunlap's American Daily Advertiser (Philadelphia), March 24, 1792.

Cabot's letters to the commissioners were chilling. Construction progress, reported Cabot, was vital to sustain support of the Residence Act. Without construction progress, he reported, the Residence Act was in jeopardy.

The commissioners had tried to move forward aggressively with construction of the needed public buildings. For example, at their meeting on September 24, 1791, they resolved that their planner-engineer Major L'Enfant be instructed to hire 150 laborers to "throw up clay" for the brick needed and to excavate for the foundations for the President's House and the Capitol.[3] These excavations never began.[4] By January 1792, the winter had advanced so that digging was no longer feasible. The commissioners halted all work throughout the new city and released the workers.[5] It was an inauspicious beginning for the new federal city.

DESIGN COMPETITIONS
FOR THE PRESIDENT'S HOUSE
AND THE CAPITOL

In the spring of 1792, the commissioners authorized advertisements for plans for the design of the President's House and the Capitol and had them placed in the various newspapers. A competition for the design of the two buildings was Jefferson's idea. President Washington had asked him the best way to develop plans for the buildings,[6] and Jefferson replied, "By advertisement of a medal or other reward for the best plan—see a sketch or specimen of advertisement."[7]

Jefferson's recommendation was innovative. At the time design competitions had rarely been used in the United States. In a letter of March 6, 1792, Jefferson told the commissioners of his recommendation and included a suggested announcement of the design competition for the President's House to be placed in newspapers. With both Washington and Jefferson pushing hard for action, the commissioners ordered an announcement of the design competition for the President's House to

James Hoban, ca. 1800, by John Christian Rauschner. Born in Ireland and having studied in Dublin, Hoban immigrated to the United States after the American Revolution and practiced architecture in Philadelphia. He moved to South Carolina, where he met President Washington. Washington introduced him by letter to the building commissioners. In 1792 his design for the President's House was selected by the commissioners and he was hired to oversee its construction. He later oversaw the reconstruction of the building after the capital was burned in 1814 by the British. White House Collection / White House Historical Association.

be advertised in various newspapers.[8] Eight days later, on March 14, 1792, Jefferson drafted a similar advertisement for a design competition for the Capitol building.[9] It was published later that month.

By July 15, 1792, the design competition plans for the President's House had been received. On July 16–17, 1792, the commissioners and President George Washington met to examine the design submissions. Irish born architect James Hoban's entry was selected as the winner: "James Hoben's plan of the Palace being approved by the President, he is intitled to the reward published and chuses a gold medal of 8 or 10 Guineas value [$22–$28] — The Ballance in money." The advertisement for the design competitions was not explicit regarding who was to supervise the construction of the winning design, but Jefferson's intent seemed clear: that the winner of the competition would be hired to build the design he submitted. So, on July 18, 1792, Hoban was hired to supervise the construction of the President's House.[10]

On August 2, 1792, George Washington and commissioners Carroll and Stuart determined the exact site of the President's House. Washington decided to put the building on the line established by L'Enfant for the north facade.[11] On October 13, 1792, the cornerstone of the President's House was laid. The *Records of the District of Columbia Commissioners* reported that ceremony:[12]

OCTOBER 13TH 1792 —

THE COMMRS OF THE CITY (OF) WASHINGTON

The Commissioners accompanied by a numerous collection of Free-Masons, Architects and of the Inhibitants [sic] of the City of Washington and George Town went in procession and laid the first corner stone of the Presidents house.[13]

The cornerstone was not a stone but an engraved plate. It is probably still in the wall of the President's House (later called the White House) but has never been found, despite the extensive renovations done to the building in 1815–1817, 1949–1952, and at other times.

The cornerstone for the President's House was laid late in the building season. It was intended to generate interest in the purchase of lots, but it did not generate much interest. The 1792 building season was rapidly coming to a close. Since the passage of the Residence Act, two full building seasons, out of the ten-year schedule, had gone by with no construction taking place. However, all looked in readiness for the 1793 building season to begin in earnest on March 1, the traditional beginning of the construction season. At the beginning of 1793, Samuel Blodget, one of the design competitors for the Capitol, was appointed by the commissioners for a year to be supervisor of the buildings. He was to give "attention to the Economical and executive part advising, or making in [the commissioners'] Absence the contracts for Materials, Labour and the like."[14]

At the beginning of 1793, there was an approved plan for the President's House but none for the Capitol. The competition for the Capitol had begun well enough, and a number of submissions had been received by the July 15, 1792, due date.[15]

Washington's intention, "to draw plans from skillful architects," was not realized in the Capitol design competition. Most of the submitters were not professional architects or experienced master-builders. Only one submitter, Stephen Hallet, was professionally trained. A few of the others were accomplished master-builders, including James Diamond, Samuel Dobie, Joseph Clark, and Samuel McIntire. Other submitters would be considered builders or carpenter-builders, such as Samuel Blodget, Abraham Faw (Abram Faw[s]), Leonard Harbaugh, Phillip Hart. Robert Goin Lanphier, and Jacob Small. One submitter, Charles Wintersmith (Charles Winterschmidt), was believed to have been a military engineer. Others, such as Andrew Mayfield Carshore, George Turner, and William Thornton, were amateurs.[16]

The commissioners were not totally pleased with any of the designs submitted for the Capitol.[17] Of the designs submitted, Washington preferred those of Hallet and Turner.[18] In late August 1792, Hallet and Turner met with Washington and the three commissioners (Johnson, Stuart and Carroll), but no decision was reached on a final design competition winner.[19] Nevertheless, Hallett's design seemed to be favored. (See plate 17.) Hallet was engaged by the commissioners to revise his design and resubmit it.

Dr. William Thornton, a physician and amateur architect then living in Tortola (British West Indies), learned of the design competition when it was too late to submit an entry. On July 14, 1792, the day before the competition closed, he sent a letter to the commissioners asking if he could still submit a design.[20] He then started working on a design for the Capitol building, one that was subsequently called "The Tortola scheme" (see plate 18). He left Tortola with his plan and arrived in Philadelphia on November 1, 1792.[21] He again wrote the commissioners, asking to submit his Tortola scheme. The commissioners responded on December 4, 1792, informing him that a design for the Capitol had not yet been selected and that they would consider his design.[22]

Once in Philadelphia, Thornton consulted with his friend George Turner, an earlier competitor in the competition. Turner probably informed Thornton of the expectations and thoughts of Washington, Jefferson, and the commissioners. Based on that information, Thornton decided not to submit his Tortola scheme; instead, he began working on a new design. The new design was for a three-part building whose center section was modeled on the shallow dome and portico of the Roman Pantheon (see plate 19). In developing this design, he was most likely assisted by his friend John Jacob Ulrich Rivardi, a Swiss military engineer and artillerist who had served in the Russian army.[23] On January 29, 1793, Thornton submitted this new design. His submittal consisted of only two drawings: the east elevation and a floor plan of the principal floor.[24] The drawings were quite stunning and were very much what Washington, Jefferson, and the commissioners wanted.

Washington, having been immediately taken with Thornton's design, lavished high praise on it for its "Grandeur, Simplicity, and Beauty."[25] Jefferson, too, was impressed by Thornton's design. He wrote commissioner Carroll (in part): "Dr. Thornton's plan of a capitol has been produced, and has so captivated the eyes and judgment of all as to leave no doubt you will prefer it when it shall be exhibited to you; as no doubt exists here of its preference over all which have been produced, and among its admirers no one is more delighted than him [President Washington] whose decision is most important. It is simple, noble, beautiful, excellently distributed, and moderate in size."[26]

The commissioners, who had been working with Hallet on modifications of his design for the Capitol, were taken aback. On February 7, 1793, they responded to Jefferson (in part): "Tho' we are much pleased that we shall at length be furnished with the Plan of a Capitol so highly satisfactory to the President, and all who have seen it, we feel sensibly for poor Hallet, and shall do everything in our power to sooth him, we hope he may be usefully employed notwithstanding."[27] On March 13, 1793, they informed a disappointed Hallet that Thornton's design had been selected for the Capitol but that they would award him "a Compensation for every Thing to this Time, 100 £ being the Value of a Lot and 500 Dollars."[28]

On April 5, 1793, the commissioners (Johnson, Stuart, and Carroll) notified Thornton that his design had been accepted. Their letter to Thornton asked him to put his business affairs in order, as it was their plan to initiate construction that autumn (implying that he was to be hired to supervise the work).[29]

On April 10, 1793, Thornton submitted to the commissioners a written report. Either he had previously given them three additional drawings of his design for the Capitol, or he provided these drawings at this time. They included an elevation of the west front and two plans for the full ground floor and the grand story above. These drawings, like the drawing of the east elevation submitted in January 1793, have not been found and are presumed lost. It has been supposed that after his design was heavily criticized, Thornton withdrew these drawings and replaced them with revised drawings in 1795 and subsequently.[30] Although the original competition drawings are lost, they have been described in Thornton's report to the commissioners.[31]

HALLET'S CRITIQUE OF THORNTON'S PLAN FOR THE CAPITOL

The commissioners, eager to soothe Hallet's wounded feelings, employed him to critique Thornton's design. Asking Hallet, a competition loser, to critique a competitor's winner's design, was a formula for trouble. Hallet delivered his critique of Thornton's design to the commissioners in June 1793. It was devastating. Hallet's critique, now presumed lost, was extensive. It took up "five manuscript volumes in folio" and was described by Thornton as containing "voluminous objections."[32] Although Hallet's five manuscript folios of objections have not been found, we have Jefferson's summary of the objections:

1 The intercolonations of the western and central peristyles are too wide for the support of their architraves of Stone: so are those of the doors in the wings.
2 The colonnade passing through the middle of the Conference room has an ill effect to the eye, and will obstruct the view of the members: and if taken away, the cieling is too wide to support itself.
3 The floor of the central peristyle is too wide to support itself.
4 The stairways on each side of the Conference room want head room.
5 The windows are in some important instances masked by the galleries.
6 Many parts of the building want light and air in a degree which renders them unfit for their purposes. This is remarkably the case with some of the most important apartments, to wit, the chambers of the Executive and the Senate, the anti-chambers of the Senate and Representatives, the Stairways etc. Other objections were made which were surmountable, but those preceding were thought not so, without an alteration of the plan.[33]

On hearing of Hallet's criticisms of Thornton's design, George Washington complained that the Capitol's design competition did not include an investigation of each submission prior to the selection of a preferred design. If such an investigation had been conducted, Washington mused, the design defects in Thornton's design would probably have been discovered before the final selection, and the time lost in wrangling over the design of the Capitol would have been avoided. In addition, no estimate of the cost of Thornton's design had been developed. Washington wrote to Jefferson:

> It is unlucky that this investigation of Doctor Thornton's plan, and estimate of the cost had not preceded the adoption of it: but knowing the impatience of the Carrollsburg interest and the anxiety of the Public to see both buildings progressing; and supposing the plan to be correct, it was adjudged best to avoid delay. It is better, however, to correct the error, though late, than to proceed in a ruinous measure, in the adoption of which I do not hesitate to confess I was governed by the beauty of the exterior and the distribution of the apartments, declaring then, as I do now, that I had no knowledge in the rules or principles of Architecture, and was equally unable to count the cost. But, if there be such material defects as are represented, and such immense time and cost to complete the buildings, it would be folly in the extreme to proceed on the Plan which has been adopted.[34]

To rectify this problem, Washington proposed that Thornton be given an opportunity to correct any defects in the plan: "It has appeared to me proper, however, that before it [Thornton's plan] is laid aside, Justice, and respect to Doctor Thornton, requires, that the objections should be made known to him and an opportunity afforded to explain and obviate them, if he can." Washington wrote Jefferson that a panel of experts should be convened to examine the validity of Hallet's criticisms of Thornton's design and, if possible, resolve the problems. What was important, wrote Washington, was that "a Plan must be adopted and good, or bad, it must be entered upon."[35] Hallet's critique was forwarded to Thornton, including Washington's letters.[36] Thornton attempted to respond to Hallet's objections in a letter to Jefferson on July 8–12, 1793,[37] but it seemed evident to Washington, Jefferson, and the commissioners that no matter how Thornton tried to explain away Hallet's objections, there were major problems with Thornton's plan for the Capitol.

THE JEFFERSON COMPROMISE

In July 1793, Jefferson convened a meeting of Thornton, Hallet, Hoban, and two Philadelphia builders selected by Thornton: Thomas Carstairs and Colonel William Williams. On July 17, Jefferson reported to Washington that the meeting had arranged a compromise using Hallet's floor plan and Thornton's exterior elevation. Jefferson wrote to Washington (in part): "This alteration has in fact been made by mr. Hallet in the plan drawn by him, wherein he has preserved the most valuable ideas of the original and rendered them susceptible of execution; so that it is considered as Dr. Thornton's plan reduced into practicable form."[38]

The compromise would permit both Thornton and Hallet to be referred to as the architect of the Capitol: Thornton as the architect of the exterior and Hallet as the architect of the interior. One major problem remained. To provide light and air to the interior of the building, Hallet proposed recessing instead of projecting the central section of the building—a major change in Thornton's original design.[39] This problem was set aside for the time being.

After receiving Jefferson's report, Washington prepared a letter to the commissioners, stating that the conferees agreed that Hallet's criticisms of Thornton's design were valid.[40] He went on to explain the compromise worked out by Jefferson: "The plan produced by Mr. Hallet, altho' preserving the original ideas of Doct. Thornton, and such as might upon the whole be considered as his plan, was free from those objections, and was pronounced by the Gentlemen on the part of Doctr. Thornton, as the one which they, as practical Architects would chuse to execute."[41]

In addition, Washington explained that Hallet's modifications to Thornton's design would save half the cost of the building: "Besides which, you will see that, in the opinion of the Gentlemen, the plan executed according to Mr. Hallet's ideas would not cost more than one half of what it would it would if executed according to Doctor's Thornton's."[42]

The issue of the design of the center section remained, to be resolved at a later date: "But whether the portico or the recess should be finally concluded upon, will make no difference in the commencement of the foundations of the building, except in that particular part—and Mr. Hallet is directed to make sketches of the Portico, before the work will be affected by it."[43] The Jefferson compromise allowed the excavations of the Capitol foundations to begin in July 1793. At least a half a year of work had been lost in wrangling over the Capitol's design, but its construction now seemed to be back on track. Meanwhile, work on the President's House had begun.

For Thornton the conference of July 1793 and the Jefferson compromise was humiliating. He had publicly asserted himself not only as an architect but as a great architect, and yet virtually all of Hallet's objections to Thornton's design were sustained by the conference participants, including the two Philadelphia builders selected by Thornton himself. Given Hallet's critique, the appointment of Thornton as architect in charge of the Capitol did not seem appropriate. Over the next year, another compromise was gradually worked out. Hallet would supervise the construction of the Capitol, in accordance with the conference plan worked out in Jefferson's office; and when a vacancy arose, Thornton would provide general oversight by being appointed a commissioner of public buildings.

On September 18, 1793, the cornerstone of the Capitol was laid, with full Masonic pomp and ceremony. The newspaper account described the silver engraved plate that was placed in the work. It included the names of the commissioners, listed James Hoban and Stephen Hallet as architect, Collen Williamson as master mason, and Joseph Clark as Right Worshipful Grand Master pro tempore, but it omitted any mention of Thornton.[44] Thornton was offended. It is probably this omission, along with Hallet's criticisms of his design, that fueled Thornton's antipathy to the professional architects employed to work on the Capitol—Hallet, and later George Hadfield and Benjamin Latrobe. The amateur gentleman architect was now arrayed against professional architect-engineers, in a battle that continued through the first public building campaign and into the second. The dispute between Thornton and Hallet later expanded to disputes between all of the commissioners and the other architect-engineers, and eventually the workers. These disputes resulted in work slowdowns, work stoppages, and general labor unrest throughout the first public building campaign.

NEED FOR WORKERS

At the beginning of 1793, the first need was for laborers who would dig the foundations for the buildings and perform other tasks. Initially the commissioners considering bringing laborers from Europe, but they found that laborers could be recruited locally and augmented with slaves. Although the pay was low, available payrolls (the first ones available are from 1795) indicated that turnover was low— few laborers left the commissioners' employ. Virtually all of these laborers were illiterate, as indicated by their marks on the monthly payrolls.[45]

As excavation continued, masons were required for building both the foundations and the walls. If laborers were at the lowest rung of the social structure of the building site, masons were at the highest end. The highest-paid masons were the stone carvers. Down just slightly from the stone carvers in status were the stone cutters, the men who finished the rough-cut stone delivered from the Aquia quarries and set that stone in the walls. Stone cutters were divided into two groups: the stone setters who worked on the walls received higher wages, and the stone cutters proper, who worked in the stonecutting shed and on the grounds surrounding the President's House and the Capitol, received lower wages.[46]

Carpenters were somewhere between masons and laborers in the building social hierarchy. They were readily available in the timber-rich country and easy to recruit locally. Hoban brought many of the carpenters with him from Charleston, including several slaves. The bricklayers were also somewhere between masons and laborers in the construction hierarchy scale. They were fewer in number than the masons.[47]

Also on the construction site were wagons, horses, and men delivering materials to the site. One example is Mr. Joseph Dove, who, on March 31, 1800, wrote to the commissioners stating that he had never received less than $4 per day for a wagon and $2 per day for a horse and cart.[48] Food was provided by the commissioners, so cooks and other workers were also needed.

Toward the end of the construction, smaller numbers of workmen from other building trades were hired, including cabinetmakers, plasterers, glaziers, roofers, and others. The President's House, for example, was painted by Lewis Clepham, foreman, with assistance from journeymen John D. Lourey and Thomas Monney in 1798. Doors and similar wooden furnishings were constructed and installed in 1798 by Joseph Middleton, foreman of the cabinetmakers; William Middleton, presumably his brother; and Thomas Brown and Charles Boon(e), apprentices.[49]

The time spent arguing over the Capitol's design was not a total loss. Work had begun on the President's House, and the needed workers' sheds, workers' housing, a hospital for the workers, and workshops for the various building trades had been built. For example, on April 12, 1792, the commissioners entered into a contract with William Knowles to build a 24-by-50-foot workers' shed near the President's House.[50] Housing for the workers, also an urgent requirement, was built in the immediate vicinity of the President's House and the Capitol. The accommodations that the commissioners erected for the workers were spartan, their rooms just 10 feet square.[51]

By the end of 1793, it was obvious that the federal construction program was at a point where great progress was needed and perhaps could be made. On September 23, 1793, five days after the cornerstone ceremony for the Capitol, the commissioners decided to substitute stone-faced brick walls for stone walls in both the President's House and the Capitol, as a way to speed the work and reduce its cost.[52] This decision was made — as was typical — without an assessment of its cost or time impact.

On the same date, the commissioners directed Hoban to assume the general superintendence of the construction of the Capitol, as well as of the President's House.[53] Hallet was hired to develop workable plans and supervise the construction of the Capitol.

YELLOW FEVER IN PHILADELPHIA

While this was going on, there still remained the threat that Congress would reverse the Residence Act and leave the federal capital in Philadelphia. Then in the autumn of 1793, yellow fever ravaged Philadelphia. It was one of the most severe outbreaks of yellow fever in US history.[54] Between August 1 and November 9, 1793, 5,000 or more people died in Philadelphia, a city of 50,000 population, most of them from yellow fever. Another 20,000 fled the city, closing shops, newspapers, schools, and other institutions. The yellow fever outbreak was especially frightening because neither physicians nor the population had any idea of its cause. Intuitively, however, they sensed that proximity to swamps and standing water was a factor. Washington, well drained and with few swamps, was seen as more healthful than Philadelphia. The outbreak of this disease in Philadelphia had the effect of convincing many of the national legislators that the transfer of the national capital from Philadelphia to "healthful" Washington was necessary. After the outbreak, there was little talk of keeping the national capital in Philadelphia.

1794: MORE DISSENSION THAN BUILDING PROGRESS

PROGRESS ON THE PRESIDENT'S HOUSE IN 1794

By August 1794, significant progress had been made in construction of the President's House. Collen Williamson, the stone mason in charge of masonry at the President's House, reported that he and his workers had completed the walls of the first floor.[55] Work on the Capitol was much slower, but delivery of foundation stone, using stone from John Mason's quarries along the Potomac River, had begun.[56]

HALLET'S DRAWINGS FOR THE CAPITOL

In accordance with the meeting in Jefferson's office in July 1793, Hallet was supposed to be working on the drawings of the Capitol, to make Thornton's design workable. These drawings were to be submitted to the commissioners for their approval. But, without submitting drawings to the commissioners, Hallet ordered the building's foundations to begin, including the disputed center section. Thornton and his friend Rivardi visited the site and became convinced that Hallet had substituted his design for Thornton's for the center section.[57] Probably at the instigation of Thornton, on June 24, 1794, the commissioners demanded that Hallet turn over his drawings to them for their review.[58]

This order produced a quandary for Hallet. He probably thought that if he turned over his drawings, the commissioners would direct him to change them to reflect Thornton's design. If he did not turn over the drawings, it was an act of insubordination. He chose not to. The commissioners wrote to him on June 26, 1794, that he did not have permission to vary from Thornton's plan without the President's or the commissioners' approval. The commissioners went on to instruct Hallet that he was to be subordinate to James Hoban. Further, they asked him to respond to them by letter that he understood these instructions.[59] The commissioners did not give Hallet much time to respond, as on the following day, June 27, they wrote him they wanted an answer within one day.[60] On the next day, June 28, Hallet angrily responded that the plan of the Capitol was his and not Thornton's.[61] Hallet was dismissed by the commissioners that day, the third architect or engineer (after Pierre L'Enfant and Isaac Roberdeau) whom the commissioners dismissed or forced to resign. He was not the last. And soon the commissioners began firing foremen and workers.

Hallet worked until November 1794, as provided by his contract. He remained in Washington until August 1796. Not much is known of the remainder of his career. He was in Havana in 1800 and New York in 1812, where he died in 1825. It is believed that he built nothing after being dismissed by the commissioners.[62]

HIRING WORKERS FOR THE NEW FEDERAL CITY

Beginning late in 1794, the commissioners, led by the new commissioners Scott and Thornton, initiated a series of actions to overhaul the public construction program. They began by increasing the number of slaves engaged in the public works.

For the tasks of construction, the commissioners also needed skilled workers. The principal issue before them at the outset was compensation for these skilled workers—not just how much to pay but in what manner: equitable compensation involved not only just payment for services but also the ideas of equity, equality, and justice that were embodied in the building trades of the time. A flat wage for hours worked was not the only method or even the preferred method of compensation. Rather, "work by the measure," where, for example, bricklayers were paid by linear feet of brick laid, was seen by most workers as a better method. That system meant that faster or more capable workers would receive more money than slower or less capable workers.

This choice, hourly rate versus payment by the measure (or piece), first arose among the stone cutters. After much debate, in October 1793 the commissioners authorized architect Hoban to pay the stone cutters at Aquia quarry "by the piece" in lieu of an hourly wage. The commissioners' record read, "Mr. Hoban is to agree for the cutting the Stone works, by the piece agreeable to the Stone Cutters proposals if he thinks them reasonable."[63]

Although aimed at the workers at the Aquia quarry and intended to increase stone production, this change was immediately noticed at the construction sites in the new federal city. At the beginning of the 1794 construction season, Cornelius McDermot Roe, a friend of Hoban's, submitted a proposal to the commissioners to supervise the masons at the Capitol and to pay them "by the piece," at six shilling the statute perch ($0.30) for regular work and six shillings six pence for oval and circular work ($0.325), "such as may be found in arches and groins."[64] On May 9, 1794, the commissioners decided to accept Roe's offer to work and invited Collen Williamson to superintend the work.[65] Having decided to pay Roe's masons "by the piece," ten days later the commissioners turned down a petition for a wage increase, made by the masons then working on the Capitol.[66]

THE FIRST SIGNIFICANT LABOR DISTURBANCE

Just as construction was starting to move forward, these decisions set the stage for the first significant labor disturbance at the construction sites in the new city. Colonel Benjamin Stoddert wrote to commissioner Carroll on May 26, 1794, that 12 to 15 stone masons had left the city and 20 more were expected to leave, publishing a warning to other masons that their usage by the commissioners had been "very bad."[67]

The basic issue in this dispute was that the commissioners had asked the stone masons to work under Cornelius McDermot Roe, a contractor who had not been trained as a mason through an apprenticeship and therefore was not considered a member of that building trade. The commissioners apparently were unaware of this prohibition among building tradesmen. The masons wrote the commissioners, "No man who is a tradesmen will Submit to work under those who are not so." They continued, "We therefore cannot work under McDermot Roe and cannot help think it very hard that we Should be told we must work under him or be discharged After having worked so long for the Publick without complaint against us." They proposed to the commissioners their own plan of working "by the measure" rather than day wages.[68]

The commissioners (Carroll, Johnson, and Stuart), were local landowners little equipped to understand or accommodate the issues in this labor dispute. The conflict was initially resolved when the masons offered to work under supervision other than Cornelius McDermot Roe's and for a rate considerably less than that which the commissioners contracted with Roe, 4 shillings 6 pence per statute perch ($0.225). The commissioners accepted this offer.[69]

The commissioners now had two groups of stone masons employed under different pay systems. The immediate labor problems in the new federal city were thus temporarily suspended, but other problems arose. Payment by the piece assumed that adequate stone would be delivered to the site. But the Aquia quarry on Government Island could not provide enough stone to keep both groups of masons employed. The commissioners handled this as best they could, by dividing the available stone equally between the two groups.[70] In effect, this policy meant that the masons who had contracted with the commissioners would receive less money no matter how hard or fast they worked "by the piece." The masons reacted by making open threats against Hoban and Roe.[71] The commissioners responded by discharging several masons. Masons Robert Brown and James Maitland were rehired by the commissioners, as indicated in the payrolls of the masons who worked on the President's House for most of the remainder of the construction period. Mason John Delahunty was discharged and not rehired, and in 1796 he was reported in a newspaper advertisement as working at the Seneca Quarry, north of the new federal city.[72] But generally at this time the commissioners did not harshly retaliate against the workers. That came later.

THE DISPUTE BETWEEN JAMES HOBAN AND COLLEN WILLIAMSON

The commissioners had entrusted architectural oversight at the Capitol to James Hoban and masonry oversight to Collen Williamson—conflicting authorities that would try the best of friends. But in this case, ethnic rivalry between the Irish architect and the Scottish mason further soured their relationship. On June 5, 1794, Williamson wrote to the commissioners demanding full authority over the stone cutters and the masons.[73] On June 28, the commissioners tried to resolve the dispute by affirming that Collen Williamson was to be the sole supervisor of the masonry work at the Capitol.[74] Nevertheless, the dispute continued and affected construction progress.

The year 1794 had begun very optimistically, but construction results were disappointing. Construction that year was characterized by workmen walking off the project, workmen publicly criticizing the commissioners in newspapers, workmen making threats against supervisors, supervisors fighting between themselves, and the commissioners adopting various and different methods of payment, further confusing the work relationships. In addition, there were more traditional problems, such as too little building stone being delivered from the Aquia quarry. In this environment it is not surprising that little construction progress was made.

REPLACING THE ORIGINAL COMMISSIONERS

It was plain to all that building progress in 1794 was a disaster. It seemed that the federal city would never be completed unless something drastic was done. What was done was to replace the original commissioners, all gentry whose principal qualifications for the post were that they were friends of George Washington and were local landowners. They were replaced with younger men, who were handsomely paid. The last meeting of the three original commissioners (Carroll, Johnson, and Stuart) was held July 27–31, 1794. Gustavus Scott, an attorney from Somerset County, Maryland, was appointed to replace Thomas Johnson on August 23, 1794, and continued in that position until his death on December 25, 1800. William Thornton, the physician who won the Capitol design competition, was appointed on September 12, 1794, to replace Dr. David Stuart. Daniel Carroll continued in office, but only until the spring of 1795. He was then replaced by Alexander White, an attorney from Frederick County, Virginia, who was a former congressman, on May 18, 1795. Scott, Thornton, and White oversaw most of the remainder of the first public building campaign. These new commissioners initiated a series of actions to overhaul the public construction program to provide for better accountability. In 1795 the new commissioners moved to place the federal construction program on a more businesslike basis. On February 18, 1795, they adopted a series of rules that became the basis of their work for the next five years. These rules established a paper system to track the payments made for goods and services related to the federal construction programs.[75] In one respect, these rules were commendable. The rules clearly sought to establish accountability in the federal construction program. Written payrolls, for example, were required—many from 1795 to 1800 are still in existence at the National Archives. Likewise, numerous receipts and other material are still in existence. The problems of pay "by the piece" were avoided as the new commissioners decided that all workmen were to be paid a daily salary. But the difficulty with these rules was that they imposed a paper system on top of the building process rather than dealing with the building process itself.

The rules adopted were rules one would expect from lawyers and accountants, not men involved in the building process.

CONSTRUCTION PROGRESS IN 1795 AND 1796

One of the criticisms of both old and new commissioners was that they rarely left Georgetown to supervise the construction work. In the lack of oversight, substandard work had been introduced into the buildings. On June 26, 1795, the commissioners reported to Secretary of State Edmund Randolph that "bad work had been put into the Walls [of the Capitol]; and in some parts prudence requires they should be taken down."[76] The commissioners found that the foundations of the south wing (the House of Representatives wing) had to be totally reconstructed. The foundations on the north wing (the Senate wing) were also defective but could be repaired and secured by laying large bond stones. The new commissioners concluded that the bad work was deliberate and that this work had been the responsibility of Cornelius McDermot Roe.[77] Roe tried to object, claiming that there was "malaise of a party against him," but he was discharged by the commissioners.[78]

The failure of the commissioners, particularly Scott and Thornton, to visit the construction sites was a frequent criticism. With the discovery of the substandard work put in place, this criticism took on a more ominous tone. To defend themselves, the commissioners explained to Secretary of State Edmund Randolph that they were not responsible for the bad work and that the press of business had kept them from making more frequent visits to the construction sites. The commissioners also told Randolph that it was very difficult to deal with the contractors and workmen, whom they referred to as "the Motley Set."[79]

The new commissioners never did resolve the problems of defective work on the Capitol. Years later Benjamin Latrobe, investigating the adequacy of the Capitol's foundations, found that the defective workmanship discovered in 1795 had never been adequately corrected and that the foundations had to be torn down and rebuilt.

Another factor in the poor work that had been done was the conflict between Williamson, the supervisor of masonry at the Capitol, and Hoban, the architect overseeing construction at the Capitol. In an effort to clear the lines of authority, on May 28, 1795, the commissioners fired Williamson.[80] Williamson vigorously objected, writing to the commissioners, "I am not the man that was represented to you."[81] But to no avail.

It was about this time that a new mason, George Blagden, was hired for the works. In 1794 Blagden, a stone cutter from England, had begun work for the commissioners in the new city, having been recommended to the commissioners in July 1792 by Philadelphia stone merchant Adam Traquair. Blagden continued in this work for the next thirty-two years.[82] He continued working on the Capitol until he was killed by the cave-in of an embankment there on June 4, 1826.[83]

The finding of defective work emphasized that the commissioners needed a responsible architect-engineer to oversee the work. At the end of 1794, artist John Trumbull, then secretary to John Jay's delegation, recommended to the commissioners his friend George Hadfield, who was an architect, for employment. On December 18, 1794, the commissioners responded that there were no vacancies. The following spring they reversed themselves and offered Hadfield a position to supervise the construction of the Capitol. He accepted on March 7, 1795, and arrived in Washington in October 1795.[84] By that time Thornton was one of the three commissioners, having been appointed a year earlier, September 12, 1794. Hadfield knew he, like Hallet before him, would have to deal with Thornton and the problems of Thornton's design of the Capitol.

Upon arriving in Washington, Hadfield immediately began studying Hallet's and Thornton's plans for the Capitol. They did not match. He found other problems and wrote the commissioners: "I find the building begun, but I do not find the necessary plans to carry on a work of this importance, and I think there are defects that are not warrantable, in most of the branches that constitute the profession of an architect, Stability—Economy—Convenience—Beauty. There will be material inconvenience in the apartments, deformity in the rooms, chimneys and windows placed without symmetry."[85]

Hadfield began searching for means to rectify the design problems: "I think the design may be improved by omitting the basement throughout the building, by this means, expense would be saved; the Legislative Body ought to occupy the principal part of the Building instead of the basement; the Portico would not be useless, and grandeur and propriety would be increased from the Order beginning from the ground."[86]

Hadfield offered two solutions to resolve the design problems he found, and both solutions moved the principal rooms from the basement to the first floor and reduced the height of the building. Thornton, who had been a commissioner for more than a year, was furious. On November 2, 1795, he sent a long letter to George Washington complaining about Hadfield's proposed changes. He suggested to Washington that Hadfield was only attempting to enhance his own reputation by "innovating throughout," thus earning him fame for redesigning America's Capitol.[87]

Shortly afterward, Hadfield and Hoban were called to Philadelphia to confer with Washington about Hadfield's proposed changes to the Capitol design. In a letter to the commissioners of November 9, 1795, Washington summarized the results of that meeting:

> From the explanation of the former [Hadfield], it would seem as if he had not been perfectly understood—or in other words—that *now* he means no change

in the interior of the building, of the least importance; nor any elsewhere, that will occasion delay, or add to the expense — but the contrary; whilst the exterior will, in his opinion, assume a better appearance, & the portico be found more convenient than on the present plan. As far as I understand the matter, the difference lies simply in discarding the basement, & adding an attic Story, if the latter should be found necessary; but this (the attic) he thinks may be dispensed with, as sufficient elevation may be obtained, in the manner he explained it, without — and to add a dome over the open or circular area or lobby, which in my judgment is a most desirable thing, & what I always expected was part of the original design, until otherwise informed in my late visit to the City, if strength can be given to it & sufficient light obtained.[88]

Washington was exasperated by having again become involved in a matter that he thought had been resolved. As president he was trying to keep the new and unprepared nation out of the European wars that followed in the wake of the French Revolution. He was also dealing with a threat by the New England states to dissolve the Union. In addition, military actions against the Indians in the Northwest Territory were going very badly. Washington responded to the commissioners that they, not he, should handle this problem:

However proper it may have been in you, to render the decision of the objection, of Mr. Hatfield [sic — Hadfield] to the Executive — I shall give no final opinion thereon. 1. Because I have not Sufficient knowledge of the subject, to judge with precision. 2. because the means of acquiring it, are not within my reach. — 3. If they were, pressed as I am with other matters, particularly at the eve of an approaching, perhaps an interesting session of Congress, I could not avail myself of them: — but above all, because I have not the precise knowledge of the characters you have to deal with — the knowledge of all the facts before you — nor perhaps the same view you can take of the consequences of a decision for or against Mr. Hatfield's proposed alterations, or of his abilities to carry them into execution if adopted. . . . you can decide with more safety than I am enabled to do, on the measures proper to be pursued under the embarrassment which has arisen from this diversity of opinion.[89]

In the same letter he chided them that he was aware, perhaps from Thornton's letter of November 2, 1795, that a major part of the problem was who was to assume credit for the design of the Capitol: "I should have no objection, as he conceives his character as an architect is in some measure at Stake — and in short, as the present plan is nobody's, but a compound of everybody's, to the proposed changes."[90]

This threw the decision on Hadfield's design modifications of the Capitol back to the commissioners. Of the three commissioners, Thornton was staunchly opposed to any changes to the design of the Capitol. Scott was a strong ally of Thornton. Thornton was also aided by architect Hoban, who reported that if Hadfield could not build the original design, he would do it himself.[91] The commissioners ruled against Hadfield and his changes. On June 24, 1796, giving three months' notice as required by his contract, Hadfield resigned after being on the job only nine months. He reconsidered and withdrew his resignation several days later.[92]

Construction proceeded in 1795 and 1796, but because of workers' slowdowns it went at a slow pace. The commissioners were aware of this organized slow pace, and their awareness influenced their later actions. In comparison to the turmoil of 1794, the new rules adopted by the new commissioners achieved a semblance of labor peace on the work sites throughout 1795, 1796, and most of 1797.

By the end of 1795, progress was reported to the Congress on the construction of the President's House, including that "the roof [was] considered as half done."[93] "The roof considered as half done" meant that the symbolic end of the building of the exterior of the structure had almost been reached and that only interior finishing would be required. This achievement was celebrated by the workmen with "a raising," a banquet celebration of their completed work.[94]

In contrast, little progress had been made on the construction of the north wing of the Capitol, the Senate wing.[95] And no work had been initiated on the south wing, the House wing, beyond the foundation. Work on the center section had been halted since the dismissal of Hallet.

SUSPENSION OF WORK ON THE HOUSE OF REPRESENTATIVES WING
By 1795 the commissioners were nearly out of money. On January 29, 1795, they asked Washington whether it would be wiser to "forgo carrying on more of that building [the Capitol] than the immediate accommodation of Congress may require."[96] They were suggesting that the Capitol be constructed in three stages: first the north wing (Senate), then the south wing (House of Representatives), and finally the center section. Immediate work would be concentrated on the north wing, and work on the south wing would be suspended. This approach was adopted. By 1800, when the Congress was scheduled to move to Washington, only the north wing was mostly finished. The entire Capitol was not completed until 1827—thirty-two years after work had begun.

SEEKING A CONGRESSIONALLY GUARANTEED LOAN
Financially, the situation got worse in the next several months of 1795. On May 1, 1795, the Greenleaf, Morris, Nicholson real estate syndicate missed their second installment payment for the 6,000 lots that they had contracted for. Nor did the syndicate make any subsequent payments or provide the one-thousand-pound-per-month loan to the commissioners that they had previously agreed upon. Further, the commissioners could not resell the 6,000 lots that were to be transferred to

the Greenleaf, Morris, and Nicholson syndicate, because the agreement with the syndicate had clouded title to these lots.

The commissioners then turned to Congress, asking for a congressionally guaranteed loan. Commissioner White was sent to Philadelphia to arrange congressional authorization for a $500,000 loan backed by the US Treasury—similar to the approach that L'Enfant had originally proposed. White was successful in achieving congressional authorization, but borrowing the money was difficult, owing to the European Wars. European capital that might normally have been available to the United States was now diverted into taxes or armies or held in a state of uncertainty. Through much effort the commissioners were able to borrow only a small portion of the $500,000 authorized.

The European wars not only made borrowing difficult but resulted in cost-of-living inflation in the United States and Europe. There was agitation among the workmen for higher wages to meet the higher living costs. On June 6, 1796, some of the carpenters working at the Capitol petitioned the commissioners for an increase in their wages.[97] The increase was not approved. Construction continued. On March 6, 1797, the stone carvers at the Capitol also appealed for higher wages,[98] and again, no action was taken. Future, more strident demands for wage increases followed.

THE 1797 CONSTRUCTION SEASON

At the beginning of 1797, work on the President's House was quite advanced, but work on the Capitol languished. This was the exact opposite of what President Washington intended. He believed that the Capitol, the House of the People, should be more advanced in construction than the President's House, referred to by its detractors as "The Palace." Washington urged the commissioners to concentrate on completing the Capitol.[99] The commissioners concurred, and priority continued to be the north wing of the Capitol.

NOT MEETING PAYROLLS: OCTOBER 1797

Labor relations were quiet at the construction sites for most of 1797. But in October of that year, an event occurred that shattered the labor peace: the commissioners could not meet their payroll. On October 3, 1797, they wrote to their architects James Hoban and George Hadfield, "The present situation of our funds renders it impossible to pay the time-Role due yesterday, or to discharge the arrears of last month."[100]

The commissioners had spent all of the original grant monies from the Maryland and Virginia state legislatures, were not able to collect the money owed them for lot sales to the Greenleaf syndicate, and had spent all of the money they were able to raise from the 1795 loan authorization. They could not meet their September payroll. From the workers' viewpoint, failure to meet payroll was sinister—a major breach of trust. The difference between the standard of living of the commissioners, from the gentlemen class, and that of the workers was huge. It would have been difficult for the workers to comprehend that money was not available for their pay when they observed the commissioners living in the style of gentlemen.

Failure to meet the payroll rapidly spread discontent among the workers. They already shared a certain inherent dislike of the commissioners anyway, particularly of Scott and Thornton. They ridiculed the commissioners through pamphlets circulated among themselves, as well as through verbal comments and jokes. These pamphlets and comments were reported back to the commissioners. There is no record that the commissioners hired paid informers, but there was probably no need to do so. Although the commissioners spent little time at either of the two primary construction sites, they undoubtedly received intelligence from several of the workmen who either sought special favors or disliked those who were spreading the ridicule. There is not much in the records of the commissioners to indicate what they were being told, but from the actions they took, we can infer what they were being informed about.

THE FIRING OF PIERCE PURCELL AND MICHAEL DOWLING

Toward the end of 1797, the commissioners issued a series of unprecedented orders, focusing first on the carpenters at the President's House. They ordered the firing of the foreman of the carpenters, Pierce Purcell, and the carpenter Michael Dowling, who was the keeper of the public stores at the President's House. The commissioners ordered Purcell to vacate his house. They gave no reason why these two men were fired, nor were any hearings held.[101] The Proceedings for November 15 provides a glimpse of the accusations against the two men: they were suspected of stealing from the public stores. After firing Purcell and Dowling, the commissioners ordered that an inventory of those stores be conducted.[102]

The record suggests that the firing for theft of public property was done without first determining whether any of the public property had actually been stolen. The Proceedings of December 5, 1797, suggests that the informant was Joseph Middleton, the master cabinetmaker, since he is the individual who was assigned Purcell's house and Purcell's former duty to take care of the public stores. Later Middleton became embroiled in a major dispute with Hoban, a friend and former partner of Purcell.[103]

This is most of what we know of these firings, with the exception of two letters written by Pierce Purcell to the commissioners approximately a year later. He wrote with the hope of clearing his name and regaining employment: "Had I been in possession of these documents when cited to the Board; I am convinced you would have assumed me innocent of the charges brought against me."[104] His enclosures have not survived the two centuries that separate his grievance from our day. However, in Purcell's second letter, written a week later, September 20, 1798, he provided affidavits from other workmen and other information clearing him of the charges brought against him by Joseph Middleton.[105]

There is no record suggesting that the commissioners considered the evidence that Pierce Purcell presented them. He was not rehired. Of the many workmen fired by the commissioners, few were ever rehired.

PROHIBITING APPRENTICES AND NEGRO CARPENTERS
FROM WORKING ON THE PUBLIC BUILDINGS

The commissioners were on a tear. They did not stop at firing Purcell and Dowling. After being informed that the carpenters at the President's House performed little work, that the apprentices and Negroes did all of the carpentry work, the commissioners issued an order on November 15, 1797, the day after they had fired Purcell and Dowling, that no Negro carpenters or apprentices would be allowed in the public works unless specifically authorized by the commissioners.[106] A few exceptions were allowed, and on December 5, 1797, they authorized a few apprentices.[107]

Pierce Purcell was a longtime friend and partner of James Hoban, his supervisor. Hoban's letter of the time to the commissioners clearly indicated that Purcell was discharged without consultation with him.[108] Discharging the slave carpenters was another action against Hoban, since he owned the largest number of carpenter slaves. The commissioners instructed Hoban to appoint a new foreman carpenter to replace Purcell,[109] and Hoban appointed John Lenox as foreman. Although Hoban was not discharged at this time, he was clearly put on warning.

CONFLICT BETWEEN COMMISSIONERS
AND WORKERS IN 1798

These events, in the last several months of 1797, set the tone for the 1798 construction year. It was 1798 that became the major year of conflict between the commissioners and the workers. Early that year the commissioners, particularly Scott and Thornton, set the stage for what one writer described as "America's first serious labor dispute."[110]

THE SECOND CONGRESSIONALLY GUARANTEED LOAN

The year 1798 was a critical year for the commissioners, because the government was to move to the new federal city two years hence. Construction of the Capitol had not progressed very far, although good progress had been made on the President's House. Further, the commissioners were very close to being broke. In 1798 they again sent commissioner White to Congress in Philadelphia to lobby for a grant or a loan. If Congress authorized a loan, the commissioners proposed to use their land holdings as security, to be sold after the government had moved to Washington so that they could bring a higher price than was possible in 1798.

White was again successful. On March 27, 1798, with a secured loan of one hundred thousand dollars in hand from Congress and with Alexander White about to come back to Washington from Philadelphia, Scott and Thornton wrote telling him that they intended to take action against what they viewed as the high wages being paid to the workmen in the federal city. They said, "We are very desirous of being fully ascertained of the prices of all wages at Philadelphia as it is high time to take up the subject here."[111]

The tone of this letter was ominous for the workers. It suggests that Scott and Thornton had thought wage reduction on the federal works had been needed for some time. Probably they also felt the workers deserved retribution for their mocking flyers circulated in Washington at the commissioners' expense.

While in Philadelphia, commissioner White consulted with a Philadelphia stone cutter, John Miller, who informed White that Philadelphia carpenters and stone cutters could be hired "from a Dollar to a French crown per day,"[112] much less than what the commissioners were paying their workmen. It is likely that the workers in Washington may indeed have been paid more than their counterparts elsewhere. There were reasons. There was the traditional extra payment that workmen received for working away from their native city. There was an additional extra payment for the longer days worked during the high construction season (March 1 through November 1). It is also likely that John Miller may have quoted low wages to White for his own financial benefit, acting as a labor jobber. Miller had told commissioner White that stone cutters could be had from one dollar per day in Philadelphia.[113] Since stone cutters traditionally received more than carpenters, his statement suggested that carpenters in Philadelphia were making less than one dollar per day, much less than what carpenters were being paid in the new federal city.

White's letter was the information that Scott and Thornton needed and wanted. The commissioners decided to forgo the normal wage increase for stone carvers and stone cutters during the traditional high construction season, when they worked longer hours in the increased daylight. The commissioners' decision not to increase those wages was in effect a wage cut.

The commissioners' new wage scale would have paid all stone cutters 10 shillings per day ($1.33), a significant reduction from the previous year. For stone cutters working at the President's House, such as Robert Tolmie, Hugh Sommervelle, James Sommervelle, Alexander Wilson, and Alexander Reid, the new wage scale represented a 30 percent decrease from the previous year. In 1797 they were paid 10 shillings per day ($1.33) during the winter months and 13 shillings per day ($1.73) during the high construction season. For lesser stone cutters, such as William Symington, William Bond, Francis Bond, James McIntosh, John Williamson, James Dougherty, and Thomas Allen, it represented a 10 percent decrease. During the previous year, their rate per day had been 10 shillings ($1.33) during the winter months and

TABLE 7.1. Days worked by stone cutters at the President's House in 1796 for selected months

	JULY	AUGUST	NOVEMBER	DECEMBER
Bernard Crook (foreman)	29	31½	25	25
Hugh Sommervelle	29	31½	25	22
James Sommervelle	29	31½	25	22
Alexander Reid	29	31	25	25
Alexander Wilson	29	31	24	24
Francis Brown	28¼	31	24	24
William Symington	28½	29⁷⁄₁₀	22	22

Source: Miscellaneous Treasury Records, Payroll for July, August, November, and December 1796, for the Stone Cutters at the President's House, RG 217, NARA, box 3, no. 524 (July); box 3, no. 600 (August); box 3, no. 826 (November); box 3, no. 888 (December).

11 shillings ($1.47) during the high construction season. Stone carvers engaged at the President's House in 1797, such as William Temmener, John Davidson, and John Hogg, faced similar wage cuts.

The carpenters and joiners were traditionally not given a wage increase during the high construction months. For them, on April 17, 1798, the commissioners dictated a per-day wage decrease to no more than 9 shillings ($1.20); $2.00 for the foremen.[114] Carpenter Joseph Hoban, probably a younger brother to James Hoban, had his pay cut from 12 shillings 6 pence per day ($1.67) to 9 shillings ($1.20). Journeyman carpenters such as Peter Lenox, Robert Aull, William Warrington, John Nixon, Richard Wright, and George Sandyford had their wages cut from the 1797 level of 10 shillings per day ($1.33) to 9 shillings ($1.20). Even the foreman, John Lenox, had his wages per day cut from 17 shillings 6 pence ($2.33) to 15 shillings ($2.00). Carpenters such as Peter Haley, Dan Coffry, Thomas Sandyford, George Thomson, Patrick Healey, Dan Caffry, Samuel Wright, and others were hired for the 1798 construction season at 10 shillings per day ($1.33) and found their daily wages reduced to 9 shillings ($1.20).

Compounding the difficulty of these wage cuts was the suddenness of the commissioners' actions. Coming just before high construction season, the reductions were announced too late in the year for the workmen to leave the new city and find employment elsewhere. Further, the wage cuts came as the cost of living was increasing.

The total effect of the wage cuts on the workers' annual wage was much more substantial than the daily percentage decrease would indicate. Construction at the time was highly seasonal. Although most of the affected workmen were provided with work throughout the year, including the winter months, it is apparent from the payrolls of the time that during the winter months, few workmen worked a full month or were paid for a full month. In contrast, during the high construction season, the workmen would usually work on all workdays, and sometimes on Sundays too (such as during the month of August). Table 7.1 indicates a representative number of days worked (and thus paid) for certain stone cutters at the President's House in selected months.

The seasonality of the building trades greatly affected monthly take-home pay. Hugh Sommervelle, James Sommervelle, Alexander Reid, and Alexander Wilson, for example, would make close to £19 ($50) during the month of July, and even more during August. But they would make only £11 or £12 ($29 to $32 per month) during the month of December. For these men, the commissioners' decision not to advance their wages during the high construction season represented a 36 percent to 42 percent cut in their yearly take home pay.

REACTION TO THE COMMISSIONERS' PAY CUTS

The workmen reacted strongly to these pay cuts. The stone carvers and stone cutters took concerted action against them. The stone carvers at the Capitol, the same men who had complained to the commissioners the previous year about their wages, were the first to protest the actions of the commissioners. On April 16, 1798, they sent a letter to the commissioners stating that the wage cuts were an outrageous act that the workers expected the commissioners to reverse.[115] The stone carvers were joined the same day by the stone cutters at the Capitol. The stone cutters went further by threatening to walk off the project unless their previous summer wages were reinstated.[116] The following day the commissioners received a letter from the stone cutters at the President's House, also threatening to walk off the job.[117]

The commissioners were thus threatened with a massive work stoppage unless they restored wages to the previous summer's levels, The workers did not intend to strike but to abandon the work in the federal city. The commissioners felt that these men were their social inferiors The reaction of the commissioners was uncompromising. In their response of April 16, 1798, the commissioners stated, "We have no intention of raising your wages," and instructed the workers to "quit the public employment at the end of the present month and all the buildings you now occupy." A similar letter was sent to the stone cutters at the President's House.[118]

Having, in essence, fired the entire workforce of stone cutters (but not the more difficult-to-replace carvers), the commissioners decided to communicate to the carpenters their decision of the previous day to cut their wages also.[119] Table 7.2 indicates the magnitude of cuts that this order effected among the carpenters and joiners.

The commissioners understood that their actions were offensive to the workers and that many, perhaps all, might quit. The commissioners believed it would be

TABLE 7.2. Daily pay cuts of carpenters and joiners for April and May 1798

	APRIL 1798	MAY 1798	DIFFERENCE
John Lenox (foreman)	17s 6p ($2.33)	15s ($2.00)	−14%
Joseph Hoban	12s 6p ($1.67)	9s ($1.20)	−28%
Robert Aull	10s ($1.33)	9s ($1.20)	−10%
Peter Lenox	10s ($1.33)	9s ($1.20)	−10%
William Warrington	10s ($1.33)	9s ($1.20)	−10%

Source: Miscellaneous Treasury Records, Payroll for April and May 1798, for the Carpenters and Joiners at the President's House, RG 217, NARA, box 5, nos. 306, 307.

easy to replace any of the carpenters or joiners. But replacing the numerous stone cutters would be more difficult. The commissioners immediately initiated action to obtain replacement stone cutters from Baltimore and Philadelphia. On April 18, 1798, commissioner Scott wrote to Robert Stewart,[120] a master stone cutter in Baltimore, asking his assistance in recruiting new stone cutters.[121] On the same day commissioner White wrote to John Miller in Philadelphia regarding carpenters and stone cutters.[122] Miller replied that he could supply the needed stone cutters. In Miller's letter he refers to the stone cutters "making a stand" and to "the shameful advantage that they have taken." Miller's language captured, no doubt, the indignation the commissioners felt about the workers' threats to walk off the public building sites.[123]

Although the commissioners had told them to quit, the stone cutters remained. The standoff between the commissioners and the stone cutters continued. The commissioners had the advantages of owning the houses that the workmen lived in and of employing them in a region that offered little alternative construction work. On the other hand, the stone cutters worked in concert with each other, understanding that it would be difficult for the commissioners to replace them en masse.

Construction was halted. The standoff between workers and commissioners horrified supporters of the new city. The federal government was due to move down from Philadelphia in only two years, and none of the public buildings were ready. Some began to look for means to resolve this dispute.

In April the commissioners, still intending to replace the fired stone cutters, asked George Blagden, the master mason, how many stone cutters at the Capitol and the President's House were required. On April 21, 1798 he responded: "I have considered the present stage of the Stone Cutting department at the Capitol, and at the Prest. House And am of opinion that to finish the Exterior of the former this Season 16 stone Cutters and setters is necessary, and for the latter 12."[124] This is a

TABLE 7.3. Daily wages for stone cutters, 1797–1798 (in shillings and pence, Maryland currency)

	MAY 1797	APRIL 1798	ADDITIONAL	MAY 1798
John Williamson	11s/day ($1.47)	10s/day ($1.33)	6p/day ($0.06)	10s6p/day ($1.40)
Alexander Wilson	13s/day ($1.73)	10s/day ($1.33)	2s6p/day ($0.33)	12s6p/day ($1.67)
Francis Brown	11s/day ($1.47)	10s/day ($1.33)	6p/day ($0.06)	10s6p/day ($1.40)
Andrew Shields	n/a	10s/day ($1.33)	2s6p/day ($0.33)	12s6p/day ($1.67)
James McIntosh	11s/day ($1.47)	10s/day ($1.33)	6p/day ($0.06)	10s6p/day ($1.40)
James Reid	n/a	10s/day ($1.33)	6p/day ($0.06)	10s6p/day ($1.40)
Hugh Sommervelle	13s/day ($1.73)	10s/day ($1.33)	2s6p/day ($0.33)	12s6p/day ($1.67)
William Bond	11s/day ($1.47)	10s/day ($1.33)	6p/day ($0.06)	10s6p/day ($1.40)
Alexander Reid	13s/day ($1.73)	10s/day ($1.33)	2s6p/day ($0.33)	12s6p/day ($1.67)
William Symington	11s/day ($1.47)	10s/day ($1.33)	6p/day ($0.06)	10s6p/day ($1.40)
James White	n/a	10s/day ($1.33)	6p/day ($0.06)	10s6p/day ($1.40)

Source: Miscellaneous Treasury Records, Payrolls, RG 217, NARA, May 1797, box 4, no. 321; April 1798, box 5, no. 206; additional Wages for April 1798, box 5, no. 307; May 1798, box 5, no. 306.

strange estimate because the exterior of the President's House by 1798 was mostly completed, and the masonry work that still needed to be accomplished was at the north wing of the Capitol. The confrontation between the commissioners and the stone carvers and stone cutters continued for the remainder of April.

RECONCILIATION OF WORKERS AND COMMISSIONERS
It was George Hadfield who proposed the reconciliation between the stone cutters and the commissioners. On April 30, 1798, he wrote the commissioners: "Some of the stone cutters at the Capitol have been with me and expressed a wish to come to a reconciliation with the Commissioners. . . . Perhaps it would be better if the progress of the department was to proceed without interruption."[125]

The accommodation that was arranged between stone cutters and commissioners was a supplemental wage for April 1798 that appeared in the *Records of the Commissioners*. This is shown in table 7.3 for the stone cutters working at the President's House in comparison to the previous year's summer wages. As the table indicates, the stone carvers and stone cutters had accomplished a victory over the commissioners by gaining a wage increase for the 1798 high construction season. It was not a total victory, because their summer 1798 daily pay was not completely restored to the summer 1797 levels. Nonetheless, it was a victory achieved through concerted action. But it was not a victory that could be sustained over time. By the winter season of 1798–1799, the commissioners reduced the workers' wages to 9 shillings per day (as compared to 10 shillings the previous winter). Daily wages were further reduced in 1799.

In announcing these wage increases, the commissioners would not acknowledge that the stone cutters had gained a victory.[126] Some historians have incorrectly stated that the resolution of this labor dispute was a capitulation of the workers to the commissioners.[127] The reason for this interpretation is that the commissioners depicted the resolution to the outside world in this way. For example, on May 7, 1798, commissioner White wrote to John Miller, the stone cutter in Philadelphia from whom he requested assistance in recruiting replacements, that the stone cutters had "come to our terms."[128]

Although the stone carvers and stone cutters took concerted action against the commissioners' proposed pay cut, the carpenters and joiners — the only other large work group at the public works (with the exception of the laborers, who were not in a position to bring concerted action) — did not. Perhaps the group organized earlier for carpenters, the Architect and Carpenters Society, was defunct by 1798. Or perhaps, by including architects and master builders, it was not in a position to organize against the commissioners' pay cuts. The only recorded action taken by the carpenters and joiners against the pay cut, from 10 shillings to 9 shillings per day, was a single letter from a single journeyman carpenter working at the President's' House, Peter Lenox (brother of John Lenox, foreman of the carpenters and joiners at the President's House). Peter Lenox proposed to be paid by "merit" at a time when all of the carpenters were having their wages reduced. He wrote, "Gentlemen of you would let any part of the work at either building by the piece, I would be very glad to contract and will give any security that is requiste for the performance."[129] This is a single artisan negotiating for higher wages. Lenox's appeal to the commissioners' reason was not accepted.

The dispute between the carpenters and joiners and the commissioners was particularly troubled. As if to make the point that many carpenters could be found to take the place of the carpenters presently employed, on April 16, 1798, the commissioners authorized George Hadfield to hire two additional carpenters who had been referred to him.[130] The commissioners also went out of their way to insult the carpenters during this dispute. In their letter of May 4, 1798, to George Hadfield and James Hoban, for example, they suggested that if the carpenters were to be paid by ability, then some were worth no more than half of what they were currently being paid.[131] Shortly after this letter, on May 12, 1798, the commissioners fired all the carpenters at the President's House.[132] Most of these men had been with Hoban since he arrived in the new federal city. Moreover, some had bought lots and owed money on the lots. Discharge at the beginning of the new construction season, at a location remote from any population center, was a major catastrophe for these men and their families. Given the realities of this situation, some of the carpenters asked to be included in the workforce at the Capitol. On May 23, 1798, the commissioners selected the ones they wanted and informed Hadfield whom he was to hire from the workers at the President's House.[133]

The only substantial work that was to continue at the President's House in 1798 was the slate roofing, through a contract with Orlando Cook. But even this work languished, because of the absence of the contractor.[134]

On May 18, 1798, the commissioners fired George Hadfield, giving him three months' notice. They told him, "Your conduct of late has rendered it proper that your occupation as Superintendent of the Capitol should cease as soon as the time for previous notice, required by your contract shall have expired."[135] President Adams was informed of the board's action, dismissing "a young man of taste" who regrettably was also "deficient in practical knowledge of architecture."[136] Years later, architect Benjamin Latrobe wrote, "He [Hadfield] was no match for the rogues then employed in the construction of the public buildings, or for the charlatans in architecture who had designed them. . . . He waged a long war against the ignorance, and the dishonesty of the Commissioners and of the workmen. But the latter prevailed."[137]

With the close-down of most of the work at the President's House, the commissioners asked Hoban to supervise work at the Capitol. They also wanted him to hire Peter Lenox as his foreman. Hoban informed the commissioners on June 22, 1798, that Lenox was not acceptable to him. The following day the commissioners informed Hoban that Lenox was not hired for work on the Capitol.[138]

The record suggests that, in response to the commissioners' unilateral pay cut of April 17, 1798, the carpenters at the President's House had stopped work, or engaged in a work slowdown, or both. For example, in their letter of July 2, 1798, to Hoban, the commissioners explained that Thomas Sandyford, a carpenter who had begun working at the President's House that year, should be reemployed, "as his indisposition and confinement were the only causes of his not working at the time of the removal of those engaged."[139] In response to these work stoppages and slowdowns, the commissioners took action against the carpenters.

At the same time they discharged the carpenters at the President's House, May 25, 1798, the commissioners warned Redmund Purcell, the foreman of the carpenters and joiners at the Capitol, to watch for a possible work slowdown by the carpenters assigned to the Capitol.[140] This warning did not avert the slowdown that continued at the Capitol. The commissioners' records also indicated other labor problems at the Capitol. For example, Samuel Smallwood, the overseer of the laborers, reported an attack on him by Robert Aull, one of the carpenters reassigned from the President's House. Before attacking Smallwood, Aull had told him, "Thire was a scheme in his Coming which pleased Mr. Berry much." What Aull was probably saying was that there was an agreement between the commissioners and certain workmen to replace other workmen, including Aull. "Mr. Berry" may have been William Barry, whom the commissioners had recommended to Hoban for employment in May 1798, the month before the attack. Smallwood expressed less concern over Aull's attack than over the possibility that "a Certain Clas of people may entice even the blackis to Commit Depridations on me."[141] Robert Auld, probably the man who attacked Smallwood, was fired by the commissioners a year later.[142] Meanwhile, construction progress suffered.

Ousting discharged workers and their families from the free housing provided by the commissioners was standard practice, and this practice lent another dimension to the problems of being fired. In Aull's (or Auld's) case, he was given a short grace period before he vacated his house. The record read, "The Board in consideration of the situation of your wife have no objection to your continuing your occupation of the Temporary Building you live in for 2 or 3 weeks."[143]

THE CASE OF RICHARD GRIDLEY, BLACKSMITH

The harshest action by the commissioners against workers was taken against Richard Gridley, the contract blacksmith at the President's House and a close friend of many of the carpenters working there. On June 13, 1798, Gridley received the following discharge message from the commissioners: "The Board having agreed with Benjamin Bacon to do the Smith work of the public Buildings, you will please give him immediate possession of the shop you occupy."[144]

Whereas the carpenters had received almost a month's notice of their separation from the public works, Gridley received none. Further, Gridley had borrowed money to buy equipment to service his contract with the commissioners. In addition, in the previous year he had invested in building lots; he had bought three lots at the public auction of August 20, 1797.[145] He needed the income from his contract with the commissioners to maintain his investments and his business. When first confronted with charges against him, Gridley had taken somewhat of a philosophical approach. On June 11, 1798, he wrote the commissioners that he suspected his dismissal was because "I am too Saucy, and that I am considered by many the author of a Ridiculous Publication." This is a reference to one of the pamphlets being circulated among the workmen ridiculing the commissioners, particularly Thornton and Scott. The commissioners' actions against Gridley and the President's House carpenters were due, in part, to the disrespect these individuals had expressed for the commissioners, both verbally and in print.[146]

Two weeks later, Gridley was less philosophical. Writing to the commissioners on June 27, 1798, he referred to them as "you who move in a higher sphere than myself" and asked, "I request of you, the reasons, of my sudden dismission without the least Notice."[147] On the same day the commissioners responded to Gridley that he was being dismissed because he had overcharged them in the amount of $14.33, according to one example they gave.[148] This accusation from the commissioners brought an explosion from Gridley that the charge was trumped up by them. On June 28, 1798 he wrote the commissioners: "Your letter of yesterday I have received and am as-

☞ TAKE NOTICE.

IN Confequence of ungenerous treat. ment received from the Commiffioners of this City; I am determined to leave it in a few days. To enable me to pay my debts, I fhall offer at PUBLIC SALE, on Thurfday, 10th day of January, 1799, at my Houfe, on Greenleaf's Point,

A VARIETY OF VALUABLE ARTICLES, among which are

Feather Beds and common bedfteads,
Mahogany Defk, Bureau, Tables and Bedftead,
Windfor and common chairs,
Looking Glaffes,
Brafs Andirons, Shovel and Tongs,
A Bath Stove, and a number of articles neceffary in a kitchen.

ALSO,

Three complete Sets of Blackfmith's tools, with an excellent large Grindftone, hung on friction rolers, one good MILCH COW, a HORSE, a handy BOA:—new laft Spring, a WAGGON, a likely NE-GRO GIRL, about 14 years old.

They will be fold without referve and delivered when the cafh is paid.

RICHARD GRIDLEY.

City of Wafhington, 31ft Dec. 1798.

The departure of Richard Gridley, blacksmith, after receiving "ungenerous treatment" at the hands of the commissioners. *Centinel of Liberty and Georgetown (DC) Advertiser,* January 4, 1799.

tonished; that such a trifling excuse should induce you to make up your minds to dismiss me from the publik employ, pray now dare you to pass and pay my acct. if they were so varied from my contract as you complain."[149] Across the top of this letter, Richard Gridley had written, "No. 2 and the Last." To this one of the commissioners or their secretary had added, "(Glad of it Richard)" and, in a different hand, "Briefer."

A comment by Gridley, "Although this is directed to the Board, yet it may be considered as addressed only to them which signed Yours to me," reflects what the workmen thought of Thornton and Scott. White was generally liked by the workers, but Thornton and Scott (who had both signed the letter to Gridley) were despised. These two commissioners were widely ridiculed by the workmen for their lack of knowledge of building (particularly in Thornton's case), their lack of interest in building (particularly in Scott's case), and their pretension to possess building knowledge. The firing of Gridley placed all other contractors on notice that such action could be used against them. It had a chilling effect on the workforce all across the city.

The last view that we have of Richard Gridley comes just before he leaves the city, in the sale of his possessions on January 10, 1799.[150]

THE 1798 AND 1799 CONSTRUCTION SEASONS

The year 1798 began with construction almost completely halted at the President's House. The carpenters and joiners who had worked there had been reassigned to the Capitol under Hoban's direction. The next year, 1799, began with the commissioners firing two more carpenters, Robert Actkeuaud and Robert Oliver.[151]

Hoban did not escape the commissioners' wrath. They charged that he had allowed the carpenters and others to slack off, and they said it was due to his negligent conduct that "fourteen carpenters for a period of Seven weeks or more did not do one third of the work justly to be expected from them, that thirteen Sawyers for a whole year played the same game." But Hoban was continued in their employ, since the commissioners' letter of April 11, 1799, to Hoban dismissed these charges.[152]

THE FIRING OF REDMUND PURCELL
At approximately the same time, the commissioners were investigating charges against Redmund Purcell, the foreman of the carpenters at the Capitol under Hoban. As a result of the investigations, they fired him for being "guilty of neglect in the execution of his duty." He was "unfit for the service to which he has been appointed by the Superintendent." Four other carpenters working for Hoban at the Capitol were also fired: John Dickey, Thomas Watkins, William Keif, and James Tompkins. As foreman of the carpenters, the commissioners appointed Peter Lenox, the same man Hoban had refused to hire the previous year.[153]

The commissioners were not in a position to fire Hoban, since his dismissal might further shake confidence in the federal city. With George Hadfield fired, firing Hoban would leave the commissioners without a qualified architect or architect-engineer at a time when the federal government was scheduled to move to the new capital only a year hence. Thus Hoban survived the purges of 1798 and 1799.

WORK SLOWDOWN AT THE CONSTRUCTION SITES
Although the commissioners could discharge workmen, reduce their daily wages, and discharge individual contractors, the workmen could also retaliate. And they did so, through a work slowdown. The commissioners responded to the slowdown by firing the workers engaged in the most obvious cases. The situation was described by James Tompkins (one of the four carpenters fired) in April 1799, in a letter to the commissioners asking them to reinstate him: "I do acknowledge that there might been more work done than was but it was so general that every man in the Yard took Whatever Liberty almost he pleased as the Main thing was to keep time." Tompkins's request for reinstatement was on the grounds that he worked slowly because all of the other workmen were working slowly. Tompkins said that, in keeping with the workers' sense of justice, he hoped he would be given a hearing.[154] The commissioners did not respond to Tompkins's request. This is probably the same James Tompkins who was referred to George Hadfield by the commissioners on May 4, 1798, indicating that the carpenters on the project could enforce the work slowdown on even the newer workers.

Two days later, April 11, 1799, the commissioners received a letter from another discharged carpenter, John Dickey, also asking to be reinstated. Wrote Dickey, "I John Dickey humbley Prayeth that you will take In Consideration my Present situasion as Viz. that I have not got no work since my Discharge; that I have got a wife

TO THE CITIZENS OF COLUMBIA.

Fellow-Citizens,

I have been employed at the Capitol in the city of Washington from April 11th, 1795, to the morning of the 5th of April, 1799; and have acquitted myself as foreman of the carpenters and joiners department at the Capitol, I believe, with integrity and abilities. If I am deficient in either; I never had a fair opportunity afforded me of defending myself, as the following letter and affidavit will shew. Two days were occupied in taking depositions, &c. against me; but the only time *I* was allowed for preparing my defence was from three o'clock, in the afternoon, till nine, next morning.

City of Washington, April 3. 1799.
Gentlemen,

Your justice in granting Mr. Hoban a public investigation of charges brought against his character, shews how worthy you are of the confidence reposed in you by the executive, who appointed you guardians red from some of these rights of natural equity, and shall have no longer time than this evening to prepare my defence. No legal proof has as yet been exhibited against me: I will beg leave to ask the board, for among them I see lawyers, whether the bare assertion of dismissed and disappointed workmen, is sufficient to condemn my character; let them be sworn legally, otherwise no proof can be stamped on their bare assertions; how do I know but these witnesses were subborned, how do I know but some of my judges may be the principal plaintiff; for you must recollect how Mr. Scott endeavored to defeat the exertions I made to hinder the embezzlement of some thousands of dollars through the mismanagement of his favorite Hadfield; set forth in my letter of yesterday and how the said Mr. Scott exerted himself against me, for having acted so. And will appeal to Dr. Thornton as a voucher for these facts.—— However gentlemen I do not shrink from having the business brought forward, even at so short a period as to-morrow, on condition however of having three men of abilities in the building line : I will appoint men, who are or have been in your employment, Messrs. Harbaugh, Lovering, and Stephenson, for as I am conscious of rectitude, I will again open day. You will exhibit my principal plaintiff, I will compare the titles we both claim to honesty, it will appear who is the honest man, and this in the presence of reputable and substantial witnesses; I will have recourse to this and my former letter on a future day, in order to ascertain by the public judgment of my fellow-citizens whether I have acted in open day.

I am Gentlemen,
Your obedient servant,
REDMUND PURCELL.

The Commissioners of the }
city of Washington. }
True Copy.

The discharge of Redmund Purcell.

Centinel of Liberty and Georgetown (DC) Advertiser,
April 4, 1799.

and family that never offendd."[155] Dickey, like all others who were fired, was not re-hired or given a hearing.

Redmund Purcell, after being fired as the foreman of the carpenters at the Capitol, captured the workmen's indignation with the commissioners in his letter of April 3, 1799. He asked that "the charges fabricated against me should be discussed openly," that his case be heard by "three men of known abilities in the building line" and that they let him know "who the principal plaintiff against me is." Purcell referred to these requests as "rights of natural equity."[156] He was not granted his hearing.

The labor conflict hindered completion of the public buildings. By the end of the 1799 construction season, there was a strong concern among the commissioners that the public buildings would not be ready a year hence, when the federal government would move from Philadelphia.

WASHINGTON, 1st January, 1800.

WANTED, for the prefent year, at the public buildings in this City, twenty-five good labourers; alfo, three thoufand bufhels of good ftone lime, and fix thoufand pounds of well-fatted pork, for which cafh will be paid by the fubfcriber.
SAMUEL N. SMALLWOOD.

Laborers wanted. *Centinel of Liberty and Georgetown (DC) Advertiser,* January 3, 1800.

PLASTERING.

THE Subfcriber having undertaken the plaftering of the prefidents houfe in the city of Wafhington wifhes to engage twenty good hands to whom he will give generous wages to commence front the firft of March.
HUGH DENSLEY.
February 18, 1800.   -14.—

Plasterers wanted. *Centinel of Liberty and Georgetown (DC) Advertiser,* February 21, 1800.

THE 1800 CONSTRUCTION SEASON

The year 1800 began with an almost new construction crew working on both the Capitol and the President's House, although at the lower wage level. The commissioners had been successful in replacing most of the needed laborers, plasters, and carpenters locally but still needed skilled joiners. Hoban was instructed to proceed to Baltimore and hire "six or eight good joiners."[157]

The commissioners paid the new carpenters 9 shillings per day ($1.20), although a few, such as Thomas Dickey and George Sanford, were paid 9 shillings 6 pence per day ($1.27). This is the same rate to which the carpenters working on the President's House had been reduced by the commissioners two years earlier. The commissioners paid this rate even though there is some evidence that carpenters in established centers were receiving more. Patrick Healey and William Knowles were paid 11 shillings per day ($1.47); undoubtedly, they were the more skilled carpenter-joiners.

At its beginning, 1800 was relatively quiet in terms of labor relations with the workers engaged at the public buildings. Then, toward the end of the year, with the deadline approaching, on October 20, 1800, the commissioners ordered the workers to vacate the temporary buildings within ten days.[158] Although most of the carpenters at the President's House were new to Washington and to the work on the public buildings, they took as militant a stand as previous workers had in facing the commissioners. They refused to vacate the buildings. They wrote the commissioners, "You cannot be ignorant of the utter Impossibility of Procuring houses for the Married, or lodging for the Unmarried Carpenters, employed at the Presidents House," and "If you Persevere in taking them down we shall every man leave the employ on the return of the Bearer by whom we shall expect your Answer in Writing."[159]

In addition, several of the carpenters stopped work on the same day. This alarmed the secretary of state and the secretary of the navy, as the deadline for the relocation to Washington by the federal government was close upon them. Although their letters to the commissioners are not included in the commissioners' records, we can detect some sense of their concern by the commissioners' response of October 24, 1800: "We shall in the meantime increase the Hands by every means in our power and shall have six additional Carpenters today at twelve O'clock."[160]

On October 22, 1800, the commissioners ordered three carpenters to leave their temporary buildings: "Orderd. that Notice be given by Mr. Belt to Thomas Knowles Wm. McMiner and Patrick Cuirrite to quite [quit] the temporary buildings occupied by them by Monday next."[161]

To calm matters, on the same day they ordered a pay increase for the remaining carpenters: "Orderd. that one Shilling & expenses per day increased Wages be allowed to the Carpenters at the Presidents house to commense from this day."[162] To this the commissioners added that the carpenters could remain in their temporary buildings on President's Square. But with these concessions to the carpenters, the commissioners also terminated them. And for good measure, they also fired James Hoban.[163]

It has not been recorded whether the workmen expected to be released so early, but certainly James Hoban did not. In his letter to the commissioners of October 29, 1800, Hoban indicated his displeasure.[164] This letter received some, although not much, satisfaction from the commissioners, who extended Hoban's employment to January 1.[165]

While the commissioners were dealing with the workers at the President's House and the Capitol, they also began construction of two executive office buildings to house the government clerks who would accompany the government to Washington. The first of these was the Executive Office Building, for the Department of the Treasury, to be erected immediately east of the President's House. On May 7, 1798, the commissioners advertised for this building,[166] which was to be built of brick, 148 feet by 57 feet 6 inches deep.[167] The building would have three floors: ground (containing 14 rooms), first (also containing 14 rooms) and a roof or attic floor (containing 8 rooms). Generally the rooms were 16 feet by 20 feet. The foundations were 30 inches thick, reduced to 23 inches in the first floor and again reduced to 8 inches in the second floor.[168] On June 23, 1798, the commissioners contracted with Leonard Harbaugh for the erection of the building for $39,511.[169]

In the following year, on July 23, 1799, they asked James Hoban to prepare an estimate for the construction of a similar building west of the President's House.[170] On August 6, 1799, the commissioners again contracted with Harbaugh, to build the War Department and other executive departments.[171] The Treasury Building was completed by the time the government relocated from Philadelphia, but not the other building.[172]

THE ARRIVAL OF THE GOVERNMENT IN THE NEW FEDERAL CITY

On November 1, 1800, President Adams moved into the partially completed President's House. Journalists commented, "On Saturday last the PRESIDENT of the United States arrived in this city and took up residence in the house appropriated to him by the Commissioners. Though not entirely finished, the part which is completed will afford ample accomodations."[173]

Mrs. Adams's impression of the President's House has been much repeated.

Describing the President's House, she said: "The house is made habitable, but there is not a single apartment finished. We have not the least fence, yard or other convenience without, and the great unfinished audience room I make a drying room of, to hang up the clothes in. The principal stairs are not up and will not be this winter."

Mrs. Adams expressed the opinion that if the twelve years' work had been

Washington Commissioners' Office,
7th May, 1798.

THE Commissioners will receive proposals until the 20th of June next, for building in the City of Washington, one of the executive offices for the United States, of the following external dimensions—148 feet in length and 57 feet, 6 inches in breadth—cellar walls, 30 inches—first story 23 inches, and second story 8 inches—partition walls averaging 15 inches, to contain on the ground floor, 14 Rooms, same number on the second story, and in the roof, 8 Rooms with a passage—The whole external of the building to be of stock brick;—the inside walls of hard burnt brick—cellars of best foundation stone, to the height of the girders—the outside walls as far as they shew above ground, to the plinth, to be of plain ashlar Free-Stone, soles of windows, sills of doors and string course, of Free Stone—The house to be covered with cypress shingles—the rooms in general to be 16 feet by 20, finished in a plain neat manner, of the best materials—six small rooms to be groined.

A plan and elevation of said building, and bill of particulars are lodged in the office for the inspection of those who may wish to contract;—also a copy of said bill, at the office of CLEMENT BIDDLE, Esq. at Philadelphia.

Proposals sealed up, will be received until the 20th of June next, on which day, the board will proceed to contract with such persons as shall appear, under all circumstances, to offer the best terms.
Per order of Commissioners,
THOMAS MUNROE, Clk.

Building proposals invited for the executive building. *Centinel of Liberty and Georgetown (DC) Advertiser*, June 15, 1798.

Treasury Building, 1804. Constructed beginning 1798 immediately to the east of the President's House. Leonard Harbaugh, contractor. Of the two administrative buildings on either side of the President's House, it was the Treasury Building that had been completed by the time the government moved from Philadelphia. *Harper's Monthly Magazine*, 1871.

going on in New England it would have been better managed and completion much nearer its end; but she saw the possibilities of the new city, for she continued: "It is a beautiful spot, capable of any improvement, and the more I view it the more I am delighted with it."[174]

The first building program was characterized by intense conflict between the commissioners, who were gentleman amateurs (particularly Gustavus Scott, Dr. William Thornton, and Alexander White), and their architect-engineers and the skilled workers. Latrobe characterized it as "warfare." The conflict began with criticisms of the inadequacy of Thornton's design for the Capitol and spread into widespread

BUILDING CAMPAIGNS

ridicule of the commissioners among the workers, leading to work stoppages and slowdowns. The commissioners retaliated with firings, pay cuts, and ousting workers from their houses. The results of the conflict were delay in completing the buildings, substandard construction, and high costs.

As Congress and the rest of the government moved to Washington in November 1800, it became obvious to all that the commissioners had spent much money but had not completed the needed government buildings. Dissatisfaction with the commissioners grew, and on June 1, 1802, Congress abolished the office of Commissioners of Public Buildings, replacing the three-man commission with a single individual, Thomas Munroe, the former secretary to the commissioners.[175]

8

The Second Public Building Campaign (1803–1811)
The President and the Architect

In a contest, similar to that in which I am engaged, first with Mr. Hallet, then with Mr. Hatfield [Hadfield], Doctor Thornton was victorious. Both these men, men of knowledge, talents, integrity and amiable manners were ruined. . . . If I felt the slightest respect for the talents of the original designer as an architect, I should be fearless as to myself, but placed as I am on the very spot from which Hallet and Hatfield fell, attacked by the same weapons, and with the same activity, nothing but a very resolute defense can save me.

BENJAMIN LATROBE
to President Thomas Jefferson, February 28, 1804

He [Latrobe] infamously endeavours to lessen my character, supposing thereby that less credit will be attached to my declarations, but I should lament exceedingly if my character for veracity was only upon the level of Mr. Latrobe's, especially "with those that have known him for 15 years."

WILLIAM THORNTON
to Jonathan Smith Findlay, editor of the *Washington Federalist*, May 1, 1808

THE NATIONAL CAPITAL IN NOVEMBER 1800

In November 1800, at the time of the removal of the federal government from Philadelphia to the new federal city, the President's House was mostly finished, but the Capitol was far from completed. Only the north wing, to house the Senate and the Supreme Court, was mostly completed. The south wing, for the House of Representatives, was only a foundation. The magnificent Pantheon-like dome and portico designed as the center section of the Capitol had not yet been started, nor was it finished for twenty-four more years. Only one of the administrative buildings, the Treasury Building, immediately to the east of the President's House, had been completed.

The few buildings that were mostly finished still had numerous problems. On July 9, 1800, the commissioners wrote John Emory, one of the contractors, asking him to correct the leaks in the President's House, to fix the resulting damage to the plaster, and to deal with related problems, and they threatened a lawsuit if he did not.[1] More problems with the public buildings were discovered later.

Because the permanent south wing of the Capitol was clearly not going to be built for some time, a temporary solution was needed to provide meeting space for the House of Representatives. On May 27, 1801, the commissioners asked James Hoban to design a temporary House chamber to be built on the foundations of the south wing that were constructed in 1793–1796. Hoban developed three alternative plans. The first two plans were to build a portion of the south wing's permanent structure: one plan would construct the walls to the height of one story, while the other would construct them to two stories high. Hoban's third plan was to put up an inexpensive wooden building that would be removed once the south wing was begun. President Jefferson was to choose which of the three designs would be built.[2] By June 1, 1801, the three plans had been completed, and the commissioners asked President Jefferson his preference.[3] Jefferson answered the following day that he preferred the alternative in which some of the construction cost of the temporary building could be used for the permanent one.[4]

On June 10, 1801, the commissioners directed that advertisements be placed in local newspapers for bids "for an elliptical Room in the south Wing of the Capitol."[5] The commissioners accepted William Lovering and William Dyer's bid of $4,789 and entered into a contract with them on June 20, 1801, stipulating that the room must be finished by November 1, 1801.[6]

Construction moved quickly on the temporary building. By the end of August 1801, the commissioners informed Jefferson that the structure was ready to be roofed.[7] Although the temporary building was not completed by the stipulated November 1 date, by December 14 Hoban was able to report that the building was essentially completed.[8]

The congressmen who were to use the building were less than impressed. The building was quickly labeled "the oven," partly because it looked like a Dutch oven and partly because it was warm and stuffy.[9] It became a symbol of the failure of the commissioners to provide adequate accommodation for the government.

CONGRESSIONAL INVESTIGATION OF THE COMMISSIONERS

With the arrival of Congress in Washington in November 1800, senators and congressmen saw the uncompleted buildings and began to raise questions about how much had been spent on the public buildings and about the performance of the commissioners over the previous decade. To head off these questions, in January 1801, only two months after Congress arrived in Washington, the commissioners (Thornton, White, and Cranch, who had replaced the deceased Scott) submitted a report to Congress, recounting the financial history of the commissioners from 1792 to 1800. The report was a defense of their actions over the decade.[10] Their explanation was found by Congress not to be satisfactory, and in the following month Congress appointed a committee to "inquire into the expenditure of money made by the commissioners of the city of Washington, the disposition of public property made by them, and generally into all the transactions at the commissioners."[11]

The committee, chaired by Congressman Roger Griswold of Connecticut, issued its report later that month, on February 27, 1801. According to the report, the commissioners had "expended more than one million dollars on various subjects," and "the accounts of the commissioners [had] not been regularly audited by any public officer, but [had] rested on their own statements." The report stated, "Whether those expenditures have been made with economy, or not, it is not necessary for the committee to decide, as the House will possess the same information which the committee possess on this point." Although the committee chose not to pass judgment on the performance of the commissioners, its principal recommendation was that the Board of the Commissioners of Public Buildings be abolished.[12]

Congress took no action on that recommendation at the time, but in the following year, on February 5, 1802, Congress directed one of its committees "to inquire into the expediency of discontinuing the office of the said commissioners." This committee, chaired by Joseph Nicholson of Chestertown, Maryland, reported on February 12, 1802: "The offices of two of the commissioners of the city of Washington ought to be discontinued, and thereafter the powers now vested in the board of commissioners ought to be vested in one only."[13]

ABOLISHMENT OF THE BOARD OF COMMISSIONERS

On May 1, 1802, Congress passed "an act to abolish the board of commissioners in the city of Washington and for other purposes,"[14] The office of the three commissioners was abolished June 1, 1802. In its place, Congress established the position of superintendent. A principal responsibility of the new superintendent was to sell the unsold lots in the city, to repay the three loans, totaling $250,000, from the State of Maryland.

Abolishment of the commission left William Thornton without a job. The wording of the Nicholson report gave Thornton hope that one of the existing commissioners would be appointed to the new position of superintendent and that he would be the one chosen. He was very disappointed when President Jefferson appointed Thomas Munroe, the secretary-clerk of the commissioners, to the position.[15]

ELECTION OF THOMAS JEFFERSON AS PRESIDENT

With the election of Thomas Jefferson to the presidency, things changed in the new federal city. Jefferson was aware of the problems of the first building campaign, and instead of appointing a commission to oversee the uncompleted work, he took on that responsibility himself, despite his other responsibilities as president. His first action was to hire the most experienced and proficient architect-engineer in the United States, Benjamin Latrobe. The patchwork funding of the first public building campaign was replaced with yearly congressional appropriations.

Congress appropriated $50,000 for work on the public buildings of Washington. It was the first congressional appropriation to support construction in the federal city. Jefferson, because of his intense interest in architecture, took a direct role in the expenditure of this and subsequent appropriations and in overseeing the completion of all the buildings of the city. On March 6, 1803, he offered the position of surveyor of the public buildings, formerly held by James Hoban under the dissolved Commissioners of Public Buildings, to architect-engineer Benjamin Latrobe; the position paid $1,700 per year.[16] In a private letter accompanying his March 6, 1803, letter appointing Latrobe, Jefferson told him that, unlike Washington, he intended to take a hands-on approach to these building projects. Jefferson wrote that, of the $50,000 appropriation, he thought $5,000 to $10,000 would be used to cover work on the north wing of the Capitol and the remainder to begin construction of the walls of the south wing.[17] He told Latrobe that he thought the south wing of the Capitol could be completed in two years for $90,000 to $95,000. But this was not to be.

BENJAMIN LATROBE AS SURVEYOR OF PUBLIC BUILDINGS

Latrobe quickly accepted Jefferson's offer.[18] Arriving in Washington on March 21, 1803, he assessed the existing foundations of the south wing of the Capitol and the general plan of the Capitol, by then an amalgam of Thornton's original design, Hallet's modifications (made by the July 1793 conference in Jefferson's office), and Hadfield's alterations. He sent his findings to Jefferson on April 4, 1803, in his "Report on the U.S. Capitol," the first of many reports.[19]

Latrobe's first report on the US Capitol was an indictment of its poor design and poor workmanship. It was also an indictment of those who oversaw its design and construction, especially James Hoban, the former surveyor of the public buildings, and William Thornton, the original designer and a former commissioner of public buildings. By uncovering these deficiencies, Latrobe generated an intense enmity toward himself on the part of Hoban and Thornton. That enmity later surfaced in a war waged against Latrobe by both Hoban and Thornton, using the local newspapers.

FOUNDATION PROBLEMS AT THE
SOUTH WING OF THE CAPITOL

Latrobe had heard from the workmen that the foundations of the Capitol were poorly constructed. One of his first acts after he arrived in Washington was to dig exploratory pits to investigate the adequacy of the foundations of the south wing. His investigation found poor enough workmanship that the foundations needed to be torn down and rebuilt.[20]

Next, Latrobe inspected the adequacy of the foundations under the temporary House of Representatives building, the oven. He found that these foundations were also inadequate. Latrobe recommended that this temporary House of Representatives building, designed by Hoban only fifteen months earlier, be torn down, because Hoban had not used "counter arches" in the foundation (i.e., upside-down arches designed to spread the building load over a wide area) and because "the brick work, above, has been forced from 2 to 6 inches out of perpendicular by the weight of the roof."[21]

The foundation of the south wing of the Capitol not only had to be rebuilt; in addition, new foundation stone had to be ordered, since Latrobe also found that the foundation stones previously used in the foundations could not be reused. This created another problem, because the commissioners' wharves that had been built for the first public building campaign had been permitted to deteriorate, as well as the road and the bridge between the upper wharf and the Capitol. The wharves, the road, and the bridge had to be repaired before building materials could be delivered to the Capitol.[22]

DRY ROT AND WATER LEAKAGE IN THE
NORTH WING OF THE CAPITOL

Next Latrobe examined the north wing of the Capitol. He found that the basement had not been properly ventilated and that dry rot was likely to develop. He predicted that the dry rot would spread throughout the timbers of the building. He also found that poor workmanship and bad design had resulted in water leakage into the building, causing subsequent wall and ceiling damage.[23]

Besides faulty construction, Latrobe had much to criticize regarding the plans for the Capitol. His first criticism, architectural in nature, targeted the decision to lower the two principal rooms of the Capitol, as agreed in the July 1793 conference in Thomas Jefferson's office, where Hallet and others attempted to correct errors in Thornton's design without modifying the exterior. Latrobe found the resulting monumental stairs planned for the building to be nonsensical, since senators and congressmen would have to ascend those stairs only to descend another flight of stairs into their respective chambers.[24]

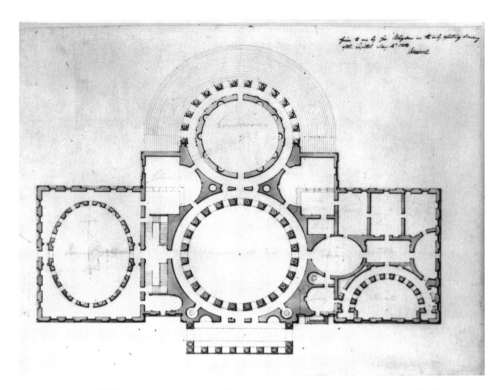

Floor plan of the US Capitol, probably by William Thornton, between 1793 and 1800. The inscription in the upper right-hand corner reads, "Given to me [Latrobe] by Geo. Blagden as the only existing drawing of the Capitol May 4th 1803. B. H. Latrobe." The elliptical hall of the House of Representatives designed by Thornton is shown on the left. Latrobe found that the foundations begun did not match this plan. He also found other problems with this design. Courtesy of the Library of Congress.

CORRECTING THE PROBLEMS OF THE
SOUTH WING OF THE CAPITOL

Since the north wing had already been constructed, there was nothing Latrobe could do to remedy faults in the building, at least at this time.[25] The south wing was a different story. Unlike the north wing, which contained a multiplicity of rooms besides the Senate chamber, the south wing had been envisioned as containing only one large room, the meeting hall of the House of Representatives. Latrobe wrote to Jefferson that there was an absolute need for additional rooms in the south wing and that these rooms "should be closely attached to the legislative Hall." What were needed were committee rooms, offices for the Speaker and the clerk, and other rooms such as toilets, record storage rooms, a lobby, and a gallery. Latrobe pointed out that there had been very little accommodation for these functions elsewhere in the previous plans.[26]

These were Latrobe's architectural criticisms of the south-wing plans. He also had a major engineering criticism of those plans—the building, if it were built, would be structurally unstable. The south wing, as planned, consisted of two enclosures, one inside the other: the rectilinear exterior wall that was to be 65 feet high and the interior enclosure of the large hall for the House of Representatives, an elliptical space defined by thirty-two columns, 25 feet high, topped by a domed roof 90 by 120 feet. The problem with the design was that the domed roof structure would exert a lateral thrust (from the weight of the roof and the live load of ice and snow in cold weather) that the structure could not resist, partially because the tall elliptical interior could not be effectively tied into the external walls and partially because the walls themselves were thin and pierced with numerous windows and therefore ill suited to resist lateral thrust. Latrobe sensed the design's structural instability, based on his training and experience.[27] Thornton had not drawn a section of the south wing, but had such a drawing been prepared, probably most architect-engineers would have reacted as Latrobe had.[28] Latrobe had much else to criticize in the plans of the Capitol, including the cost of the building as proposed and the "absolute want of light in all that part of the building which lies between the Wings and the center of the building."[29]

From the date of this report, sent to Jefferson on April 4, 1803, Latrobe had approximately six months to show Congress construction progress using the $50,000 appropriation before Congress came back to Washington and considered a second appropriation for the public buildings. To assist him in the work, Latrobe hired John Lenthall to serve as his superintendent or clerk of the works and to oversee the work when Latrobe was not there.[30] At the time Latrobe was still working as chief engineer on the Chesapeake and Delaware Canal, and that work may have delayed the preparation of his first report to Congress, which was not completed until February 20, 1804.

LATROBE'S FIRST REPORT TO CONGRESS

In this first report to Congress as surveyor of public buildings, dated February 20, 1804, Latrobe reported what he had done the previous year to fix the problems with the Capitol building. For the north wing, he had solved the problem of dry rot and inadequate ventilation in the basement; repaired the leaking roof, which was damaging the interior finishes; and provided adequate heating for the building. On April 4, 1803, Latrobe had reported on the deficient foundation of the south wing. Now he informed Congress that the substandard foundation had been torn out, a new foundation had been laid, and the external walls had been built to half the height of the ground floor.[31]

Latrobe was also responsible for correcting the problems found in the President's House. In a report not found and presumed lost, Latrobe had previously reported to Congress on the water leakage and other problems in this building. In his February 20, 1804 report to Congress, he wrote that the heavy slate roof had to be removed and replaced with a much lighter sheet-iron roof. The weight of the slate roof had caused the walls to spread. The walls needed to be tied to the roof structure. Also, the building's gutters had to be taken down and rebuilt. All of this work, except replacing the slate roof, had been finished.[32]

THE 1804 CONSTRUCTION SEASON

CHANGES IN THE PLAN FOR THE
SOUTH WING OF THE CAPITOL

To complete the Capitol's south wing, Latrobe was convinced that many changes would have to be made to Thornton's design. These changes would bring Latrobe into full confrontation with Thornton, the nominal designer of the Capitol. Thornton had defined his persona as an architect, and that persona was providing a high standard of living for him and his family. He had previously tried private practice in medicine but had found it was not to his liking. His estate in the West Indies did not provide enough income for their support, nor could his wife's family provide adequate income. It was through winning the Capitol competition that Thornton came to be appointed one of the three commissioners of public buildings, with a salary of sixteen hundred dollars per year at a time when a skilled worked made a dollar a day. In attempting to build the south wing, Latrobe, like Hallet and Hadfield before him, was challenging the self-established architectural persona that Thornton had created for himself.

Official Washington, beginning with George Washington, had little interest in who was acclaimed as the architect of the Capitol. They also did not want additional changes to the architectural plans. President Jefferson wrote Latrobe, "In order to get along with any public undertaking it is necessary that some stability of plan be observed. Nothing impedes progress so much as perpetual changes of design."[33] But Latrobe convinced President Jefferson that the changes he was proposing were essential.[34] Faced with the problems in Thornton's original plan, Jefferson again developed a compromise approach: approve whatever changes Latrobe needed to make in order to build the structure and make it usable, but keep as much as possible of Thornton's design, which had been approved by President Washington. For that purpose, Jefferson urged Latrobe to discuss his proposed changes with Thornton. A meeting between Thornton and Latrobe was arranged.

The meeting between Thornton and Latrobe went quite badly. Latrobe reported to Jefferson, "In a few preemptory words, he [Thornton], in fact, told me, that no difficulties existed in his plan, but such as were made by those who were too ignorant to remove them."[35] Latrobe was extremely discouraged by Thornton's reaction and briefly considered resigning. However, he pushed on.

On February 28, 1804, Latrobe submitted a report to the chairman of the committee of the House of Representatives, Philip R. Thompson, to whom his report of February 20, 1804, had been referred. In this document, Latrobe recounted the history of the development of the plans for the US Capitol. Without mentioning Thornton's name, he reported, "The evidence of the books of the office proves that it [Thornton's design] was not considered practicable" and that "its author [Thornton] was not a professional man."[36] Latrobe's report was an open government document in a small town, and Thornton would have quickly learned of its contents and of Latrobe's criticisms of his design. This particular passage, however accurate, could only serve to further infuriate Thornton — and it did. He responded to Latrobe later, on April 23, 1804.

Meanwhile, in February 1804, Latrobe met with the congressional committee tasked to review his objections to the plans of the Capitol.[37] The outcome of the meeting was favorable for Latrobe, who wrote, "As the result of the meeting it was understood, that an appropriation of 50,000 Dollars should be recommended."[38]

Thornton and others continued to strongly argue that the original plan approved by President Washington should not be modified by Latrobe. This brought up the question of exactly what plan had been approved by President Washington. On February 28, 1804, Latrobe submitted a report to Philip Thompson, chairman of the committee of the House of Representatives reviewing the Capitol issue. Latrobe reported, "Of the plan approved by General Washington [in 1793] no drawing can at present be found among the papers belonging to the office."[39] Apparently the original drawings for the competition had been lost or destroyed.

On March 13, 1804, the House of Representatives passed the appropriation of $50,000 for the public buildings by a vote of 57 to 23. On March 24 the Senate deadlocked twelve to twelve on the same appropriation. The bill was sent to conference, where an effort to postpone it for a year was defeated. It was enacted on March 27, 1804.[40] This was a substantial victory for President Jefferson and Latrobe. The second year of construction funded with congressional appropriations could then begin.

NEW PLANS FOR THE SOUTH WING OF THE CAPITOL

With the passage of the $50,000 appropriation, Latrobe had to develop new plans for the south wing. These drawings must have already been in an advanced stage, as he forwarded a roll of the drawings to Jefferson on March 29, 1804, only two days after the appropriation passed Congress.[41]

In the drawings he provided, he included an exact copy of Thornton's floor plan for the ground floor. Latrobe found many problems with Thornton's plan, especially the elliptical-shaped enclosure of columns for the hall of the House of Representatives. His objections included that the space inside the elliptical wall could not be adequately lit, that the doors leading into the space inside the elliptical wall were useless, that the rooms could not be furnished with fireplaces, and that a staircase to the Speaker's chair could not be built for lack of space.[42]

Among Latrobe's other objections to Thornton's elliptical enclosure of columns for the hall of the House of Representatives, not the least was the cost of cutting stones for the architraves — the main beams resting on the top of the columns. An elliptical pattern for the required architraves was more difficult and therefore more expensive than cutting stones in a semicircular shape. Latrobe recommended converting the elliptical enclosure of the Representatives' hall into "two semicircles abutting upon a parallelogram."[43] That is, he proposed that the enclosure of columns be composed of two straight lines along the long axis of the hall, terminating in a semicircle at each end.

Like others before him, Latrobe criticized Thornton's plan for the Capitol for its lack of light. A particular problem was lighting the spaces between the wings of the building and the portico or center section — called by Latrobe "the recesses." Latrobe wrote Jefferson: "The total want of light in those parts of the Building, which lie behind the Recesses, and between the Corps de logis[44] and the wings, has produced all the bad arrangement, and the waste of room which is found in the plan of the North wing." Latrobe's plan modified this portion of the south wing by designing a centrally located vestibule to provide access to the hall of the House of Representatives from the portico or north side of the south wing.[45]

JEFFERSON'S PRIORITIES FOR
CONSTRUCTION IN 1804

Because of the press of other business, Jefferson did not react immediately to Latrobe's drawings. He did send Latrobe his ideas on what needed to be accomplished in 1804. On the south wing, the walls were to be completed, and work on the roof structure and interior carpentry was to be pushed forward so that the roof could be readily installed at the beginning of 1805. For the north wing, Jefferson wanted the roof to be thoroughly repaired that year. For the President's House, Jefferson wanted the new sheet-metal roof to be completed, to have the glazing of the windows finished, and to have the floor of the great room and the rooms above it completed.[46] Jefferson concluded this letter to Latrobe with the statement that he would be back to supervise the construction of the kiln for seasoning the floor boards. Probably no other American President has ever exerted such hands-on oversight as supervising the construction of a kiln for seasoning floorboards.

The drawings that Latrobe sent to President Jefferson on March 29, 1804, were received by Jefferson on April 6, 1804. On April 9, Jefferson responded to Latrobe that he generally approved of Latrobe's proposed drawings, except for the work that Latrobe proposed for the recess between the south wing and the portico. Jefferson pointed out to Latrobe that the purpose of the appropriation was to work on the south wing; therefore, they must not initiate construction on the recess north of the north wall of the south wing.[47]

THORNTON VS. LATROBE

Progress seemed at hand on the construction of the south wing of the Capitol. But the enmity between Thornton and Latrobe had not been diminished. On April 23, 1804, reacting to Latrobe's report to Congress of February 28, 1804, Thornton wrote Latrobe (in part), "I am sorry to be obliged to declare that your Letter to the Committee is, as it respects me, not only ungentlemanly but false."[48] Latrobe, willing to give as good as he got, wrote back to Thornton five days later, on April 28 (in part), "I now stand on the Ground from which you drove Hallet, and Hatfield [sic — Hadfield] to ruin. You may prove victorious against me also; but the contest will not be without spectators. The public shall attend and judge."[49]

This was a serious matter. "Call you to the field," wrote Latrobe to Thornton, referring to the field of honor. Both Latrobe and Thornton indicated that they were quite close to engaging in a duel with the other. On June 27, Thornton wrote to Latrobe, perhaps stepping back a bit from a possible duel, "You mention some Instances of incivility on my part. I remember none ever intended, nor any act that could be so construed."[50] If Thornton intended to smooth matters over with Latrobe, nonetheless he also indicated, "I am thus, and shall always be prepared to repel any attack from any Quarter."[51]

Perhaps it would have been diplomatic of Latrobe to respond to Thornton in a more conciliatory manner. But Latrobe was having none of it. On July 21, 1804, he wrote Thornton (in part), "It is impossible that you should even be on a level with me, excepting in your own opinion, and equally so that I should revert to the ignorance of the art with which I began to study 25 Years ago, in order to descend to your scale of knowledge."[52] And there the matter remained until 1808, when the dispute between the two men again exploded with public charges and a resulting lawsuit.

LATROBE'S SECOND REPORT TO CONGRESS

By December 1804, Latrobe was ready to report construction progress to Congress. He began by discussing the problems with water leakage at the President's House, "Of the inconveniences attending the house, the greatest was the leakiness of the roof." This leakage was a serious problem, wrote Latrobe. It ruined the furniture and ceilings and was due to faulty gutters and bad slating. Latrobe also found quite a bit wrong with the building's rain-shedding systems: the gutters did not have an adequate slope to discharge rainwater, the openings into the downspouts in the walls were too small, the troughs in the walls were badly constructed, and the lead work was faulty.[53] Consequently, rainwater drained into the building.

Another cause of the water leakage at the President's House was the slate roof: Latrobe found that the slates were of inferior quality and quite small, particularly near the ridge. The entire roof had to be replaced because it leaked and because the weight of the slate roof had overstressed the roof trusses and forced the front and back walls of the building outward.[54]

BUILDING THE SOUTH WING OF THE CAPITOL

It was the south wing of the Capitol, not the President's House, that was Latrobe's principal construction responsibility. Regarding the Capitol building, Latrobe reported to Congress, "Various causes have conspired to prevent our carrying up, this Season [1804] as large a Mass of building as was expected." Latrobe gave several reasons for not achieving as much as had been desired. First, before construction on the south wing could commence, the bad foundations erected ten years earlier had to be torn out.[55] He wrote, "Bad as the workmanship appeared before the Walls were taken down, the measure of removing them was still more justified by the State in which they were found to be on their demolition."[56] Other reasons he gave for lack of construction progress were the lateness in the season when the appropriation had been made by Congress, the unusual wetness of the season, and sickness of the workmen.[57]

Another problem was the location of the temporary building for the House of Representatives, "the oven." Because the oven was within the perimeter of the external walls of the new building, the interior walls could not be built in concert with the exterior ones.[58] So Latrobe decided to build the external walls of the south wing, later to be laterally tied into the internal walls yet to be built.[59]

By the end of 1804, Latrobe was able to report: "The work [i.e., the external walls] [has] been raised to the level of the Selles [sills] of the Attic windows externally, and by far the most tedious and expensive part of the work in freestone has been completed, excepting the Cornice and the Capitals of the Pilasters."[60] For the interior of the south wing, Latrobe reported: "Of the interior parts of the building all the foundations are laid, and brought up to the floor of the Cellar story on the North side, and although they do not appear to view, the work done in them is very considerable. The whole south half of the Cellar story is Vaulted, and ready to receive the Walls of the Basement or Office story."[61]

THE 1805 CONSTRUCTION SEASON

APPROPRIATION FOR CONSTRUCTION IN 1805

On December 21, 1804, Congressman Philip Thompson, in a letter to Latrobe, asked what money he would need for the coming year, 1805. Latrobe answered:

For the south wing of the Capitol	$109,100
For the Recess of the south wing of the Capitol	25,200
	$134,300 [62]

On January 25, 1805, an appropriation for both the north and south wings was enacted, providing Latrobe almost everything he asked for:

For the south wing of the Capitol	$110,000
For "necessary alterations and repairs" on the north wing of the Capitol and on "other public buildings at the City of Washington"	20,000
	$130,000 [63]

In addition to his work on the Capitol in 1805, Latrobe developed a plan for colonnades on either side of the President's House, connecting the house to the two flanking administration buildings, and a plan for a fireproof building for the US Treasury.[64]

The 1805 construction effort was the largest to date in the new federal city. But construction progress depended on the smooth flow of construction materials to the work site. The most critical material was building stone, and the largest potential bottleneck for delivery of that stone to the building site was unloading the stone at the wharves. On June 5, 1805, Latrobe suggested that walking wheels, also called treadmill cranes, be used. An ancient device dating back to Roman times, the walking wheel was a large wheel inside of which men or horses walked, driving the wheel, which hoisted the rope in the crane.[65] Latrobe was familiar with the walking wheels then used on the London docks (they continued in use into the twentieth century). There is no indication that Latrobe's recommendation was accepted, and no walking wheels were used on the Washington, DC, wharves.

DRY ROT IN THE NORTH WING OF THE CAPITOL
At the end of August 1805, Latrobe reported to Jefferson the status of the work. On the north wing of the Capitol, completed only five years earlier, Latrobe had found dry rot in the trussed girder above the ceiling of the room that the Senate had used in its previous session. Failure of this girder would have collapsed the ceiling of the Senate meeting room and would also have collapsed the floor of the room above, where the House of Representatives met.[66] Latrobe was able to strengthen the defective girder, but the incident placed the architect-engineer and his builders on warning that more defective work might be found in the north wing.

A DOMED ROOF FOR THE SOUTH WING OF THE CAPITOL
With respect to the south wing, Latrobe told Jefferson that the exterior walls were nearing completion.[67] Thus, the type of roof for the building needed to be decided

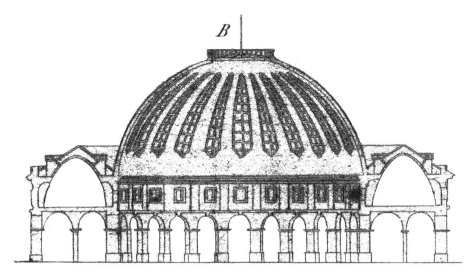

Section, the Halles au Blé, Paris. The first de l'Orme dome of the structure consisted of a system of ribs of short wooden struts, with glazing installed in between. The domed roof on the Halles au Blé had many of the same problems discussed by Latrobe. It was subject to expansion, contraction, and deformation (and fire, which destroyed it in 1802). The second domed structure was an iron-ribbed dome covered in copper sheathing. Author's collection.

upon. This roof would also serve as the ceiling of the principal room in the main floor of the south wing, the hall of the House of Representatives and its colonnaded enclosure. It was on the question of the south-wing roof that Latrobe and President Jefferson had their biggest disagreement.

While in Paris, Jefferson had been much taken with the light framed, wooden, domed roof of the municipal grain market, the Halles au Blé. The grain market was an octagonal structure surmounted by a hemispherical roof 120 feet (approximately 38 meters) in diameter.[68] It was so much admired that one guidebook wrote, "All who have seen this dome say, that it is the most beautiful and magnificent object they have ever beheld."[69] Developed by architects J. G. Legrand and J. Molinos, the domed roof was built in the manner of sixteenth-century architect Philibert de l'Orme and consisted of ribs made up of short laminated wooden sections.[70] What differed from de l'Orme's earlier designs was the glazing inserted between the ribs that functioned as skylights in the Halles au Blé. This feature brought a great deal of light into the building. Although the first dome no longer existed by 1805,[71] Jefferson remembered it well and wanted such a domed roof for the House of Representatives wing of the Capitol building.

Latrobe had great difficulties with the idea of using a roof similar to that of the Halles au Blé on the new south wing. He wrote to Jefferson that such a roof would leak, that the glass panes would be subject to breakage, that condensation would be a problem, and that the glazing would be expensive.[72] In fact, the domed roof on the Halles au Blé had such problems. The dome in Paris was subject to shrinking, expansion, contraction, and deformation. Repairs were frequent.[73] Few of Latrobe's' arguments swayed Jefferson. Jefferson wanted a domed roof like that of the Halles au Blé to cover the new House of Representatives. He wrote Latrobe, "I cannot express to you the regret I feel on the subject of renouncing the Halle au bled lights of the Capitol dome." In the end, Jefferson left the decision up to Latrobe.[74] From Latrobe's point of view, this was the worst outcome. He very much wanted to please his sponsor, but he had little faith in the roof structure of the Halles au Blé. Eventually, he came up with his own approach.

LATROBE'S THIRD ANNUAL REPORT

Just before Christmas 1805, Latrobe submitted his third annual report on the status of the construction of the public buildings in Washington. He began with an apology for not having completed the House of Representatives: "The hopes which were entertained that it might be possible to compleat the Hall of Representatives in time for their occupance during the present Session have been disappointed." Latrobe gave two reasons for not having finished the south wing of the Capitol: other projects were competing for workmen, and it had been difficult to secure large and fine blocks of stone from the Aquia quarry.[75]

After recounting the status of the work on the south wing and the problems encountered on the north wing, Latrobe turned to the need for an additional appropriation, the fourth one, "to meet any eventual deficiency, and to provide for fitting up the house [i.e., the House of Representatives] when finished."[76] His request to Congress included these items:

1	To meet any eventual deficiency in the appropriation for the South wing, etc.	$25,000
2	To erect that part of the building which is to contain the communications of the offices with the house of Representatives	13,000
3	To render the building accessible, by removing earth and rubbish,to fill up, on the South front to the Gallery doors, and to restore thecommunication with the North wing	2,000
	Total	$40,000[77]

This amount was appropriated by Congress on April 21, 1806.[78]

FINISHING THE SOUTH WING OF THE CAPITOL

The year 1806 was the fourth year of the construction of the south wing of the Capitol. President Jefferson, who had initially estimated that the south wing would require only two years to complete, was determined that the House of Representatives would occupy a completed south wing by the end of 1806. He wrote Latrobe asking to be provided with periodic construction status reports.[79]

Jefferson was particularly concerned with the progress of the masonry work in the south wing. He asked Latrobe to recruit six additional stone cutters from Philadelphia to facilitate the work.[80] This was easier said than done. Two days later, July 3, 1806, Latrobe informed his clerk of the works, John Lenthall, that stone cutters were not readily available in Philadelphia.[81] Latrobe was able to recruit two stone cutters in Philadelphia, Daniel Hunt and Patrick Barr,[82] but this did not satisfy Jefferson. On July 17, 1806, he again wrote to Latrobe, asking him to hire not just six additional stone cutters, but twelve.[83]

Jefferson also commented on the need for stone of extraordinary size and expressed his exasperation that Latrobe was in Delaware dealing with the Chesapeake and Delaware Canal, not in Washington to resolve the stone-delivery problems from Aquia.[84] Latrobe had stayed in Delaware primarily for his work as chief engineer of the Chesapeake and Delaware Canal, leaving day-to-day supervision at the Capitol to John Lenthall. In 1806 work on the canal was halted for financial reasons (not to be resumed for twenty years), eliminating the primary reason for Latrobe not to spend more time in Washington.

August 1806, as usual, was the high construction month of the year. Typically, workers would work all thirty-one days of that month. After August, they reverted to six-day weeks, and the days, sunrise to sunset, became appreciably shorter. If construction progress could not be made in August, it would seem unlikely that the building could be finished by the end of the year.

On August 15, 1806, Latrobe summarized his stone cutter recruitment efforts for Jefferson, "I engaged five and in Baltimore two, all of whom arrived here. Two of them are run away again, after having their expences of travelling paid, and six remain. In all we have however only 25 stone cutters." Latrobe was continuing his efforts to recruit stone cutters in Philadelphia and New York, but, as he informed Jefferson, most of the stone cutters in Philadelphia had left, to work either on the new City Hall in New York or on the new State House in Albany.[85]

The lack of stone cutters was not the only problem. There were also difficulties in getting stone quarried and delivered from the Aquia quarries. Because Latrobe was no longer working for the Chesapeake and Delaware Canal and was now in Washington, he began addressing the stone problem at Aquia. The stone that was most needed was largely for the entablature of the parallelogram with semicircular ends — the massive horizontal beams sitting on top of the columns, consisting of cornice, frieze, and architrave — that, with the columns, would define the space to be occupied by the hall of the House of Representatives. This stone was not the ordinary ashlar used in walls but blocks of massive dimensions requiring special cutting, handling, and installation. By mid-August all but five of the great stones had been delivered.[86]

By the end of August 1806, it began to appear that the south wing would again not be completed that year for the House of Representatives. Latrobe reported to President Jefferson:

> We have now *all* our most heavy stone in the Yard, and shall not probably be again at a stand for stone. The architrave is set round the East semicircle, and by Friday evening [i.e., by August 29, 1806] the three Stones which form the East angle of the straight part behind the speaker's chair will be set so as to compleat the whole of the Architrave. The frieze of which one half has been for some time set on the East side will then be compleated without delay, including the Eagle of which a considerable part is finished.[87]

Despite this seemingly optimistic report on construction progress, Latrobe felt compelled to warn that the south wing might not be completed by the end of the year so that it could be used by the House of Representatives: "I beg to assure you, that nothing that was within the compass of our means has been neglected to forward the work, and if after all we are disappointed, it will only be because it was *impossible* to be otherwise. I have still hopes however that Congress though in a very unfinished state may sit in the house during the next Session."[88]

THE DOME FOR THE SOUTH WING
OF THE CAPITOL REVISITED

Jefferson was not pleased that once again the south wing would not be ready for Congress. Lack of construction progress was not the only source of friction between the president and his surveyor of public buildings. Another one was their continuing debate on how the hall of Representatives in the south wing of the Capitol was to be lighted. There were three alternatives. The method strongly preferred by President Jefferson was to have the interior of this cavernous room lit in the manner of the Halles au Blé in Paris, by glass panes installed between the dome's ribs. Latrobe argued against this approach, stating that there would be "2400 [joints between the glazing and framing], being so many chances of leakage."[89] Latrobe did not discuss the structural behavior of domes with President Jefferson at this time, but domes tend to "flatten" when live loads, such as snow and ice, are imposed upon them. Such flattening is controlled by holding the ribs in place by tension rings surrounding the dome or by substantial abutments at their base, or by both measures.[90] When the load is removed, the structure, being relatively elastic, resumes its normal shape. The glass between the ribs, however, tends to be inelastic. Changes in the dimensions of the dome would tend to crack the glazing. This problem is usually resolved by the design of the joints mounting the glazing to the structure. If well designed, the joints would take up the expansion and contraction of the structure, thereby preventing the glass from cracking. But, Latrobe told Jefferson, these joints were difficult to design and install in a way that would avoid cracking of the glass. The difficulty arose partially because of mounting rectilinear plates of glass in the geometry of the hemispherical dome and partially because of the poor quality of the building caulking that was available at the time. Leakage would occur after fine-line cracks developed in the joints, water entered, and freezing and thawing enlarged the cracks. Thermal contraction and expansion would produce the same result.

Neither President Jefferson or Latrobe discussed whether the roof of the Halles au Blé leaked, but it seems likely that it did. After the original roof burned in 1802, it was replaced by an iron-ribbed dome covered in (presumably watertight) copper sheathing. Some leaks might be tolerated in the Halles au Blé, since it was a warehouse for grains. Leaks in the roof of the south wing of the Capitol would be more troublesome, as it was essentially an auditorium where congressmen would have to spend most of their day.

The roof of the south wing of the Capitol had to be designed to let light into the assembly hall. Latrobe's proposal was to install a lantern (sometimes called a cupola or monitor) on top of the dome. Lanterns had the advantage of having vertically mounted glazing in a wooden frame designed to easily shed water and therefore avoid leakage. President Jefferson disliked lanterns. He wanted a dome like the dome on the Halles au Blé.

Eager to please his patron, Latrobe began to work on designing a leakproof dome in the manner of the Halles au Blé during the autumn of 1805. He came up with the idea of "panel lights," framed skylights to be mounted between the structural ribs of the dome. Five sizes would be fabricated and a total of one hundred produced. From below, the skylights would appear to be ceiling coffers, although their upper surface would be glazed. Venetian blinds would control the sunlight admitted. (See plate 26). President Jefferson agreed with the plan.

After sending President Jefferson the plan for the panel lights, Latrobe had second thoughts regarding the effectiveness of his skylight design. He was especially concerned that condensation would drip on the members of Congress during their proceedings. He shared these concerns with Jefferson.[91] Despite his own misgivings, Latrobe decided to proceed with the plan and ordered glass from Hamburg on December 6, 1805. He wrote: "I was ashamed to condemn my own plan, lest I should appear capricious, and shrunk from a recantation. I proceeded therefore to the execution of the project, and on the 28th. of Novr. 1805, I gave to Mr. Lenthall the working plans of the whole roof at large with all the detail of the lights, their fittings, and their mouldings." Latrobe continued to explain his reasons for vacillating: "Thus have I gone on, though entirely contrary to my decided opinion, to the present hour. For I am convinced by the evidence of my senses in numberless cases, by all my professional experience for near 20 Years, and by all my reasonings on the subject, that the pannel lights, if made, must inevitably be again destroyed; partly on account of their most unpleasant light upon the desk, but chiefly on account of their droppings the condensed vapor, every morning in winter, and always on every change from warmer to colder weather, over the whole floor of the house."[92]

Latrobe saw other problems with the panel lights, including expense. The one hundred panel lights would be expensive, costing $3,000, partly because the frame sashes were not rectilinear but of a complicated geometry, "the sides of truncated spherical Sectors"[93]—they were more difficult to make than rectilineal sashes—and partly because one-eighth-inch-thick pane glass was expensive, $12 each pane. So Latrobe decided to substitute a "Lanthorn" (a lantern or cupola) for the panel lights. Unfortunately, he did not inform President Jefferson of this change in plans; nor had he received Jefferson's assent.

President Jefferson discovered the substitution of the lantern for the panel lights sometime in October 1806. He wrote John Lenthall expressing his strong displeasure about the substitution.[94] Latrobe was mortified at President Jefferson's reaction to the abandonment of the panel lights in the dome and the adoption of a lantern in their stead. On October 29, 1806 he wrote to the president asking him to reconsider his decision to insist upon the installation of a glass domed ceiling. Jefferson refused.[95] And that was that. Washington was to have its example of a glass domed roof in the manner of the Halles au Blé.

LATROBE'S FOURTH ANNUAL REPORT

In November 1806, Latrobe turned to writing his fourth annual report to Congress, just before its members would assemble in Washington. Latrobe knew that he had much explaining to do. The south wing of the Capitol was not yet complete, and he would be asking for more money to complete it. He decided to submit two reports to Congress: the official report of the surveyor of public buildings, provided to President Jefferson on November 26, 1806, and to Congress on December 15, 1806,[96] and a longer report, printed as a pamphlet and given to each member of Congress. It was entitled *A Private Letter to the Individual Members of Congress, on the Subject of the Public Buildings of the United States of Washington, from B. Henry Latrobe Surveyor of the Public Buildings* and dated November 28, 1806.[97] Latrobe asked the doorkeepers of the Senate and the House of Representatives not to distribute *A Private Letter* until President Jefferson had submitted to Congress the official report prepared by Latrobe.[98]

President Jefferson felt compelled to add an introductory paragraph to Latrobe's annual report explaining why the south wing was not yet finished: the basic problem was that the Aquia quarries could not deliver enough stone.[99] This explanation was echoed by Latrobe in the opening paragraph of his report.[100] Latrobe was quick to point out, however, that this was the only substantial problem encountered in finishing the building. "In every other branch of labour and of materials we have been in sufficient forwardness."[101] And even the stone problem had already been resolved. Wrote Latrobe, "All the building stone which will be wanted is on the spot."[102] What was needed, reported Latrobe, was more money; that additional money would guarantee completion of the south wing by the next session of Congress.[103]

In addition to this construction, money was also needed to furnish the south wing. The north wing needed work as well. Latrobe reminded the congressmen that "towards the close of the [last] session, a large part of the ceiling of the central lobby fell down [in the north wing]" and that "many attempts have been made during the last session to prevent the leakage of the gutters and skylights." In addition, he said, "The plaister's work is universally bad, and scarcely adheres even to the brick walls, and the carpenter's work is not only rotten, but injudiciously and insecurely put together." While correcting these and other deficiencies in the north wing, Latrobe proposed to rearrange the interior to better accommodate the Senate and the Supreme Court. The Senate would be moved up one floor, and the Supreme Court would be housed below. As Latrobe wrote, "The expense of the proposed arrangements will therefore be comparitively small, when the necessary expense of repair is deducted." This would be done in two parts: first the eastern portion of the north wing would be completed, then the western part. He predicted, "In 1809–1810 the whole wing will be completed."[104] For this work on the wings of the Capitol, Latrobe's cost estimate was $100,000.

Estimate of expenditures proposed to finish the south wing
for the occupancy of Congress, previous to the next session[a]

independently of the money in hand	25,000
Furnishing the same	20,000
Towards altering the east side of the north wing	50,000
Contingencies	5,000
Total	100,000

I am, sir,

    With high respect,

        Yours faithfully,   B. Henry Latrobe

[a] I must observe that the finishing of the capitals of the columns of the House of Representatives will be the work of a few years to come, the time of finishing them will depend on the number of artists which can be procured.[105]

TABLE 8.1. Money spent on public buildings in Washington

An account stating the several sums received from the treasury of the United States, and expended on the Capitol, the President's house, the Public offices, and other objects of Public expense, within the city of Washington.

THE MONIES RECEIVED FOR THESE OBJECTS FROM THE TREASURY OF THE UNITED STATES, HAVE BEEN UNDER THE AUTHORITY OF APPROPRIATIONS MADE BY THE FOLLOWING ACTS OF CONGRESS:			RECEIVED FROM THE TREASURY.	
	Dollars.	cents.	Dollars.	cents.
Act of 6th May, 1796	200,000			
18th April, 1798	100,000			
24th April, 1800	10,000			
			310,000	
3d March, 1801 *(general appropriation act)*	5,122			
3d May, 1802	8,000			
2d March, 1803 *(general appropriation act)*	3,702	66		
3d March, 1803	50,000			
27th March, 1804	50,000			
25th January, 1805	110,000			
Ditto	20,000			
1st March, 1805	9,000			
21st April, 1806	40,000		295,824	66
			605,824	66

THESE MONIES HAVE BEEN EXPENDED AS FOLLOWS:

ON THE CAPITOL

		Dollars.	cents.	Expenditures.
Act of 3d March, 1803. *Of the 50,000 dollars, appropriated by this act, there were expended on the capitol*		37,342	75	
27th March, 1804. *Of the 50,000 dollars, appropriated by this act, there were expended on the capitol*		36,896	04	
25th January, 1805. *An appropriation by this act, was expended on south wing*	110,000			
And of the $20,000 further appropriated by the same act, were expended on the north wing	1,130 89			
		111,130	89	

	Dollars.	cents.	Expenditures.
25th April, 1806.			
Of the appropriation by this act, there have been expended on south wing	35,327	96	
Remaining to be expended	4,672	04	
		40,000	
			225,369 68

OTHER OBJECTS OF PUBLIC EXPENSE.

1796, May 6th. *Of a loan of 200,000 dollars, by the state of Maryland, guaranteed by the United States, $120,000 have been paid, and $80,000 are still due, (both sums exclusive of interest) paid from time to time*			200,000
1798, April 18th. *By this act the United States loaned the city this sum, which, with the other monies borrowed under the guarantee of the United States, was expended promiscuously, amongst the monies which arose out of the funds of the city, on the public buildings, and other objects of expense*			100,000

Source: The President's house, the Public offices, and other objects of Public expence, within the city of Washington, in *DH*, 125.

Congress was concerned about the amount of money being spent on the public buildings. On the day President Jefferson transmitted Latrobe's report to Congress, Congressman John Randolph of Virginia offered a resolution to the House of Representatives calling on President Jefferson to provide a complete accounting of all of the moneys that had been spent on the Capitol, the President's House, the public offices, the Navy Yard, and the Marine Barracks. Congressman Randolph considered these expenditures on public buildings to be "a sink of expense, of increasing expense." Furthermore, "he believed [in] the case of the building in which they sat, the conduct of persons employed on it [had] always fallen short of the promise made."[106] The resolution was passed, and on December 23, 1806, President Jefferson presented his accounting to the House (table 8.1).

LATROBE'S *PRIVATE LETTER TO THE INDIVIDUAL MEMBERS OF CONGRESS*

Latrobe would have anticipated such an adverse reaction to his expenditures on the public buildings. He prepared his *Private Letter* to put forth his arguments to the congressmen to counter their criticisms. Addressing the problem of building stone, Latrobe discussed the problems of cutting, handling, and shipping stone from the various quarries at Aquia. He suggested that it was the annual appropriation of Congress that was largely the cause of high stone costs and stone-delivery delays. Because the appropriations were made annually, argued Latrobe, the public quarries were not immediately opened, and stone had to be purchased at a higher cost from private firms. What Latrobe meant was that if a single appropriation had been passed to cover multiple years, instead of annual appropriations, he would have been able to more efficiently and effectively organize the public quarries for the shipping of stone from the quarries through such means as the construction of a railroad or tramway.[107]

There were other problems with the annual appropriations made by Congress, wrote Latrobe. They came late in the construction year, thereby increasing construction costs, particularly labor costs, and delaying construction. In addition, he argued, the annual nature of the congressional appropriation was linked to the congressional yearly time frame and not the construction season. Appropriations passed by Congress in the spring acted against the goal of keeping good workers over the winter months. The uncertainty of passage of an annual appropriation was compounded by the continued uncertainty of whether Washington would remain the national capital. The result, wrote Latrobe, was that it was difficult to persuade workers to move to Washington, and, when they did move, higher salaries had to be paid to them. The system of annual appropriations also adversely affected the supply of brick and building stone. Latrobe pointed out, "We have often been obliged in the early part of the season, to discharge for a time most of our masons and brick-layers for want of a supply of building stone and bricks."[108]

Other sources of delay and additional costs were attributable to a lack of a master plan for Capitol Hill. "The Oven"—the temporary building that housed the House of Representatives—could not be promptly demolished because there was uncertainty that an appropriation would be passed for a replacement. Therefore, work on the south wing was delayed because it was necessary to work in the narrow confines between the Oven and the new walls.

Latrobe also had much to say about the initial design of the south wing, particularly its lack of structural stability to resist lateral forces. Because almost the entire south wing was planned for a single great hall for the House of Representatives, there were few interior walls to provide lateral bracing for the forty-foot-high external masonry walls. The north wall of the south wing had already been begun, and Latrobe felt compelled to continue that construction and reinforce it by building the recess to the north.[109]

Latrobe explained why the south wing of the Capitol was not ready to receive the House of Representatives: "That the House has not been completed, has been simply owing to this, that its completion was impossible in itself,—for it depended on circumstances which could not be controuled." The principal circumstance that could not be controlled was "the situation of the quarries ... the failure of stone has been in a great measure the cause of our not having effected more." But it was not the only controlling circumstance: "The difficulty in procuring stone was great:—had it been less, we should have labored under another equally so, that of procuring stone cutters."[110] To these reasons for the south wing not being completed he added another: that it was designed by an amateur he did not name—William Thornton—as a result of a design competition. Design competitions, wrote Latrobe, attracted amateur architects with little knowledge or understanding of building.[111]

After offering these reasons for not having completed the south wing, Latrobe provided details of the plan for that wing. The great hall for the House of Representatives was to be especially impressive when complete: "The House is surrounded by a plain wall seven feet high. The 24 Corinthian columns which rise upon this wall and support the dome, are 26 feet 8 inches in height, the entablature is 6 feet high, the blocking course 1 foot 6 inches, and the dome rises 12 feet 6 inches, in all 53 feet 8 inches. The area within the wall is 85 feet 6 inches long and 60 feet 6 inches wide. The space within the external walls is 110 feet by 86 feet."[112]

THE APPROPRIATION FOR 1807

The *Private Letter to the Individual Members of Congress* was intended to get Latrobe out from under the many criticisms that had been leveled at him by Congressman John Randolph and others. It was also intended to gain acceptance of his ideas for the Capitol and assure passage of an appropriation for completing the south wing and starting renovations on the north wing in 1807. It worked. On February 13, 1807, the House of Representatives began debate on an appropriation of $25,000 to finish the south wing. A separate appropriation of $20,000 was debated to buy new furniture. By the time the votes were taken, the full $25,000 was appropriated to finish the south wing, and $17,000 was voted to furnish it.[113] An additional $55,000 was sought, and approved, to work on the east side of the north wing.

THORNTON'S REACTION TO LATROBE'S *PRIVATE LETTER*

Latrobe's appropriations victory was not without cost. On March 3, 1807, the last day of the legislative session, William Thornton distributed to all members of Congress an anonymous, malicious satire entitled *Index to My Private Letter etc*, which was purported to have been written by Latrobe. The introductory paragraph to this work read:

In most elaborate performances of authors of high reputation, who have exerted their utmost powers to enlighten the world, and especially in works which contain a variety of useful matter, there has been some attention paid to an Index; and, as without vanity, it will be admitted that my private letter to you contains much important matter, mingled with many brilliant thoughts, I have imagined it might be more extensively beneficial, if supplied with such a Table as would point out its principal beauties, without descending to those minor elegancies, to which the refined understanding will be directed in a perusal of the whole work.[114]

A heavy handed work, Thornton's *Index to My Private Letter etc* further increased the enmity between Thornton, the amateur architect, and Latrobe, the professional architect-engineer—an enmity that reached a court of law in the coming years.

THE 1807 CONSTRUCTION SEASON

March 1 was traditionally the first day of the new construction season. On March 1, 1807, with appropriation in hand, Latrobe began his fifth year of construction on the Capitol. There were problems. His previous order for glass for the panel lights, sent to Hamburg, never arrived, and he found he could not replace the order in the United States. Even before the passage of the appropriation for 1807, Latrobe had sent an order to England for glass for the lights in the domed ceiling of the House of Representatives. He ordered "double the quantity required, to provide against accidents."[115]

It took months for this glass to be delivered to Washington. In the meantime, work on the dome and the installation of the sash for the one hundred skylights continued. On April 14, 1807, Latrobe informed President Jefferson that heavy rains entering the building through the joints of the skylights' sash had run down the ribs of the dome and had damaged the recent plasterwork. Latrobe proposed sealing the skylights with lead. But that would cost one thousand dollars, and the lead would have to be removed later in the construction season when the glass arrived. Again, as an alternative, Latrobe recommended the installation of a lantern. Previously, in November 1806, Latrobe had sent President Jefferson a perspective of how the Capitol would look with cupolas on both the Senate and House of Representatives domes (see plate 27).[116]

President Jefferson responded to Latrobe's ideas eight days later, on April 22, 1807. He wrote: "The idea of spending 1000. D. for the temporary purpose of covering the pannel lights over the representatives chamber [with sheet lead], merely that the roof be plaistered before the roof is closed, is totally inadmissible." Jefferson offered a commonsense answer to this problem: "I do not see why that particular part of the plaistering should not be postponed until the pannel lights are glazed." As for Latrobe's continuing suggestions that lanterns be used on top of the domes of the Senate and House of Representatives wings, Jefferson wrote Latrobe, "I confess they are most offensive to my eye, and a particular observation has strengthened my disgust at them."[117]

By mid-August 1807, Latrobe was reporting to President Jefferson that the roof on the south wing still leaked.[118] Latrobe found that part of this water leakage problem was due to incorrectly joined iron sheets. But most of it was related to the deteriorated putty used on the roof. Once again, Latrobe recommended the use of a lantern.[119] And once again President Jefferson responded that he did not want to use a lantern; he wanted to use skylights in the dome similar to the glass used in the Halles au Blé.[120]

On August 21, 1807, Latrobe wrote to President Jefferson stating that he had abandoned the plan to shingle between the skylight panels and had undertaken to reputty the entire roof. In the same letter, he reported that he had just received an invoice for the glass for the skylight panels that he had ordered from England, "The amount of the Glass is enormous, and will cripple our fund exceedingly." Including estimated import duties, the glass cost $4,130.10, or approximately $20 for each of the one hundred panel lights (twice as much glass as was needed was ordered, the remainder to be used later in the north wing).[121]

Latrobe also reported to Jefferson that because visitors to the Capitol had been hindering the work, he was going to restrict their entrance to the building.[122]

As Latrobe pushed to have the south wing ready for the reception of the House of Representatives by December, he began to have labor problems. On September 5, 1807, he informed President Jefferson that the carpenters were threatening to strike. The following week Latrobe reported, "I have fortunately arranged everything with both [th]e parties to my perfect satisfaction."[123] The second public building campaign had very few labor problems compared to the first public building campaign.

By September 17, 1807, Latrobe was able to report that the scaffolding in the great hall of the south wing was nearly down. "At the Capitol all goes on well," wrote Latrobe.[124]

Meanwhile Latrobe had begun fixing the roof of the north wing. On the same date, Latrobe provided Jefferson with an update on the north wing roofing work and an explanation of why the whole roof was not being replaced that season. He explained that not all of the timbers of the north wing roof were rotten. The portion of the roof that was rotten had been removed and replaced by interior vaulting, "and so roofed as to be forever safe."[125]

CAPITOL of the UNITED STATES.

No Person whatever can be admitted into the south Wing of the Capitol, excepting those actually engaged in the work

It is hoped that the citizens will see the necessity of this restriction being general The injury done already by improper Visitors has been very considerable, and in the present state of the building, which is about to be furnished, the interruption occasioned by the admission of strangers might defeat all endeavors to get the house ready at the time required

B. H. LATROBE, Surveyor
of the Public Buildings of the U. S.
September 11—tf

No visitors to the Capitol. *Washington National Intelligencer,* September 11, 1807.

The south wing was mostly but not totally complete. On Saturday, October 17, 1807, Latrobe held a banquet to celebrate its completion.[126] Paid for with public funds, the event was attended by 167 workers who had been engaged on the south wing and some workers from the Navy Yard who had also assisted. One man who did not attend was clerk of the works John Lenthall, perhaps the single man most responsible for bringing the project to a successful close. Lenthall, to Latrobe's mortification, boycotted the event because the laborers had not been invited.[127]

LATROBE'S 1807 REPORT TO CONGRESS

On October 27, 1807, Latrobe submitted his report on the House of Representatives wing to Congress.[128] The report marked Latrobe's moment of achievement. He wrote: "In the design and construction of the House of Representatives, all that I could effect by study and labor has been done to carry the desire of the President to give to the House every practical convenience and accommodation, into execution. In the permanence and solidity of the work, I hope nothing will be found deficient." Latrobe explained that the south wing was mostly, but not totally, finished: "The meeting of Congress at a period so much earlier than was expected in a very advanced part of the season, has prevented the completion of many small arrangements, which will render the house still more commodious. Every exertion is now making to complete them."[129]

There was great public interest in the newly completed House of Representatives. The editor of the *National Intelligencer*, Samuel Harrison Smith, wrote a laudatory review of the new House chamber in his newspaper.[130] Smith asked Latrobe to provide him with a description of the new Capitol wing for publication in the *National Intelligencer*, and on November 22, 1807, Latrobe did so.[131] This piece, a more detailed description of the south wing than Latrobe had given to Congress on October 27, 1807, was also a defense against criticisms of the building and of Latrobe himself.

Latrobe began his piece with a brief history of the building, starting with the design competition of 1792–1793 and Dr. William Thornton's winning design: "The design of the Capitol of the United States at Washington, as presented by Dr. Wm. Thornton, and accepted by general Washington under the law, placing at his option both the plan of the city and of the public buildings, — consisted of a central building, with two wings. The whole front from north to south extends 355 feet in length. The width of the wings is 121 6 [121 feet, 6 inches]; — but the center projected Westward beyond the wings, so that, its depth, including the Porticos, would have been near 200 feet."[132]

Latrobe felt compelled to write something conciliatory about Thornton and his original design of the Capitol: "Dr. Thornton, . . . it must be confessed, has, in his design exhibited talents which a regular professional education and a practical knowledge of architecture would have ripened into no common degree of excellence."

With regard to George Hadfield's contributions to the evolution of the design of the Capitol, Latrobe was more forthcoming: "Nor ought I on this occasion to withhold the praise due to Mr. Geo. Hatfield [Hadfield], architect, who for some time conducted the work at the Capitol in the city. His exquisite taste appears in many parts of the exterior of the North wing."[133]

Latrobe discussed the changes in the design of the south wing, such as the inclusion of committee and other rooms in the wing itself: "The original plan of the South wing underwent the alterations which will be described principally for the purpose of attaching to the hall of Representatives itself all its committee rooms and offices, and of avoiding the inconvenience of having them at the distance of near 200 feet in the North wing." He also described the changes from the original design of the hall of the House of Representatives from Thornton's elliptically shaped space to what was built, a parallelogram with semicircular ends: "The first design proposed [was] an eliptical colonnade of 36 equidistant columns standing upon an eliptical arcade in the lower story, and the present hall has a colonnade of 24 columns arranged in a different manner, and the arcade is entirely omitted. The present room is about 20 feet less in height than the first designed."[134]

Latrobe was proud of the hall in the House of Representatives wing. "If there is any degree of merit in the building, it is, I think, in the arrangements of this story." He wrote of the hall:

The Legislative Hall occupies the whole area of the wing within the walls, excepting the angles: being a parallelogram of 110 feet by 86 feet. The angles of this area are cut off, so as to reduce it to a species of oblong octagon. The two angles on the south are occupied by the gallery stairs, those on the north, by a committee room, and the door-keeper's office. Within the walls of the room, at about 10 feet distance from them, is the base wall that supports the colonnade. It consists of two semi-circles, 60 feet in diameter at the east and west ends, united by straight lines, 25 feet in length, on the north and south, so that the internal space within this wall is nearly 85 by 60. The wall itself is 7 feet in height and the top is nearly level with the selles [sills] of the principal windows.[135]

The hall was encircled by twenty-four monolithic fluted sandstone columns, each 26 feet 8 inches tall. The columns were of the Corinthian order, although by the end of 1807, only two of the capitals had been carved by Giovanni Andrei, the Italian sculptor brought over to the United States by Latrobe.[136]

Above the hall was the domed roof, with its one hundred skylights in the manner of the Halles au Blé. Latrobe singled out this feature for special note:

The space between the columns and the external wall is covered with a solid brick vault, but the internal area has a roof of timber, which forms a flat dome, rising 12 ft. 6 in. in height. This enormous roof is in thickness only 16 inches and is a very remarkable specimen of excellent carpenters work. It is constructed on the plan of Philibert de l'orme, and is pierced with square lights, five over each intercolumniation. There are 20 tiers or ribs of these lights, in all 100. Each is covered with a single square of plate glass, and the effect is very striking, especially to those who have not seen the Hallaubled [i.e., the Halles au Blé] or Corn market in Paris, which was lighted on the same principle.[137]

After describing the interior, Latrobe ended his article with, "All this scenery is lighted by the lanthorns of the two domes."[138]

LATROBE'S ANNUAL REPORT FOR 1807

Normally at the end of a year, usually in November or December, Latrobe submitted to the president an annual report, subsequently transmitted to Congress, summarizing work achieved in that year and outlining the scope and cost of work to be undertaken in the coming year. The annual report for 1807, Latrobe's fifth annual report, was not submitted to President Jefferson until March 23, 1808. Although the south wing of the Capitol was a resounding success (except for heating and acoustical problems), Latrobe had exceeded his appropriation for 1807. It was in this annual report that Latrobe felt he had to explain why more money would be required, why he had exceeded the appropriation, and how much additional money would be needed to cover the deficit. He also explained that the reason for submitting the report much later than usual was that some of the accounts could not be closed until the spring.[139]

In this report, Latrobe detailed the work that had not yet been completed in the south wing of the Capitol:

1 All the wood work require to be painted. The wood work is only primed.
2 Of the 24 Corinthian columns of the Hall of Representatives the capitals of only two are entirely finished, 8 are in a state of forwardness, and 14 are only rough hewn or bosted.
3 Only part of the moulding of the Cornice are finished.
4 The Sculpture over the entrance is incomplete.
5 The enclosure of the Lobbies is not yet finished.
6 All the Chimney pieces of the principal story and two of the Vestibules 10 in number are wanting.
7 Two small Capitals in the Circular Vestibule are still to be carved.
8 The platform on the South front giving access to the Galleries is erected upon the old scaffolding, which having been some years in use is weak and decayed. It is required by the nature of the ground, that a permanent platform on arches should be extended along this front.[140]

Much of this work, such as the stone carving, was expected to extend beyond the date of completion of the south wing. A more serious matter was that the western end of the four-foot-wide, forty-foot-high north wall of the building was settling and needed to be fixed.[141]

Besides this work, some additional moneys would also be needed to address two problems of the south wing: (1) the hall of the House of Representatives had poor acoustics, and (2) the stoves for heating the hall produced steam and excessive heat. With regard to the acoustics, Latrobe recommended large wall tapestries.[142] To improve heating, he noted that the flues were still wet from construction and that the problem would resolve itself as the flues dried and as less wood was added to the stoves.[143]

On the other side of Capitol Hill, there were more severe problems. On the north wing, Latrobe reported that there was serious wood decay arising from the leaky roof. Latrobe and his workers found that the wood rot was not confined to the roof structure but continued down to the foundation, all of which had to be replaced. What could be torn out without disrupting the operations of the Senate had been done, and the wood joists and flat ceilings had been replaced with more durable brick vaulting.[144]

Latrobe also reported on the President's House: "The work performed at the Presidents house has consisted of the covered way in front of the Offices on each Wing, of the erection of one half of the wall of enclosure, and of one of the Gates, of the levelling of the greatest part of the enclosed grounds, and of minor repairs and improvements of the house itself."[145]

In addition, Latrobe reported on the work on the roads of Washington, primarily the widening of the carriage-way of Pennsylvania Avenue and for drain improvements and related expenses.[146]

For 1807, Congress had appropriated $85,000:

South Wing of the Capitol	$25,000
North Wing of the Capitol	25,000
Presidents House	15,000
Publick Highway	3,000
Furniture of the South Wing of the Capitol	17,000
	$85,000[147]

Unfortunately, Latrobe had to report a $51,949.22 cost overrun on the 1807 appropriation, an overrun of about 61 percent. The deficit included the following items:

South Wing of the Capitol	$35,212.54
To make good the sum loaned	
to the public Offices	3,218.65

Presidents house		3,919.46
Prior Claims to 1807 do	1,737.44	5,656.90
Publick High Ways		3,644.79
Furniture fund		4,216.34
		$51,949.22[148]

The matter of the deficit was referred to a committee chaired by Congressman Richard Stanford of North Carolina. On April 5, 1808, Stanford introduced a bill to cover the deficit and to provide additional required funds. Congressional reaction to Latrobe's report of the deficit was predictable. Congressman John Randolph of Virginia, an administration critic, stated on the floor of the House, "This expense has been incurred, not by the Executive, not by the Head of a Department, but by a somebody whom we do not know."[149]

On April 8, 1808, Latrobe sent Congressman Stanford a reasoned letter defending himself and the cost overrun. After briefly explaining the background and purpose of the position of surveyor of public buildings, Latrobe pointed out that the south wing of the Capitol had cost substantially less than the north wing, even though the south wing was a more substantial structure (because of the brick vaulting used by Latrobe).[150] He went on to state that the work to correct faulty construction and deteriorated wood in the north wing would cost more than $100,000. So, once that work was completed, the north wing would have cost the US government $435,000, compared to $274,000 for the south wing.[151]

Latrobe's critics charged him with overrunning his cost estimate. Disputing that charge, Latrobe stated that in 1803 the Congress had decided to build the south wing in the same style, dimensions, and materials of the north wing and at that time it was known that the north wing had cost $313,000, that is, $39,000 more than the eventual cost of the south wing. Latrobe also pointed out that in earlier years he had underspent the appropriation for the south wing[152] and that Congress had not always appropriated the money that he had requested.[153] According to Latrobe's accounting, a total of $252,061.48 had been appropriated by Congress for the south wing—which was not as much as previously reported.[154]

Latrobe gave several reasons why more money had to be spent on the construction of the south wing. For example, the work needed to be secured for winter.[155] Price inflation was another factor,[156] as was the need to complete the building. By September 1807 Latrobe had exhausted his appropriations, and yet the south wing was still not ready for occupancy. Latrobe wrote: "All these causes exhausted the funds of the South Wing, and about the month of September I was informed by the superintendent of the city they would be greatly deficient."[157]

At this juncture Latrobe was faced with the decision to either exceed the appropriation or fail to have the south wing ready for the House of Representatives. He wrote: " In this dilemma there w[as a] choice of only two resolves—either to dismiss all the workmen and to commence the measurement and valuation of the work as far as it had been completed and thus to defeat the desire of the House of Representatives — or to apprize them of the state of the funds and to propose to them the risk of confiding in a future appropriation." The workmen agreed to take the risk of not being paid for their labor in case no additional appropriation was passed, and they completed the building: "There was no hesitation among them as is proved by the enclosed certificate of those who are now in the city, many of whom possessing capital, consented to encrease their own risk by the payment of those who could less easily afford it."[158]

On April 21, 1808, Congressman Stanford reported the findings of his committee to the House. The committee had recommended passage of the bill to cover the deficit.[159] The bill to cover the $51,400 deficit was passed, with Congressman Randolph vigorously objecting.[160] In addition to the funds to cover the deficit of 1807, Congress added $11,500 to finish the south wing of the Capitol, $25,000 for work on the north wing, and $14,000 for work around the President's House.[161]

THE NEWSPAPER WARS: THORNTON VS. LATROBE

While Congress was deliberating what to do with Latrobe's cost overrun, Dr. William Thornton leveled his first public blast at Latrobe. It appeared in a letter of April 20, 1808, from Thornton to Samuel Harrison Smith, the editor of the *National Intelligencer*. The letter was published in the *Washington Federalist* on April 26, 1808.[162] It is not clear why Thornton chose to attack Latrobe at that particular time. Perhaps it was a reaction to the acclaim Latrobe was receiving for the south wing of the Capitol — acclaim that Thornton thought was rightfully his. Perhaps it was an attempt to influence the outcome of the congressional deliberations on Latrobe's cost overrun in 1807. Wrote Thornton: "The design of the Capitol he [Latrobe] admits is grand, but the alterations made in every part have improved it, and though he knows that most of these alterations were made by me, yet he endeavors to leave an impression that they were made by others."[163]

Thornton criticized Latrobe for the fifty thousand dollars spent but not covered by appropriations. He also criticized Latrobe for not estimating the cost of materials previously provided by the commissioners (including Thornton). But in the same letter he went much further. Thornton questioned Latrobe's professional training in Europe, suggesting that Latrobe had misrepresented his education and training and that instead of receiving an education as an architect, he had been a chimney piece carver in Europe. Thornton also claimed that Latrobe had come to this country as a Moravian Missionary. In the same letter, Thornton asserted that George Washington had no confidence in Latrobe, that Latrobe had changed his name, and that he was not of English origin but of French. Thornton also attacked Latrobe's architecture, stating that the Bank of Pennsylvania, in Philadelphia, designed by Latrobe, was a mere copy.[164]

After Thornton's attack on Latrobe was published the morning of April 26, 1808, in the *Washington Federalist,* Latrobe's reaction was rapid. He sent his response to Jonathan Smith Findlay, the editor of the *Washington Federalist,* on the same day. His letter began:

> In your paper of this morning I observe a letter signed William Thornton. This letter is filled with the coarsest personal abuse of me, professedly in return for the too flattering observations with which, in a conciliatory humor, I prefaced the account of the capitol, published some time ago in the National Intelligencer. It is not my intention to pay any regard to that part of the letter which does not implicate my moral character. Those who know the Doctor, whether they understand the subject on which he writes, or not, will set no value on all he has said. But I owe it to myself and to my family to contradict those falsehoods, invented or collected by him, by which he would reduce me, in respect to veracity, to a level with himself; in the opinion of those at least, who have known me only for the last 15 years.[165]

Latrobe refuted as "utterly false" Thornton's assertion that he had been a chimney piece carver or a Moravian Missionary. He dismissed Thornton's statement that George Washington had said that he had no confidence in him as "a malicious fabrication." And he dismissed Thornton's assertion that Latrobe had changed his name as a falsehood. In response to Thornton's charge that Latrobe's Bank of Pennsylvania was merely a copy of a Greek Temple, Latrobe commented: "The grossest ignorance alone could assert that the Bank of Pennsylvania is the copy of a Greek Temple. All that is said on this subject is as absurd as it is false — Even the Porticos vary in every part of their proportions of columns and entablature from every temple in existence."[166]

Latrobe summarized his response to Thornton's charges: "The father of children who bear his name, owes it to them to place on the same record, on which the slander is enrolled, the rebuttal of the falsehoods that assail him, even if the slanderer be such a man as Dr. William Thornton; a man too feeble for personal chastisement, and too ignorant, vain, and despicable for argumentative refutation."[167]

Latrobe's letter was published in the *Washington Federalist* on April 30, 1808. The following day, Thornton sent a letter to Jonathan Smith Findlay, editor of the *Washington Federalist.* In this letter, published in the *Washington Federalist* on May 7, 1808, Thornton repeated and enlarged the charges he had made against Latrobe in his April 20 letter.[168] Two days later, Latrobe sent a second letter to Findlay, stating that he was taking legal action against Thornton for slander.[169]

On June 26, 1808, Latrobe submitted a memorandum to his lawyers, Walter Jones and John Law, on the differences between him and Thornton.[170] Thornton was represented by Francis Scott Key. Key and Thornton were able to postpone the trial by claiming that one witness, Ferdinand Fairfax, was not available. Eventually on June 24, 1813, *Latrobe v. Thornton* was decided in the Circuit Court of the District of Columbia, which ruled in favor of Latrobe. Latrobe did not press for damages, and the court awarded Latrobe one cent plus costs. It was, however, a great moral victory for Latrobe.[171]

Meanwhile, work continued in 1808 on the wall around the President's square ($14,000), on finishing the south wing of the Capitol ($11,500), and on the north wing of the Capitol ($25,000). Because of the previous year's cost overrun, Jefferson gave Latrobe specific instructions on how to limit cost overruns in the coming construction year. On the wall around the President's square, Jefferson wrote Latrobe, "no gate or lodge to be attempted till we see the state of our funds at the finishing of the wall so far." On the south wing, he told Latrobe, "The work [is] to be done *successively* paying off each before another is begun." The bulk of the construction work in 1808 was to be remedial work on the north wing of the Capitol. Jefferson instructed Latrobe: "North wing. To be begun immediately, and so pressed as to be finished this season. 1. Vault with brick the cellar story. 2. Leave the present Senate chamber exactly in it's present state. 3. Lay a floor where the gallery now is to be the floor of the future Senate chamber, open it above to the roof to give it elevation enough, leaving the present columns uninjured, until we see that, every thing else being done and paid for, there remains enough to make these columns of stone."[172]

Although Latrobe's cost overruns in 1807 were vindicated by Congress, he received an admonishment from President Jefferson:

> You see, my Dear Sir, that the object of this cautious proceeding [the appropriation for 1808] is to prevent the possibility of a deficit of a single Dollar this year. The lesson of the last year has been a serious one, is [it?] has done you great injury, & has been much felt by myself — it was so contrary [to] the principles of our Government, which make the representatives of the people the sole arbiters of the public expense and do not permit any work to be forced on them on a larger scale than their judgment deems adopted to the circumstances of the Nation.[173]

VAULTING THE NORTH WING OF THE CAPITOL

Of the work to be done in 1808, the vaulting of the north wing of the Capitol was the most important, as this masonry vaulting would replace the deteriorated wood joists. On May 23, 1808, Latrobe wrote Jefferson that the deterioration of the wooden joists and trusses was even greater than expected: "Its state is now evident to the public: the building being now surrounded with rotten girders, plates and joists, with exception only of the *Pine* timber, which is generally sound, altho' the

Oak trusses which have been enclosed within them are rotten." Latrobe found other structural problems in the north wing:

Independently of rotten Timber we have found other dangerous circumstances in the state of the building. The East Wall of the North lobby, has been cut away 40 feet high to the thickness of 9 in. and the remaining part so shattered and weak as to require great care in managing this part of the work. This 9 in. shattered wall carries 15 feet of 3 brick Wall at the top of the house, filled with rotten timber. In a few Years it is very probable that this tremendous weight held up only by decayed girders would have suddenly fallen into the Senate chamber, whenever the timber was so far decayed as to be too weak to resist its pressure.[174]

Besides reporting on the progress of construction, Latrobe used this letter to vent his frustration and rage at the criticisms he had received from Congress because of the 1807 cost overruns. Because of those criticisms, Latrobe told Jefferson, he had considered resigning. In addition to criticisms from foes of the administration, Latrobe had been routinely lambasted on the floor of the House of Representatives by members who he had thought were sympathetic, such as Virginia congressman John Eppes, a friend and ally of Jefferson.[175]

Latrobe felt that he had done what Jefferson wanted him to do—complete the south wing of the Capitol for the House of Representatives by the beginning of the 1807–1808 congressional session. He wrote to Jefferson: "Now by your letter of the 30th. of July I was expressly ordered to hire more workmen, and not on any account whatever to neglect to get the House ready by the Session."[176]

Jefferson was having none of this. On June 2, 1808 he wrote Latrobe (in part):

When the great deficit therefore happened the last year [1807], it was impossible not to consider it proceeding from a defect in your estimates, and continuing the work after the funds were exhausted. . . . It is true, as you observe, that I had urged you to employ a greater number of workmen, to ensure the completion of the S. wing for the ensuing season. But I did it on the ground, always expressed, that, the money being fixed and in hand, it would cost no more to employ 100. hands 50 days, than 50. hands 100. days. There never was a hint expressed, or a thought entertained, of going beyond the appropriation. Still I will say candidly that had it been suggested to me that the appropriation was inadequate, I should in the first place advised the doing only those things substantially necessary for the comfort of Congress, and if a moderate sum beyond even this were necessary (omitting every thing of mere ornament) I should probably have advised the going on to make the room capable of receiving them, and would in that case have taken on myself a candid explanation of the motives to Congress, and thrown ourselves on their indulgence.[177]

For Latrobe, attacked in the press, attacked on the floor of the House of Representatives, and not supported by Jefferson, this must have felt like the nadir of his career in America, a descent that had begun with losing his engineering position with the Chesapeake and Delaware Canal in 1806 and had accelerated downward with public attacks by Thornton and on the floor of the House of Representatives by members. But he was not at the nadir; worse was to come.

In June of 1808, Latrobe suffered from "bilibous fever,"[178] which left him "so weak, that I have been obliged to have recourse to bark and quiet to restore me to strength."[179] On September 2, 1808, his wife delivered a stillborn daughter, to have been named Louisa.[180]

Latrobe spent little time in Washington during the summer and early autumn of 1808. As usual, work continued under the immediate direction of Latrobe's clerk of the works, John Lenthall, particularly on the north wing of the Capitol. On September 11, 1808, Latrobe was able to report to Jefferson that all of the arches in the upper story of the north wing would soon be completed:

NORTH WING OF THE CAPITOL
Today all the arches of the upper story will be completed: 1.) That is, of the Senate chamber, a half dome of 60 feet diameter and a cylindrical extension of the same to the East Wall of 56 feet: 2.) The spandrils surrounding the same, all cylindrical arches, forming, one large lumber room to the south, and wide passages round the back, to the Westward, and 3.) Two very commodious committee rooms on the North.[181]

Latrobe reported to Jefferson the problem of the very slow drying of the mortar in the vaulting, necessitating the delay in removing the temporary framing underneath the centering.[182]

THE DEATH OF JOHN LENTHALL
Then, on September 19, 1808, disaster struck. While trying to remove the centering under the vaulting in the new court room of the north wing, John Lenthall, clerk of the work and good friend to Latrobe, was crushed to death by falling masonry. Two hours later, Latrobe penned the following note to John P. Colvin, editor of the *Monitor:* "This day, about half past 11 o'clock, the vault of the new court room in the north wing of the Capitol fell in. There were three or four men under it when it first cracked, all of whom escaped excepting Mr. John Lenthal, the clerk or superintendent of the work, who is buried in the ruins and has not yet been found. The hopes of his being alive are very faint."[183]

News of Lenthall's death spread quickly throughout Washington. To combat his enemies and detractors, it was necessary for Latrobe to get his account of what happened quickly published. On the following day, September 20, 1808, he sent

On Monday last, the vault of the new court room in the north wing of the Capitol, at Washington, fell in. Three or four men were under it, when it first cracked ; but they escaped, except Mr. John Lenthall, superintendant of the work, who was discovered after three or four hours search. Mr. L. was dead when he was found, having his scull fractured, and being otherwise wounded. [*North American.*]

After several hours industrious search the lifeless body of Mr. Lenthall was found, with the scull fractured and other parts considerably injured—under the sanction of Mr. Buzard, the coroner, his death was pronounced *accidental*, and the corpse conveyed to the dwelling of his afflicted family, there to receive the necessary preparations for the rights of burial. The regret of the whole city on the exit of Mr. L. constitutes for him an honourable eulogium, surpassing the panegyric of the pen.

The redoubled exertions of Mr. Latrobe, and the facilities which every class of our citizens are cordially disposed to render for the repair of the accident, are ample securities that the hall of the senate will be in a suitable situation for the reception of that body in November, notwithstanding the unexpected drawback upon the operations of the workmen.

[*Monitor.*]

The death of John Lenthall.
Annapolis Maryland Gazette, September 29, 1808.

this account to Samuel Harrison Smith, the editor of the *National Intelligencer*:

Yesterday the Vault of the Court Room in the North Wing of the Capitol fell down. Several workmen, under the direction of Mr. John Lenthal, the clerk of the works, were under the vault, lowering that part of the centre which still stood under it, just before it fell. A loud crack gave notice of their danger, and all of them escaped out of the windows, or under the adjoining vaults excepting Mr. Lenthal, who, to judge from the place in which his body was found, wanted only a single step to have secured his retreat also. But being under that part of the arch, the centre of which had been removed on Friday; he was suddenly buried under many tons of bricks, and must have been instantaneously deprived of sensation and life.[184]

The letter was not only an explanation of the accident, it was also Latrobe's eulogy for his good friend Lenthall:

Mr. John Lenthal was born at Chesterfield, in the county of Derby in England, and brought up to the business of a carpenter. He was in his youth much employed in the mines and had acquired a thorough knowledge of the manner of working them, and of all the machinery used in that part of England. He was not less acquainted with the cotton works, and joining an insatiable desire of knowledge to a strong memory and sound judgment, his acquaintance with the arts and sciences not immediately connected with his trade, was very extensive and by no means slight. He was also a superior draughtsman, and of his own particular business a perfect master. Though of very moderate stature his personal strength and activity was unequalled, and the command he acquired of all those who were placed under him was achieved by his own superior ability.[185]

Although some historians have written that Latrobe assumed the responsibility for this fatal accident, from this letter it seems apparent that Latrobe thought the accident was either a result of sabotage or that Lenthall had removed the centering under the vault too soon. Latrobe did not speculate who might sabotage the work or why they might want to do so. Despite mention of sabotage, Latrobe thought: "The real causes of the failure, I conceive to have been a too early removal of the centers, and perhaps a removal not altogether judicious, the early removal of the centering supporting the vault."[186]

On September 21, 1808, Latrobe provided a somewhat longer explanation of the structural failure to the editor of the *Monitor*. He again broached the subject of sabotage: "That attempts have been made in the absence of the workmen to injure this work can be legally proved."[187]

Latrobe thought a contributing cause of the failure was possibly the incorrect construction of small wooden boxes on top of the centering, intended for use in building coffers in the underside of the arch. Still, although the arch warped, Latrobe wrote that this was probably not the principal cause of the structural failure. He reported that Lenthall had proceeded in removing the centering under the arch in a slow and methodical manner. Latrobe had recommended to Lenthall that the backing on top of the vault be removed before removing the centering (thereby greatly lessening the load on the vault), but Lenthall did not think this was necessary. As the centering was removed, Lenthall carefully examined the masonry for the development of cracks that might indicate that the structure would not sustain itself. None were found. Over that weekend cracks began to develop in the arch, readily apparent by the following Monday, September 19, 1808. Lenthall took ten men with him to examine the structure. Shortly after, the arch failed, killing Lenthall but not the other men. Above this structural failure was the Senate chamber, itself surviving with little damage.[188] The vault of the Senate chamber above, with a fifty-four-foot span,[189] "is as firm as a rock." Latrobe wrote to Jefferson, "and has a most extraordinary and beautiful appearance," and "every other Wall and Arch in the Building is as sound as at the moment it was built."[190]

On September 23, 1808, in a letter to President Jefferson, Latrobe repeated his conclusions on the cause of the vault failure. He wrote that he would not publicly blame his good friend Lenthall for the failure. Latrobe explained to Jefferson that his original plan was to construct more or less continuous centering to build the vaulted ceilings of the Supreme Court and to ensure a level floor for the Senate chamber above. Instead of using this continuous system of centering, wrote Latrobe, Lenthall constructed a ten-foot-long centering structure, installed the brick arch above it, let the mortar set, and then removed the centering to repeat the process at the next ten-foot-section (thereby avoiding the necessity and cost of building multiple centering structures).[191]

Once the debris had been cleared, there was a question whether the damage could be repaired in time for the Senate to meet. Only $3,049 was left in the Capitol north wing appropriation, a sum inadequate to repair the damage. However, all of the workmen donated a week of work, and "many gentlemen have proposed

If the Surveyor will calculate the number of bricks used in the arches with the mortar between each, and cons... ... small rise he will not be surp... ... they fell and if he is no body else is. Why did he not stay at the seat of government instead of going away for weeks together when such new inventions and experiments as these arches were going on ! for he receives three thousand dollars a year—and yet sports with the public money, and the feelings of the citizens, who by his blundering works, are threatened yearly with the removal of the seat of government.

A PLAIN MAN.

The last paragraph of the newspaper notice by "A Plain Man" (James Hoban), anonymously attacking Latrobe. *Washington Federalist*, October 8, 1808.

subscribing to an amount amply sufficient to repair and render the Mischief invisible by the meeting of congress." Repair work began. Latrobe modified the design of the vaults above the Supreme Court Chamber by substituting three arches for the one great arch.[192]

MORE NEWSPAPER WARS:
THE "PLAIN MAN" VS. LATROBE

The failure of the brick vaulting and the death of John Lenthall opened Latrobe to even more criticism than he had received for the cost overrun. No more severe criticism was printed in the newspapers than that by an author identified as "A Plain Man," which appeared on October 10, 1808, in the *Washington Federalist*. The "Plain Man" wrote that Latrobe was to blame for the failure and that he had tried to pass blame on to Lenthall.[193] "A Plain Man" questioned why Latrobe had "gutted the Senate room by knocking down the whole and clearing it out"; why he "spen[t] the public money unnecessarily when this chamber would have accommodated the Senate many years?"; why he acted, "contrary to the President's directions in taking down the columns of the Senate Chamber?"; and why he had "taken down the steps of the Great Stair Case?"[194]

"A Plain Man's" most damaging accusation against Latrobe was that Latrobe had designed an inadequate arch against the east wall of the Supreme Court room and had subsequently cut into the arch for various purposes, further weakening it.[195] According to "A Plain Man," the failure of the arch and the death of Lenthall were thus the responsibility of Latrobe. Latrobe, not knowing at that time the identity of "A Plain Man," refuted these charges in a published letter on October 20, 1808, in the *Washington Federalist*. To defend himself against charges that he had unnecessarily removed sound construction from the north wing, he quoted from the Day Book written by John Lenthall that there was severe deterioration in the structure after only eight years of occupancy.[196]

"A Plain Man" charged that the arch was inadequately designed and that the remaining wall was unsafe. Latrobe responded that the wall was perfectly safe and that the cracks observed in the wall had existed for at least six years and were due to the characteristic of Aquia freestone to shrink.[197] Three days before this letter was published, on October 17, 1808, Latrobe was informed that "A Plain Man" was

James Hoban. In his letter to the editor of October 20, 1808, Latrobe pointed out that many of the defects that he had needed to correct in the north wing of the Capitol were installed by Hoban when Hoban was responsible for its construction.[198]

Hoban responded to Latrobe's letter of October 20 a week later, stating that he was not responsible for the decayed structural members found by Lenthall and Latrobe in the north wing. He went on to accuse Latrobe of falsifying the Day Book quoted in the October 20, 1808, article.[199] Hoban continued his attacks in a further letter to the editor of the *Washington Federalist*, on November 15.[200] And that was where the public denunciations ended, although the antipathy between Latrobe and Hoban continued.

LATROBE'S CONSTRUCTION REPORTS OF 1808
Latrobe ended 1808 with two reports, the annual report to Congress (his sixth annual report), submitted on November 18, 1808,[201] and a report to Senator Stephen Row Bradley, in response to a Senate resolution "to ascertain as nearly as may be, the amount which would be required to complete and finish the President's house and square, and the two wings of the capitol."[202]

In his sixth annual report, Latrobe wrote that most of the interior finish work had been completed in the south wing.[203] The bulk of the construction undertaken in 1808 was on the north wing. Stung by Hoban's recent accusations that he had wasted a large amount of public funds by arbitrarily gutting the north wing, Latrobe felt compelled to begin his report in conformity with the language of the appropriations act passed by the legislature.[204]

Latrobe stated that the principal reason that most of the interior of the north wing had needed to be removed was the deterioration of the timbers. "Scarcely a single principal girder or beam was entirely sound," wrote Latrobe. To back up these statements, Latrobe exhibited the deteriorated timbers on the grounds of the Capitol and invited all to examine them. This was not all. The brick piers that Latrobe had counted on had been weakened to such an extent that they could no longer be relied upon and had to be rebuilt. Because of the deterioration and other weakened work found, Latrobe had no other choice than to gut the interior of the building and reestablish sound walls from the basement up.[205]

In this report Latrobe had to mention Lenthall's fatal accident. He included a paragraph that was laudatory to Lenthall, lacking in specificity as to the cause of his death, and acknowledging his, Latrobe's, overall responsibility.[206]

Latrobe's annual report not only included a statement of work accomplished in 1808 but also a request for funds for the coming year, 1809. For the south wing, Latrobe asked for $6,000 to continue carving the capitals in the hall of the House of Representatives, to repair the glass in the building, and to repair defective chimney pieces. To finish the rooms of the north wing, $20,000 was needed, and an additional $25,000 for the library and judiciary offices. In addition, Latrobe again asked for funds, $18,000, to repair the northwest corner of the south wing; this request had been denied in the previous session.[207] Finally, he asked for $12,000 to finish the President's House and make various repairs to it.[208]

For the coming year, 1809, for the public building, Latrobe requested a total of $81,000:

RECAPITULATION

1	South Wing	6,000
2	North do. Senate	20,000
3	do. Library and Judiciary	25,000
4	N. W. Corner of South Wing	18,000
5	Presidents House	12,000
		$81,000[209]

Latrobe requested an additional $3,000 for roads, for a total of $84,000.

In his report to Senator Bradley (table 8.2), Latrobe amplified his comments on the functional problems at the President's House. And he explained that in addition to these problems, the President's House, like the north wing, had much decayed structural timber that had to be repaired or replaced.[210]

Latrobe's request for $84,000 in 1809 had a difficult time in Congress. Congress was primarily concerned with the Embargo Act of 1807 and its economic consequences on both the economy and the federal budget. Eventually, on March 3, 1809, Congress passed a reduced appropriation of $31,000.[211]

With the inauguration of James Madison as president, Latrobe was continued in his office as surveyor of public buildings and given some additional duties. One of them was to work with Dolley Madison to furnish the President's House.

On June 28, 1809, Congress appropriated an additional $15,000, to "defray the expenses of finishing and furnishing the permanent Senate chamber, its committee rooms and other apartments."[212] Also appropriated was $1,600 for various expenses and repairs to the temporary accommodation of the Senate.

On September 8, 1809, Latrobe reported construction progress in Washington to President Madison. In the south wing of the Capitol, carving of the column capitals continued. He estimated that another six years would be required to complete all of the capitals.[213] His report included comments on the completion of the dome of the Senate chamber.[214] The bulk of the work undertaken in 1809 was in the north wing.

On December 11, 1809, Latrobe submitted his annual report to President James Madison — his seventh annual report and his first to be submitted to President Madison. He reported on a severe hail storm in June 1809 that had broken almost all of the glass on the south elevation of the building.[215] Latrobe did not mention whether the glass panes broken were in the dome, but it is likely that they were.

Work proceeded slowly on the north wing of the Capitol. As Latrobe explained, the delay in completing this work had occurred because the Senate had been using the building, because of a scarcity of workmen, and because it was difficult to get materials. But despite these problems, the Senate chamber was almost complete, as were the offices for the Supreme Court.[216]

Latrobe's annual report for 1809 discussed the immediate need in the south wing for additional committee rooms to house the growing number of standing committees in the House of Representatives. There was an additional, immediate reason for undertaking this work now—to remedy past construction defects in the northwest corner of the building.[217] In addition, funds would be required to continue the carving of the capitals and to improve the heating system.

On the north wing, having substantially completed the work on the east side and the center of the building, Latrobe urged that work commence on the west side.[218]

Work at the President's House had also progressed,[219] but more work was needed. Latrobe began this topic of his report by summarizing the state of the house in 1800, when the government moved from Philadelphia to Washington: "On the removal of the seat of government to Washington in the year 1800, the presidents House was in a most unfinished state, and quite destitute of the conveniences required by a family. The roof and gutters leaked in such a manner, as materially to injure the Cielings and furniture. The ground surrounding the house barely enclosed by a rough fence, was covered with rubbish, with the ruins of old brick kilns, and the remains of brickyards and stone cutters sheds."[220]

Latrobe continued by summarizing the work he had undertaken on the house between 1804 and 1809: "During the presidency of Mr. Jefferson from the year 1804, annual appropriations have been made, by the aid of which several bed chambers were fitted up; the most necessary offices and cellars, which before were absolutely wanting, were constructed; a new covering to the roof was provided; a flight of stone steps and a platform built on the North side of the house; the grounds were enclosed by a wall, and a commencement was made in levelling and clearing them in such parts as could be improved at the least expense."[221]

But still much work needed to be undertaken:

But notwithstanding the endeavors of the late President, to effect as much as possible by these annual legislative grants, the building in its interior is still incomplete. It is however a duty which I owe to myself and to the public, not to conceal that the timber[s] of the presidents house are in a state of very considerable decay, especially in the Northern part of the building. The cause of decay, both in this house and in the capitol, is to be found, I presume, in the green state of the timber when first used, in its original bad quality, and in its long exposure to the weather before the buildings could be roofed. Further progress in the levelling and planting of the ground, in the coping of the wall, and in current repairs and minor improvements, are also included in the estimate submitted.[222]

TABLE 8.2. Estimate of cost and schedule for completion of the public buildings

OBJECTS OF EXPENDITURE.		TOTAL.	1809.	1810.	1811.	1812.	1813.
I. Presidents house and square.		Dollars					
a. Carriage houses, north gate, south Platform, regulation of the ground Etc	12,000						
b. S. West gate, extension of the wings, Garden and planting, South portico etc.	12,000						
c. Completion, and repair of the interior	24,000	48,000	12,000	10,000	10,000	15,000	1,000
d. Removal of Earth front of the public Offices, sewers, and road in the Square		3,000	3,000	—	—	—	—
II. Wings of the Capitol.							
SOUTH WING.							
a. Completing all the carving of the hall of Representatives, and painting all the walls of the house, including minor repairs		20,000	6,000	4,000	4,000	5,000	1,000
b. Platform along the south front, paving along the east and west front, steps, &c		8,000		5,000		3,000	
c. North west corner of the wing (see my report) containing, the door-keeper's dwelling—water closets—pumps—three committee rooms—passages and sundry store rooms		24,000	18,000	6,000			
NORTH WING.							
a. Senate chamber and committee rooms.	20,000						
Alterations after the experience of one session, probable	1,500						
		21,500	20,000	1,500			
b. West side of the house, containing the library and the offices of the judiciary, solid work and carpentry	30,000						
Finishing, book-cases, and fitting up generally	10,000						
		40,000	25,000	15,000			

OBJECTS OF EXPENDITURE.	TOTAL.	1809.	1810.	1811.	1812.	1813.
c. S. W. corner of this wing, containing the door-keeper's dwelling—pump courts—privies and water closets—back stairs and the great conference room	48,500		30,000	18,500		
	213,000	84,000	71,500	32,500	23,000	2,000

Having, in the general plan of the Capitol, upon which the above estimates are founded, comprised a detailed design of the centre part of the building, I respectfully submit to the committee, a description and estimate of its expense; as also, of regulating and planting the ground within the Capitol square, in a manner suitable, and convenient to the building. These two objects, are not included in your requisition, but the information I take the liberty to offer you, may tend to throw some light on this subject.

CENTRE OF THE CAPITOL.

a. Containing general communication of the lower stories, of the wings and principal public staircase—the great Vestibule of the whole building—ten rooms for committee rooms, or rooms for refreshment—an extensive portico on each front—on the east, a flight of steps leading to the principal story	225,000			100,000	100,000	25,000
b. Planting and regulating the ground including the necessary walling	25,000		5,000	5,000	5,000	10,000

Source: Latrobe's Five Year Estimate of the Sums and Periods required to finish the Public Buildings of the United States, 1809–1813, submitted to Senator Stephen Row Bradley by Benjamin Latrobe, "To the honorable general Bradley, chairman of the committee of the Senate of the United States, appointed on the 12th of December, 1808, on the subject of the public buildings of the United States," December 13–21, 1808, in *CL2*, 685–692, 686; and in *DH*, 151–152.

Finally, Latrobe made a request for $6,000 for "the purpose of repairing and draining the high roads around and connecting the public buildings, and for widening the bridge over the Tiber."[223]

The cost of all the work Latrobe described in his report would be $103,500, as detailed in the following list:

Estimate.

1	South wing, sculpture, warming and ventilating the house	$7,500
	South wing, north west addition	25,000
2	North wing, defraying the expense of completing the court room and the offices of the judiciary on the east side, completing the Senate chamber, and for the library	40,000
	North wing, for the platform and external access on the north side	5,000
3	President's house, offices, wall, and grounds	20,000
4	Highways	6,000
		103,500[224]

Latrobe called for an early passage of the 1810 appropriation. But it was not to happen. His report of December 11, 1809, was on December 22, 1809, referred to a committee consisting of Joseph Lewis Jr. of Virginia, Richard Stanford of North Carolina, Edward St. Loe Livermore of Massachusetts, Erastus Root of New York, and John Brown of Maryland.[225] The appropriation did not come before the House until May 1, 1810, the last day of the congressional session. It passed at 12 midnight. But it was significantly cut back. Instead of $103,500, as requested by Latrobe, only $32,250 was appropriated: $7,500 for the sculpture and heating system on the south wing; $20,000 for completing the courtroom and the judiciary offices on the east side of the north wing and for completing the Senate chamber and stopping the roof leaks; and $5,000 for repairs to the President's House.[226]

On December 28, 1810, Latrobe submitted his eighth annual report.[227] Little progress had been made in 1810, and because of the expense of furnishing the courtroom, most of the workmen had been discharged in June 1810. Some work did get accomplished, however. The Senate chamber had been mostly finished. The east wall, the portion of the Senate chamber that had collapsed, still remained to be plastered. In the south wing, twenty-two of the twenty-four capitals in the hall of the House of Representatives had been carved. Some improvements had been made in the heating system. Blodgett's Hotel had been purchased for the use of the Patent Office and other offices. But much work remained uncompleted. Latrobe again reported the need to fix the deteriorated timbers in the west side of the Senate chamber, and again he pointed out the need to fix the northwest corner of the south wing.[228] For the coming year, 1811, Latrobe asked for an appropriation of $48,932, in the following categories:

View of Washington in 1807, by Nicholas King. Blodgett's Hotel is at the right, and in the far distance is the President's House. Courtesy of the Library of Congress.

ESTIMATE.

For sculpture, being for the wages of two Italian sculptors
   engaged under a specific contract, and for wages of the
   assistants and of two laborers to attend them, scaffolding,
   utensils, tools and all expenses incident to this branch $6,000

For ventilating and warming the south wing, painting on
   the roof, repairing of glass, moving and replacing furniture
   and carpets, and all incidental work   3,500

For replacing the expenditure of fitting and furnishing the
   court room and offices, thereby covering all outstanding
   claims excepting current wages by agreement $2,432

For salaries and their arrears, and all contingent expenses   4,000

* * *

For the plaistering and residue of finishing in the senate
   chamber, minor repairs and painting, additional shelves
   and book-cases, and fitting up the room over the secretary's
   office as a deposit of books and papers not in use 2,500
For the permanent construction of the west side of the north
   wing and the north west angle of the south wing, as high as
   the same can be carried in the season of 1811 20,000
For the regulating of the ground in front and on the sides
   of the capitol and repairs of the roads 2,500
For the platforms to the north and south entrances 8,000
                                                      ———
                                                    48,932

All of which is respectfully submitted.

B. HENRY LATROBE.[229]

Congress appropriated no moneys for these purposes.

By mid-1811 Latrobe had left his post as surveyor of public buildings. The eight years he had spent in Washington were marked by newspaper attacks by Thornton and Hoban, criticisms of his cost overruns on the floor of Congress, and the death of his close friend John Lenthall. (See plate 28.) He relocated to Pittsburgh to seek the fortune that had been elusive in Washington. He returned in 1815 to rebuild the Capitol after the British burned it in August of 1814.

The Third Public Building Campaign (1815–1824) **9**
Rebuilding the Federal City

We drank tea at Mrs. Thornton's, who described to us the manner in which they conflagrated the President's H., and other buildings, — 50 men, sailors and marines, were marched by an officer, silently thro' the avenue, each carrying a long pole to which was fixed a ball about the circumference of a large plate — When arrived at the building, each man was station'd at a window, with his pole and machine of wild-fire against it, at the word of command, at the same instant the windows were broken and this wild-fire thrown in, so that an instantaneous conflagration took place and the whole building was wrapt in flames and smoke. The spectators stood in awful silence, the city was light and the heavens redden'd with the blaze.

The poor Capitol! Nothing but its blacken'd walls remained! 4 or 5 houses in the neighborhood were likewise in ruins . . . these houses were so thoroughly destroy'd as the House of Representatives and the President's House. Those beautiful pillars in that Representatives Hall were crack'd and broken, the roof, that noble dome, painted and carved with such beauty and skill, lay in ashes in the cellars beneath the smouldering ruins, were yet smoking. In the P.H. not an inch, but its crack'd and blacken'd walls remain'd.

MRS. SAMUEL HARRISON SMITH

The British forces advanced on Washington to burn the public buildings after defeating the American forces at Bladensburg on August 24, 1814. A period of great uncertainty ensued about where the nation's capital should be rebuilt, but slowly a consensus coalesced that it should remain at Washington. On February 13, 1815, Congress passed an appropriation of $500,000, "that the President of the United States cause to be repaired or rebuilt forthwith, the President's House, the Capitol, and public offices, on their present sites in the city of Washington."[1] The burning of Washington also had a terrible economic impact on the building class of Washington. Until Congress decided that Washington would continue to be the capital, there was little or no work for the building trades, and the workmen suffered severe hardships.[2]

COMMISSION FOR REBUILDING THE PUBLIC BUILDINGS

The process for reconstruction of the public buildings was very similar to the process used to build them initially. As in the first public building campaign, a commission was formed to oversee the rebuilding.

City of Washington.—We are happy to learn that the rebuilding of the Public Edifices at the seat of the General Government proceeds with unexpected rapidity ; and we cannot but add our hope that they will at last be *finished* in a style becoming the metropolis of the United States. The Capitol is under the direction of Mr. Latrobe, its old superintendant, and the President's House in the charge of Capt. Hoban, its original constructor. The damage sustained from the flames of war is found to be less than was apprehended.

Rebuilding proceeds. *Weekly Raleigh (NC) Register,* June 2, 1815.

The commissioners were appointed by President Madison on March 10, 1815: John P. Van Ness, Richard B. Lee, and Tench Ringgold. Many of the principals engaged for the reconstruction were the same as those involved in the original construction. James Hoban was employed as the architect for the rebuilding of the President's House; Benjamin Latrobe was hired as architect for the rebuilding of the Capitol; and Peter Lenox was selected as foreman of the carpenters at the President's House and clerk of the works. Lenox had worked as a carpenter on the President's House twenty years earlier. George Blagden, another old name from the first public building campaign, was hired as the inspector of stone and superintendent of stone cutters and stone setters at both the Capitol and the President's House. Robert Brown was hired to be the foreman of the stone cutters at the President's House, where he had worked twenty years earlier. Leonard Harbaugh, the contractor from Baltimore who had built the first bridge over Rock Creek, was hired as foreman of the carpenters at the Capitol.[3]

Severe damage had been done to the buildings by the fires of August 24, 1814. Although some parts of the buildings could be used, most of the restoration required totally new construction. This new construction was planned and executed along the same lines, using the same building materials, as the original construction.

BUILDING MATERIALS FOR RECONSTRUCTION

Securing adequate building materials was much less a problem for the reconstruction than it had been during the original construction. The region, sparsely inhabited twenty years before, was now fairly populous, although hardly a metropolis on the order of Philadelphia or New York. There were now building suppliers who could provide building materials, and there was a transportation infrastructure to deliver them. There were wharves where building materials could be unloaded, although many of the wharves needed repair. Bridges and roads had been built. Building materials that had previously been difficult or expensive to acquire were no longer so. For example, throughout the first public building campaign, it was a problem to secure lime at reasonable prices; now lime was available in abundance.[4]

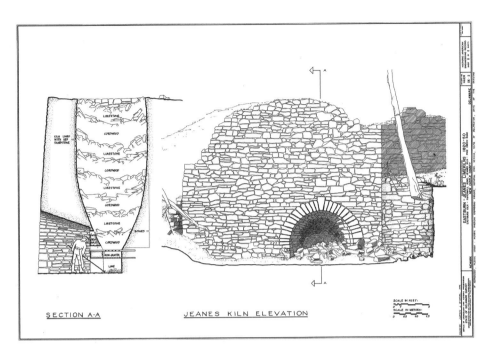

Section and elevation of Jeanes Kiln, typical of lime kilns used in the early nineteenth century. Judson J. McIntite, delineator, 1979. In the first federal construction campaign, lime was expensive, primarily owing to shipping costs. When the Potomac Canal Company opened the Great Falls bypass canal in 1802, shipping costs were greatly reduced, making lime from Loudoun County, Virginia, and Frederick County, Maryland, less expensive and more available. In 1815 Benjamin Latrobe, architect of the Capitol, proposed to Samuel Clapham that he agree to provide ten thousand bushels of lime. To do so, Clapham would have had to construct lime kilns similar to this one. The kiln was "charged" (i.e., loaded) from the top with alternate layers of limestone and fuel (wood, charcoal, or sometimes coal). As the material burned down, the lime and the clinkers were extracted from the base of the kiln and separated. The lime had to be ground into powder. Historic American Engineering Record collection, Library of Congress.

The commissioners experienced very little difficulty in obtaining the building materials necessary for the reconstruction program, except for building freestone.

For building freestone, the commissioners and Latrobe reopened the Aquia quarries on Government Island, purchased by L'Enfant twenty-four years earlier. In addition, they contracted for Aquia freestone at other quarries on or near Aquia Creek, as they had done in the second public building campaign. They also investigated other building-stone sources, including the Seneca quarries above the Great Falls on the Potomac River. Latrobe, then architect of the Capitol, visited the Seneca quarry in 1815. He reported, "The quarry is situated a mile and a half from the little Village around Seneca mills," and described the quarry for the commissioners:

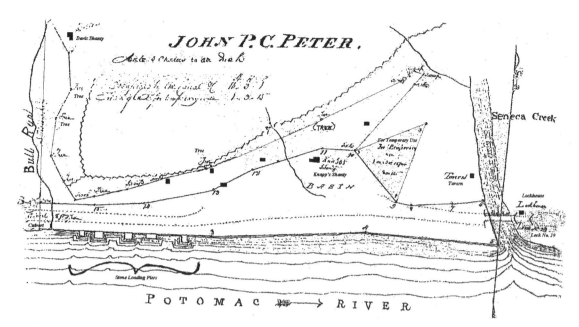

Location of Bull Run quarries of Seneca, 1830. Before 1802, when the Potomac Canal Company opened the Great Falls bypass canal, there was no way to economically deliver Seneca stone to Washington, and therefore the stone was not used in the first federal building campaign. It was not until the third building campaign that Benjamin Latrobe began to use Seneca stone for pavers, balusters, and other purposes. This map, prepared in 1830, some fifteen years after Latrobe visited the site, locates the Bull Run quarries of Seneca approximately one-half mile due west of the mouth of Seneca Creek (*right*). The hamlet of Seneca Mills was approximately a mile up Seneca Creek. The basin and the aqueduct shown would not have existed at the time of Latrobe's visit, having just been constructed by the Chesapeake and Ohio Canal Company. Lee's Quarry, which also furnished Seneca stone to the public buildings of Washington, was immediately to the left of this map. Author's collection.

On my return I called at the Quarry of Freestone above the Seneca falls. It is situated beyond a narrow flat, (being the lower part of the Sugar land bottom) and forms a perpendicular Cliff 50 or 60 feet above the bottom. The stone lies in very regular, strata inclined to the river near the bottom. The stone lies in very regular, strata inclined to the river near the top, and has a very moderate head of Earth and Rubbish upon it which is easily thrown down the Cliff out of the way. The stone is then easily got at, and will undoubtedly furnish all our steps, pavement, and Ballusters as well as cornice Mouldings of moderate highth, to greater advantage than any other quarry of freestone which I know. The stone is however harder than that of Acquia, and will cost a little more in working, which additional expense is amply concentrated by its superior quality.[5]

These quarries had been unavailable to supply stone to Washington until the Potomac Canal Company opened the Great Falls bypass canal in 1802. After 1802, Seneca sandstone was used in Washington for pavers and balustrades and other uses.[6]

But the quarries of Aquia and Seneca could not provide the fine stone needed for the columns for the House of Representatives and Senate wings of the Capitol.[7] Latrobe found a suitable stone for that purpose on the banks of the Potomac River in Maryland, just above Conrad's Ferry (present-day White's Ferry). He wrote:

> There is on the S. East of the Cotcktin [Catoctin] Mountain a very large extent of country, which abounds in immense Rocks or Marble, or Limestone *Breccia*, that is of a Stone consisting of fragments of ancient Rocks bound together by a calcareous cement, and thus becoming one solid and uniform (homogeneous) Mass of Marble. This Range of Rocks I have traced from James River to the Delaware, but it appears nowhere of a more beautiful kind than on the Potomac. A specimen will be submitted to you as soon as I can get it polished.
>
> The largest Mass of this Kind of Rock is situated on the Maryland side of the Patowmac on land the property of Samuel Clapham Esqr. It overhangs the River, and would furnish without any land carriage *all the Columns of the Capitol of one block each* if required, and of beauty not exceeded in any modern or ancient building. It is impossible for me at this moment to say more, than that it is *practicable* to obtain all our Columns there, but at what price of materials or Labor, must be determined by further investigation.[8]

He added, "The columns 2 feet 8 inches in diameter, 22 feet long, may be easily procured in one block, and conveyed by water from the quarry to the foot of the Capitol Hill."[9]

REPLACING THE COMMISSIONERS

The original three commissioners for the reconstruction of the public buildings (Van Ness, Lee, and Ringgold) served only a little over a year. On April 29, 1816, President Madison signed legislation abolishing their positions and establishing a single commissioner of Washington.[10] Samuel Lane, a War of 1812 veteran, was nominated by President Madison for that position, and his nomination was confirmed by the Senate on April 30, 1816. One of Lane's first actions was to advertise for skilled workers: stone cutters, bricklayers and masons.[11]

WORKERS TO DO THE REBUILDING

Acquiring the needed skilled labor for reconstruction was, as in the first public building campaign, a problem. Skilled labor, especially stone masons, stone cutters, and stone carvers were not available in Washington, and it was necessary to go far afield to obtain the skilled labor required. Advertisements were taken out in the various newspapers for the necessary labor.[12] Agents were dispatched to New York to negotiate with master and journeyman stone cutters. Offers were sent to qualified workers.[13] Samuel Lane also looked overseas. Foreign workers were recruited at the docks as they disembarked.[14]

PAY POLICY

After recruitment, the commissioners had to decide how to pay the workers for their labor: "by the measure" (also called "by the piece"), "by contract," or "by the day." During reconstruction a consensus emerged among the commissioners that the preferred method was to pay "by the day." In the previous federal building campaign, "by the measure" had been the mode preferred by workers, since it rewarded them for extra exertion and ingenuity in accomplishing their work.

Although the commissioners preferred "pay by the day," they were willing to consider payment "by the piece" if it would assist in hiring skilled workmen. For example, the commissioners offered to pay "by the piece" to hire Baltimore stone cutters.[15]

Supervision of the workers was provided by foremen also directly hired by the commissioners—the same type of work organization used in the first and second public building campaigns. But a new form of work organization had begun to emerge in the private sector that could be termed a modern building arrangement. By this arrangement, designs were prepared by a professional architect or engineer, who then advertised for competitive bids from contractors. The contractors worked on a lump-sum fee basis. They hired their labor by the day and were responsible for all aspects of the work, including provision of all building materials.

Other people in Washington were also rebuilding and were also employing workers. It was not unusual for one organization to hire workers from another. In 1815, for example, the army, reconstructing the arsenal on Greenleaf's Point, hired several laborers from the President's House. They received a strong response from the commissioners for this act of labor piracy, and the army responded that it did not approve of hiring workers away from others.[16]

Construction costs for rebuilding the public buildings were greater than for their original construction. Inflation was one factor, and another was the increase in wages, typically greater during rebuilding (table 9.1).

PRICE ESCALATION

The literature of the time also indicates that there had been a substantial increase in prices since the first two building campaigns, resulting in increased construction costs. An editorial of 1816, in discussing the need for a pay increase for government officials, provides comparative prices for the years 1805 and 1815 (table 9.2). The same article indicates large price increases in all aspects of life: "The list might be

TABLE 9.1. Wages paid on President's House, original construction and reconstruction

	1791–1800 WAGES	1815–1817 WAGES
Architect	$1400 per year	$1600 per year
Inspector/Superintendent of the Stone Cutters		$1500 per year
Clerk of the Works		$4.00 per day
Foreman of the Stone Cutters	$2.00 per day	$3.75 per day
Overseer	$0.57 per day*	$2.00 per day
Foreman of the Plasterers		$3.75 per day
Carpenters	$1.20 to $1.63 per day	$1.50 to $1.88 per day
Cabinet Makers	$1.11 to $2.00 per day	$1.12 to $2.50 per day
Stone Cutters	$1.34 to $1.67 per day	$1.50 to $2.75 per day
Bricklayers	$1.47 per day	$2.00 per day
Sawyers	(Usually slaves)	$1.25 to $1.63 per day
Plasterers		$2.00 to $2.50 per day
Laborers	$0.31 per day*	$0.87 to $1.00 per day

Source: Samuel Lane to Lewis Condict, January 4, 1817, in *RDCC*, Letters Sent, 182–183; and various day books submitted by Peter Lenox, Clerk of the Works of the President's House, 1816–1817, in *RDCC*, rolls 19–20.
*Plus provisions.

readily enlarged, for the increase of prices runs through all the articles of life, into the mechanical arts, etc. Among other things, house rent has risen enormously, as well as fuel, both of which are weighty objects of expenditure."[17]

DEMAND FOR PAY INCREASES

The overall increase in prices fueled the workers' demands for pay increases throughout rebuilding. Virtually every group of workers employed by the commissioners appealed their agreed-upon daily wages. This began early with the bricklayers at the President's House. On July 21, 1815, they appealed to the commissioners for a pay increase to $2.50 per day because bricklayers elsewhere were being paid more.[18] The tone of the bricklayers' request was subdued, but the tone of subsequent letters from other building trades, and the commissioners' responses, became more strident as the year progressed. In the autumn of 1815, for example, when the carpenters asked for a pay increase, they threatened that they would leave the works if they did not receive it. Their request was denied by the commis-

TABLE 9.2. Cost of Commodities in Washington, 1805 and 1815

	1805	1815
Labor, per day	$0.80	$1.00
Corn, per bushel	$0.95	$1.00
Butter, per pound	$0.21½	$0.45
Potatoes, per bushel	$0.40	$0.80
Lard, per pound	$0.14½	$0.20
Hams	$0.13½	$0.22

Source: Washington City Weekly Gazette, January 6, 1816, 4.

sioners.[19] The stone cutters at the Capitol made a similar demand for an increase in wages, with a similar threat to leave the works. The commissioners sent a similar, but much stricter, letter to the stone cutters employed at the Capitol, denying the pay increase.[20]

If the skilled workmen had walked off the job, construction would have halted. The commissioners had to weigh the difficulty of replacing these men against the demand for increased wages. Stone cutters were the most difficult to recruit. The commissioners went to great lengths to persuade these men to remain on the job. Laborers, however, were the easiest to recruit and hire. The commissioners sent a terse letter to Peter Lenox, advising him to pay the best laborers at the President's House no more than $1.00 per day and the others proportionally.[21] The laborers knew they could be easily replaced, and when they requested a pay increase, their letter was much more deferential than similar letters sent by the proud stone cutters, carpenters, or bricklayers. When the laborers at the Capitol requested a pay increase on April 15, 1816, it was very much with their hat in their hands.[22] The commissioners also denied this request.

Shortly thereafter the carpenters at the President's House had James Hoban request an increase to $2.00 per day for them.[23] The commissioners did not grant this request but did advance their pay to $1.88 per day. In addition, the commissioners adopted Hoban's suggestion to vary the pay according to the abilities of the workmen.

CONFLICT WITH A FOREMAN

Workers also took other actions to assert their independence. At the Capitol, while demands for higher wages were being made, the carpenters, joiners, and bricklayers engaged in "disorderly or abusive conduct towards Mr. Harbaugh," the foreman.

The commissioners called on Latrobe to immediately discharge the workers who were abusing Harbaugh.[24] Latrobe refused to do so. In a confidential response to the commissioners, Latrobe, who had a low opinion of the abilities of Harbaugh, explained, "Men will not obey a Man whom they think their inferior in talent or knowledge." He added that the workers could have earned more in Baltimore but did not leave the works because of "personal good liking to Mr. [Shadrach] Davis and perhaps to myself." He defended the workers: "In the department of the Centers [i.e., the wooden structure for the construction of masonry vaults and arches], I affirm, that no centers have been more cheaply and better made, and put up."[25] The commissioners did not accept Latrobe's explanation and on March 30, 1816, sent him another letter insisting that the individuals involved in abusing Harbaugh "be forthwith discharged."[26] But by then it was apparent that the three sitting commissioners would soon be replaced, and Latrobe took no action.

THE TEN-HOUR DAY

Besides the issue of daily wages, the workers disputed the length of the working day and the amount of money they were to be paid for the longer days of summer. The traditional workday was from 6:00 a.m. to 6:00 p.m. with two hours for a midday meal, a ten-hour day. But with the longer hours of daylight during the summer, more hours could be spent working. In both the first and the second public building campaigns, many of the skilled workers were paid additional amounts for these extra hours. Beginning in 1816, the commissioners established a standard day of ten hours for all of the workers, except for the unfortunate laborers.[27] The laborers did not like being excluded from the ten-hour day. On May 1, 1816, Benjamin Latrobe wrote to commissioner Lane, stating that the laborers' response to being excluded from the ten-hour day was a work slowdown and stoppage:

> There is at present a total desertion of all the Work by our Laborers, in order to force a regulation respecting their time, similar to that which has lately been made as regarding the mechanics. Formerly, with the exception of the Stone cutters, all our Mechanicks began work at Sunrise & quitted it, at Sun Set; — but by a regulation of the late Commissioners the working hours were limited to 10 hours pr. day. The laborers now demand the same indulgence, and an early decision on their claim is become of very great importance to the progress of the work.[28]

Despite the laborers' slowdown, work progressed, particularly at the President's House, the easier of the two major public buildings to reconstruct. On October 10, 1816, James Hoban requested funds for a "raising" — a traditional celebration — to mark the placement of one of the principal rafters on the building.[29]

In early 1817 commissioner Lane came under substantial pressure from the president and Congress to complete the public buildings. Lane reacted by increasing

TABLE 9.3. Daily wages for stone workers for four different days, 1816–1817

Stone Specialties	JUNE 29, 1816		JULY 3, 1816		APRIL 3, 1817		MAY 5, 1817	
	No.	Daily wage	No.	Daily wage	No.	Daily wage	No.	Daily wage
Cutters in shop	10	$ 2.50	7	$ 2.50	9	$ 2.50		$ 2.50
	1	2.375	1	2.375	3	2.375	5	2.375
	5	2.25	2	2.25	2	2.25	2	2.25
	2	2.00	2	2.00			1	1.75
Setters	2	2.75	2	2.75	2	2.75	1	2.75
Fitters	2	2.50	1	2.50	2	2.50	1	2.50
	1	2.25	1	2.37				
Cutters out	1	2.50	1	2.50				
on the Wall	1	1.50	1	2.37				

Source: Peter Lenox to the commissioners, Day Book, June 29, 1816, in *RDCC*, Letters Received, roll 19, 2342; ibid., July 3, 1816, 2350; ibid., April 13, 1817, 2448; ibid., May 5, 1817, 2464.

the number of hours to be worked each day by the skilled workers. For some of the workers, Lane also increased their pay for the additional hours worked. This brought a reaction from those workers not paid for the additional hours, such as the carpenters at the President's House and the carpenters at the Capitol.[30]

WINTER WORK

Winter work—who was to be employed and for how long—was another labor issue. Commissioner Lane's policy was to discharge the workers who were easy to replace, such as the bricklayers and the laborers, during the winter months and to retain the skilled workers who were difficult to replace but to reduce their wages, in accordance with custom. In addition, Lane varied workers' pay according to merit.[31] This policy was more or less acceptable to the skilled workers.

The policy of varying the pay of skilled workmen based on merit had been begun by the three commissioners, Van Ness, Lee, and Ringgold. On March 4, 1816, they instructed George Blagden to institute payment to stone cutters based on merit.[32] The result was a large disparity in wages paid. Tables 9.3 and 9.4 contain a few examples taken from the day rolls. The merit-based daily wages of stone cutters, for instance, varied from $1.75 to $2.50. Among carpenters, the cabinetmakers, who worked in expensive mahogany on the paneled doors, easily earned more than any of the other carpenters employed on the President's House (table 9.4). The master cabinetmaker, perhaps alone of the skilled workers, was paid from $2.00 to $2.50 per

TABLE 9.4. Daily wages for carpenters for four different days, 1816–1817

CarpenterSpecialties	JUNE 29, 1816		JULY 3, 1816		APRIL 3, 1817		MAY 5, 1817	
	No.	Daily wage	No.	Daily wage	No.	Daily wage	No.	Daily wage
Shop	6	$1.88	6	$1.88	30	$1.50	46	$ 1.63
	8	1.88	6	1.88	5	1.50	1	1.50
Girders	1	1.50	2	1.75	2	1.62	2	1.62
							1	1.50
Jobing	1	1.62	1	1.63				
Cabinet Makers	2	2.00	1	1.12	2	2.50	2	2.50

Source: Peter Lenox to the commissioners, Day Book, June 29, 1816, in *RDCC*, Letters Received, roll 19, 2342; ibid., July 3, 1816, 2350; ibid., April 13, 1817, 2448; ibid., May 5, 1817, 2464.

day. For both the stone cutters and the plasterers, commissioner correspondence indicated that they would consider payment "by the piece"; they were among the few employees paid by the measure.

WORK STOPPAGES

Available records indicate that the workers were much better paid during the reconstruction program than during the original public building programs. Labor problems between workers and commissioners still remained, although they were not as intense as during the first public building campaign. In May 1818 commissioner Lane faced a threat of work stoppage by the stone cutters and stone setters. He responded that he would dismiss them if they did not go to work on the following day.[33] They yielded. Later the same year, there was a work stoppage that lasted thirty days.[34] But despite these few disturbances, work continued.

ASSESSING THE FIRE DAMAGE

The first step in rebuilding the burned public buildings was to assess the damage they had sustained and to stabilize the ruins. Two months after the British attack, George Hadfield was instructed by the Senate to inspect all of the damaged buildings, make a report, and prepare a cost estimate for their restoration. He reported that the entire interior of the north wing of the Capitol that had been constructed of combustible materials had been destroyed, but that some of the masonry external and internal walls, as well as much of the basement, remained uninjured.

The President's House, which had been built mostly of combustible materials, was more severely damaged. In the same document, Hadfield estimated that $692,000 would be required to restore the public buildings.[35]

On October 29, 1814, Thomas Munroe, the superintendent of the city, submitted to Congress an estimate that the cost of rebuilding the public buildings would be $1,215,111:

Report from the Superintendent of the Public Buildings[36]

Rebuild north wing of the Capitol	$457,388.36
Rebuild south wing of the Capitol	329,774.92
Rebuild President's House	334,334.00½
Rebuild public offices	93,613.82
Total	$1,215,111.10½

And on February 13, 1815, Congress approved an appropriation for rebuilding, not to exceed $500,000. The money was to be borrowed at 6 percent interest or less.[37]

HOBAN'S REPORT ON DAMAGE TO THE PRESIDENT'S HOUSE

Without an appropriation, no work had been done on the President's House or the Capitol during the severe winter of 1814–1815. With appropriation in hand, on April 22, 1815, the commissioners asked James Hoban to investigate the damage to the President's House. He reported that the walls had survived the severe weather of the winter, and therefore there was no need for further protection from the elements. He had begun the preliminary arrangements for the reconstruction of the President's House: "The Stone cutters Shed is now ready for operations in that line and a Work shop for the carpenters is progressing, which will be put up between the Presidents house and the Treasury Office."[38]

LATROBE'S REPORT ON DAMAGE TO THE CAPITOL

On March 14, 1815, the new commissioners asked Benjamin Latrobe to come to Washington to interview for a position in the rebuilding the other federal buildings, especially the Capitol. Because of other work, he did not depart Pittsburgh for Washington until April 6, 1815,[39] but he was subsequently hired by the commissioners. His first task was to assess the damage to the Capitol and to recommend interim measures for protecting the work.

THE NORTH WING OF THE CAPITOL

At the time the Capitol was burned, it was two buildings connected by a colonnade. The north (Senate) wing had been mostly finished by 1800, initially with masonry bearing walls, wooden floor joists, and timber girders. The wooden members were susceptible to decay and fire. Latrobe, during the second public building campaign, had replaced many of the decayed timber joists and timber girders with brick vaults impervious to fire. Nevertheless, there was still much within the north wing that could burn. The fire of August 24, 1814, had damaged the Aquia sandstone walls and limited their ability to bear load, especially the weight of the Senate dome. Of particular anxiety to Latrobe was that the walls holding the dome were in danger of failure. He reported: "The injury which the North wing has received is more serious as to the necessity of an early exertion to prevent it becom[ing] more fatal to the whole of what remains. What I have said as to the propriety and even necessity of a temporary roof applies equally to both wings. But the circumstance I particularly consider as deserving immediate attention, is the failure of the Walls on which the arches rest, and against which the thrust of the great Dome of the Senate Chamber acts."[40]

In addition, the fire had damaged the columns and other masonry. Latrobe described that damage to Thomas Jefferson:

In the North wing the beautiful Doric Columns which surrounded the supreme Court room have shared the fate of the Corinthian Columns of the Hall of Representatives and in the Senate Chamber the Marble polished Columns of 14 feet Shafts in one block are burnt to lime and have fallen down. All but the Vaults is ruined. They stand a most magnificent ruin.

The West side containing the Library which was never vaulted burnt very fiercely and by the fall of its heavy timbers great injury has been done to the adjoining Walls and Arches, and I fear that the freestone is so much injured on the outside that part of the outer wall must be taken down.[41]

He reported to Congress:

The North wing of the Capitol was left, after the fire, in a much more ruinous State, than the South wing. The whole of the interior of the West side having been constructed of Timber, and the old Shingle roof still remaining over the greatest part of the Wing, an intensity of heat was produced, which burst the Wall most exposed to it, and being driven by the Wind into the Senate chamber, burnt the Marble Columns to Lime, cracked everything that was of Freestone, and finding vent thro, the windows, and up the private Stairs, damaged the exterior of the Wing very materially. Great efforts were made to destroy the Courtroom, which was built with uncommon solidity, by collecting in it, and setting fire to, the furniture of the adjacent rooms. By this means the Columns were cracked exceedingly, but it still stood, and the Vault was uninjured. It was however very slenderly supported and its condition was dangerous. Of the Senate Chamber, no parts were injured, but such as were of Marble or Freestone. The Vault was entire and required no repair whatever. The great Staircase was much defaced, but might have been reinstated without being taken down.[42]

THE SOUTH WING OF THE CAPITOL

The fire also did tremendous damage to the south (House) wing. Although the brick vaulting installed by Latrobe during the second public building campaign did not burn, there was much in the building that did.[43] The resulting fire was very intense, and it was that intensity that greatly damaged the Aquia freestone. Wrote Latrobe, "The ruin of the Hall of Representatives arises from the effect of fire on the Freestone of which it is built, a stone which cannot bear heat."[44] Virtually all of the Aquia freestone work was lost, as well as the woodwork. "The destruction which has been occasioned by the *burning* of the Capitol, consist in the total loss of all the Work in Freestone and in wood in the house of Representatives, almost without exception: in the loss of all the woodwork in the Clerks office and passages, the destruction of part of the freestone dressings of the Windows of the South and east sides, and the total Loss of the Roof of every part of the building."[45]

The roof of the south wing was built of wood, and when it burned, water got into the building and caused considerable water and frost damage, especially to the plasterwork. To prevent further such damage and to protect the surviving plasterwork in both wings, Latrobe recommended that both wings of the Capitol be covered with a temporary roof.[46]

REMOVING THE DAMAGED MASONRY IN THE HALL OF THE HOUSE OF REPRESENTATIVES

Once the wings had been secured from further weather damage, the next priority was to remove the damaged masonry. In the hall of the House of Representatives, this was no small task, because the hall had been surrounded by twenty-four Corinthian columns, each 26 feet 8 inches high, mounted on a 7-foot-high wall and supporting a 6-foot-high entablature (and above that, the remains of the Halles au Blé–like domed roof). All of this masonry, as well as the wood and glass roof, had been severely damaged by the intense fire. The masonry had to be removed before rebuilding could start. It was very dangerous work, and the workmen were reluctant to engage in it for fear of being crushed by falling masonry. If the damaged columns and entablature were to collapse, more than one hundred tons of masonry would fall, probably killing workmen and likely collapsing the vaulting below. To safely remove the damaged stone, the workmen would need to have a work platform almost four stories tall adjacent to the columns and the entablature. Scaffolding could not be safely used, because the necessary poles supporting the scaffolding might upset a damaged column, causing a cataclysmic collapse of the rest. Latrobe came up with a clever solution to safely remove the fire-damaged masonry. He bought five hundred cords of firewood and had his workers fill half of the hall with the firewood, from the floor up to the architrave level of the entablature. The workmen then used the piled firewood as a platform to work on as well as a means to stabilize the badly damaged columns, wall, and entablature.[47] As the damaged masonry was removed from the top down, some of the firewood was removed, thereby lowering the workers' platform. After the masonry was removed in one half of the hall, the process was repeated for the other half. Masonry removal proceeded quickly. By August 12, 1816, all of the damaged masonry had been removed from the south wing.[48] The firewood was then sold.

NEW DESIGN FOR THE HALL OF THE
HOUSE OF REPRESENTATIVES

The $500,000 appropriation by Congress for rebuilding the public buildings was passed probably with the expectation that they would be restored to their former appearance and condition. This was the course James Hoban pursued with the President's House, except for the unseasoned wood, faulty downspouts, and the heavy slate roofing installed in the first public building campaign and replaced by Latrobe in the second. However, Latrobe had no intention of rebuilding the south wing of the Capitol as it had been before the fire. On April 27, 1815, he submitted a report and a drawing of his proposed new design. He began the report:

> I have already reported to you that the Colonnade of the Hall of Representatives U.S. has been so utterly defaced, and indeed destroyed by the fire, that the whole of it must be taken down, together with the Vault which covers the space between the Columns and the external Wall.
>
> This, when done, will have the whole Area open either to receive a Structure, the exact copy of that which has been ruined, or one of a new Design.[49]

There was no question in Latrobe's mind which was the appropriate choice. A new design would eliminate the problems of the old: "It cannot be denied, that the former Hall of Representatives, notwithstanding the general Admiration which sanctioned its appearance, had numerous defects. It was agreed, that it wanted light and ventilation, and that it appeared to be ill adapted to the purposes of debate, it being difficult to speak, and to hear what was spoken, either within its area, or in the galleries." Without mentioning William Thornton's name, Latrobe wrote that these prefire defects were not the fault of his work but were the result of Thornton's imperfect design: "The design of the Hall of Representatives was indeed never a work of the choice of the Architect who erected it."[50]

In 1817, Latrobe described his new design for the hall of the House of Representatives to commissioner Lane:

The Hall of Representatives in the South wing of the Capitol may be said to consist of two parts, that which is square, and that which is semicircular.

These two portions of the Hall are separated by a range of eight Columns, the object of which is to support the ties necessary to connect the opposite Abuttments of the Vault covering the square part of the house; as well as to connect the Architectural character of the Room from side to side. . . .

The circular part of the room contains the seats of the members. It is to be enclosed first by 4 Committee rooms, and two lobbies. These rooms carry the Galleries at the heighth of 13 feet above the floor of the house. The Wall of the Galleries is of Freestone pannelled towards the house, and is surmounted by an entablature which forms the parapet of the Gallery.

Within these Rooms and Galleries, are 14 Columns which are to carry the Vault of the Room internally, and the Vault covering the Gallery externally.[51]

Latrobe provided Lane a sequence of construction that needed to be undertaken in 1817:

From the above description which will be elucidated by the plan and description in Your possession it is evident; that the following must be the course of the work, in building the Hall.

1   to carry up the square part, which is 26 ft. 8 in. high to the Entablature, and with the entablature 34 feet.
2   to set the 8 Columns and their Entablatu[res.]
3   to turn the arches between the 8 column[s] and the South Wall.
4   To let down the Chains connecting the West and East sides of the Main Vault.
5   To turn the South part of the vault.
6   To turn the interior part of the vault.
7   To carry up the semicircular front of the Committee rooms, and the Committee rooms and Lobbies themselves.

This appears to me to be ample work for the Year 1817, including externally, the elevation of the South front which covers the arches, and as much of the Roof (covering) as they admit of being put on; and also the gallery stairs which are include[d] in the Mass of square work on the East and West.[52]

By the beginning of 1817, construction costs had increased from the original reconstruction estimates. Commissioner Lane explained to Congress that the reasons for this cost increase were that the size of the Senate chamber had been expanded, causing work already done to be taken down; and that damage to the exterior walls of the President's House was found to extend deeper than was previously thought.[53] An additional $100,000 was appropriated to cover this cost increase.[54]

On April 4, 1817, President James Monroe sent specific instructions to commissioner Lane for work on the Capitol for the coming construction year. The Senate dome, specified Monroe, was to be built of brick, while the House dome was to be constructed of wood (both were eventually made of wood).[55] He also informed Lane that a General Swift and a Colonel Bomfort had found that the Potomac marble located above Conrad's Ferry, previously identified by Latrobe, was structurally adequate and that columns from that source were to be used in both the Senate and the House of Representatives chambers. Mason Robert Leckie was assigned to supervise the quarry operations, and Leckie would deliver the blocks to John Harnett to undertake the column carving. Monroe's goal was "the placing the Capitol in a state to accommodate Congress at the next session."[56]

DIFFICULTIES BETWEEN LATROBE AND COMMISSIONER LANE

Despite Monroe's direction, the rebuilding of the Capitol was not finished by the 1817–1818 congressional session. This exacerbated Latrobe's numerous difficulties with commissioner Lane, and the ensuing criticism was an influence on Latrobe's decision to resign his position. But it was probably Lane's letter of October 31, 1817, that finalized that decision. In that letter Lane informed Latrobe that he had reassigned Latrobe's clerk of the works at the Capitol, Shadrach Davis, and that he was assigning Peter Lenox, Hoban's clerk of the works at the President's House, to replace Davis. Wrote Lane, "The Clerk of the Works is an agent responsible to me and to me alone. Your sphere is a different one."[57] Reacting to this move, Latrobe wrote Baltimore architect and colleague Jacob Small, then clerk of the works for Latrobe's Baltimore Exchange, "Colonel Lane has dared to discharge Mr. Davis at three days notice. He offers him the President's house where there is only a month's work. All this is abominable."[58]

Latrobe's problems with Lane and the uncompleted Capitol building left him in a jeopardized position. In a letter to architect Charles Bulfinch, Boston merchant William Lee, for whom Latrobe had designed a house in the new city,[59] described Latrobe's situation:

[I have] good reasons for thinking that the President of the U. States will very soon after his return here displace Mr. Latrobe and the Commissioner charged with the execution of public buildings. He threatened to do it before his departure if they did not proceed with more harmony than they have done and with greater expedition with the Capitol. So far from this, things have gone on worse, and nothing has been done. I am sorry, for Latrobe, who is an amiable man, pos-

US Capitol, west front, as completed by Charles Bulfinch, architect of the capitol, in 1823. Courtesy of the Library of Congress.

sesses genius and has a large family, but in addition to the President's not being satisfied with him there is an unaccountable and I think unjust prejudice against him by many members of the Government, Senate and Congress.[60]

DEPARTURE OF BENJAMIN LATROBE AND ARRIVAL OF CHARLES BULFINCH

Latrobe resigned on November 20, 1817,[61] and was replaced as architect of the Capitol by Charles Bulfinch, appointed by President Monroe in January 1818.

On January 28, 1818, a motion was put forward in Congress to have the president of the United States report on the expenditures on the public buildings and to annually report on their progress to Congress. Bulfinch prepared the first report required by Congress, dated February 5, 1818. The Capitol had not yet been completed, and Bulfinch estimated that an additional $177,803.46 would be required to complete it.[62] An additional $3,634 was requested to build a temporary building to provide twelve committee meeting rooms.[63] An appropriation of $80,000 was passed for completing the wings of the Capitol, an additional $100,000 for beginning the foundation of the center section of the Capitol, and $3,634 for the temporary building for committee rooms. An additional $30,000 was appropriated for furnishing the House of Representatives chamber and an additional $20,000 for furnishing the Senate chamber and committee rooms.[64]

With the departure of Latrobe and the arrival of Bulfinch, the atmosphere between architect and Congress markedly improved. The battles of the first and second building campaigns between Thornton and the architects hired to work on the Capitol receded. Perhaps after twenty-five years of battle, all sides were exhausted. Construction proceeded without most of the criticism that had plagued Latrobe's years, and by 1824 the copper-clad wooden dome of the Capitol was completed.

Early in the first building campaign, architect Stephen Hallet predicted that the Capitol, as designed by William Thornton, would take thirty years to build. It was a remarkably accurate prediction. Ironically, it was the destruction of the Capitol and the President's House in 1814 that hastened their completion. During their extended construction period, extensive conflict had reigned between those appointed to oversee the work and those hired to conduct the work.

10

Later Transportation Improvements
Canals and Aqueducts

*Unquestionably, the Potomac Aqueduct is one of the most
important pieces of engineering, if not the most important, in
the entire [Chesapeake and Ohio Canal] construction. Measured
between the two river banks, it is 1,500 feet long.*

FRANZ ANTON RITTER VON GERSTNER, 1842

For the new federal city to become a major urban center, it was necessary
for the agricultural produce of the Potomac Valley to flow into the new
city, including Georgetown and Alexandria. The Potomac Canal, which
had been intended to fulfill this function, never lived up to expectations.
It was a river navigation with short bypass canals around major obstacles
like Great Falls and Little Falls. Because the Potomac Canal was a river
navigation, its traffic was heavily dependent on fluctuations in water levels. The Potomac River could be used for transport only during several
months of the year.

THE CHESAPEAKE AND OHIO CANAL

What was needed to enable the new federal city to develop as a major
commercial center was a still-water canal adjacent to the river. In 1828
work began on such a canal, the Chesapeake and Ohio Canal, successor
to the Potomac Canal. Planned on the Maryland side of the river, it was
intended to connect Georgetown with Pittsburgh. To avoid the cost of
expensive property acquisition and extensive excavation in Georgetown,
the Chesapeake and Ohio Canal Company planned the lower terminus of
the canal to be at Little Falls, approximately six miles above Georgetown.
On July 4, 1828, groundbreaking began near the US Magazine at Little
Falls three-tenths of a mile above what became Lock 6 on the Chesa-

peake and Ohio Canal. The termination of the new canal six miles from Georgetown brought protests from the merchants of Georgetown, who strenuously argued that the commerce of the new canal would bypass their town. So a compromise was reached, extending the new canal into the middle of Georgetown.

The excavation of the canal through Georgetown generated extensive spoil, for which the company planned to build a mole at the mouth of Rock Creek: "The tide lock provided in the mole which is to sustain the water of Rock Creek at this level, will connect its navigation with the wharves of Georgetown, above the basin, and those of Washington and Alexandria, below; while a breadth and structure are given to the mole, by the disposition of the surplus excavation of the Georgetown sections."[1] The mole was 840 feet in length and so designed "that vessels should lie on the river side of the mole, or dike, which forms the barrier between the basin and the river, and that the boats should lie opposite the vessels in the basin, and discharge their cargoes across the mole."[2]

Lieutenant Colonel John J. Abert and Lieutenant Colonel James Kearney give a more detailed description of the mole and the terminus of the canal:

This basin [at Georgetown] is formed by a dam thrown across the mouth of the creek, forming an extensive quay or landing place, one of its faces being on Rock Creek, and the other on the Potomac river. The length of the quay on the Potomac face is one thousand and eighty feet; two hundred feet of which is occupied by a tumbling dam, for the delivery of the surplus water of this creek, and thirty-eight feet occupied by the tide-lock leaving eight hundred and forty feet front on the Potomac river. Piles, each one foot in diameter, were driven throughout the whole extent of the river front, touching each other, and then at every three feet of the interior, to a distance of twelve feet, until they refused a pile driver of eleven hundred pounds. The whole of these piles were then connected by heavy timbers, bolted to the head of each pile, and this framework was then united by a course of hewn timbers, fitting close to each other, and five inches thick, and well secured to the frames and piles. On the front of this pile work there is a well laid dry wall, twelve feet thick, and seven feet high, including the coping. Strong and frequent ties of timber, firmly connected with the pile work, are extended under the soil of the quay. . . .

The width of the quay is one hundred and sixty feet, except at the city end, where it narrows to eighty feet. Sixty feet in width of the centre of this quay is intended for warehouses and stores, and the rest of the space is to be left open for streets and landing places.

A bridge is connected over the head of the tumbling dam, connecting the Georgetown part with the city part of the quay. The bridge is of timber on piles—a simple, but substantial structure.[3]

The Chesapeake and Ohio Canal had other notable construction features. Overcoming the descent in elevation between Great Falls and Little Falls was a major challenge. Wrote army engineer William G. McNeill:

> In the words of the late President of the company [Charles Mercer] . . . it may emphatically be remarked, that "on no canal in America, and very few, if any, in the world, will there be found, and certainly on no part of the Chesapeake and Ohio canal do there remain to be encountered, obstacles more appalling than have been overcome." True it is that he refers more especially to that particular section which embraces the descent between the Great and Little Falls of the Potomac; "a compass of eleven miles along precipices, bounding a river which has borne on its bosom ice and snow, elevated for several miles 30 feet above its ordinary height." Yet the numerous locks and culverts, the extent of rock excavation, and of outer walling which has been found necessary to the support and protection of the towpath, two expensive acqueducts, etc. etc., constitute difficulties characterizing the whole section under consideration, which, from description alone can scarcely be appreciated.[4]

To overcome these difficulties, chief engineer Benjamin Wright sought to avoid the extensive rock excavations near Great Falls that had plagued and delayed the completion of the Potomac Canal by utilizing the old riverbed, now called Widewater, immediately downriver of Great Falls Tavern. At the upriver end of the old channel, between Locks 16 (milepost 13.6) and 17 (milepost 14), a massive dry-laid stone wall needed to be built, carrying the towpath approximately fifty feet above the Potomac River.[5] Today, most visitors to the Chesapeake and Ohio Canal think this is a natural feature and not manmade. It is usually referred to as "Mary's Wall," a name of unknown origin that is not historical. Army engineers Abert and Kearney describe the construction of "Mary's Wall":

> From [lock] No. 16 to lock No. 17, the distance is about two hundred and forty yards. Here, again, the embankment had to encroach upon the river, from the action of which it is protected by a heavy slope wall of stone, continued up to the top of the embankment, and in places it is as high as fifty six feet. This wall had partially yielded to the great pressure to which it is exposed, but was immediately repaired and enlarged, and is also aided by strong and well-built buttresses.
> In these cases, where the wall is stated as being so high, rows of plank piling were driven on the inside and the canal filled with earth, and secured by a puddle work [i.e., tamped clay]. Where the deep places widen into large ponds, as in the vicinity of Bear island the inside of the embankment is judiciously and carefully secured by plank piling, and by a puddle wall.[6]

Also of substantial construction on the canal is Seneca Aqueduct (Aqueduct Number 1), at milepost 22.8. Seneca Aqueduct is 114 feet between abutments. The two intermediate piers are 7 feet wide and support three arches 40 feet on center (now flood-damaged). Army engineers Abert and Kearney also described the Seneca Aqueduct, then under construction:

Over this river [Seneca Creek] an aqueduct is constructing. The abutments and piers which rest upon a rock foundation, are completed, the centering is up, and the arches partly turned. The masonry is to be entirely of the red sandstone of Seneca.

The length of the aqueduct, from the face of one abutment to the face of the other, is one hundred and fourteen feet. It will consist of two piers and three arches. The span of each arch is thirty-three feet, and the thickness of each pier seven feet. The sheeting, as well as the ringstone, are to be cut to the proper angle, and the whole of the arch work is to be laid in cement, and grouted carefully over the extrados. The front or facing ranges of the piers and abutments are laid in cement or hydraulic mortar, and the interior of the masonry carefully grouted with cement at every range. No stretcher is admitted with a bed less than its face, and no face is less than a foot wide, and the length of each stretcher must be not less than four feet. No header is admitted that does not extend into the masonry at least four feet, and with a face one foot high and two feet long. The spandrels are to be built up with rubble stone, and grouted with cement at every range.

The stone, before being used, are subjected to a rigid inspection, and if an improper piece finds its way into the work, it is ordered out as soon as discovered. . . .

We believe that this structure will be both beautiful and enduring.[7]

Larger is the Monocacy Aqueduct (Aqueduct 2) located at milepost 42.2 near the town of Dickerson, Maryland. The total length of that structure is 516 feet, and it has seven 54-foot span arches.[8] The largest of eleven stone aqueducts on the canal, it is one of the two finest features of the Chesapeake and Ohio Canal (Paw Paw Tunnel being the other). Army engineers Abert and Kearney described the Monocacy Aqueduct, then under construction:

The next object of our examination was the aqueduct over the Great Monocacy. This structure is 438 feet long from the face of one abutment to the face of the other, and the masonry of the abutments and wing-walls extends ninety-six feet further. The whole work will consist of two abutments, six piers, and seven arches. The masonry of the abutments and piers rest upon the solid rock which forms the bed of the river, and which have been previously cleaned and prepared for the purpose.

The arches are to be fifty-four feet in the span, with a rise of nine feet. The two arches, which rest against the abutments, are conducted, within the abutments, by what is called a blind arch, down to the rock foundation. The centering of one arch is up, the masonry partly laid, and preparations were in activity for erecting other centres.

The piers and abutments are thirty-three feet four inches long, exclusive of the pilasters. The piers are ten feet wide above the water table, and fourteen feet wide, and thirty-eight feet long at the foundations, which last dimensions are preserved up to within one foot of the low water surface.[9]

Both aqueducts were designed in 1828–1829 by chief engineer Benjamin Wright as part of the eastern portion of the new canal to reach Cumberland, Maryland, approximately 184 miles from Georgetown. The middle and western sections that would have connected the canal with Pittsburgh were never built.

THE ALEXANDRIA CANAL AND THE POTOMAC AQUEDUCT

Groundbreaking on the Chesapeake and Ohio Canal added new impetus to building a canal that would extend to Alexandria. The Potomac Aqueduct and the Alexandria Canal were planned to extend the Potomac Canal (later replaced by the Chesapeake and Ohio Canal) down the Potomac River from Georgetown to Alexandria, to provide Alexandria with access to the trade from the Potomac valley.[10] This project was first authorized by Congress in 1812,[11] but work was never begun at that time, owing to the onset of the War of 1812, and the authorization expired. Following the war, the feasibility of the new canal was studied by engineer James Geddes. The key engineering problem for Geddes was where the Alexandria Canal should cross the Potomac River. The crucial issue was the depth of the river at the proposed water crossing. In a time when cofferdam dam construction for piers was rare and expensive, and water more than ten feet deep was considered deep water, it was imperative that the crossing should be as shallow and as narrow as possible. Crossing at Chain Bridge, where the river was narrow, was not feasible because of the rocky terrain on the Virginia shore. Crossing at Three Sisters, above Georgetown, had been proposed, but Geddes rejected that location owing to the deep water there. Instead, he recommended that the canal cross the Potomac at Mason's Ferry at Georgetown.[12]

In 1830 Congress again authorized the incorporation of the Alexandria Canal Company,[13] paving the way for the construction of the canal and the Potomac River crossing, called the Potomac Aqueduct. Congress appropriated $400,000 for construction.[14] Groundbreaking for the new canal and aqueduct was held July 4, 1831.[15] The Chesapeake and Ohio Canal Company agreed to build the northern abutment for the needed aqueduct. In 1830 Benjamin Wright, chief engineer of the Chesa-

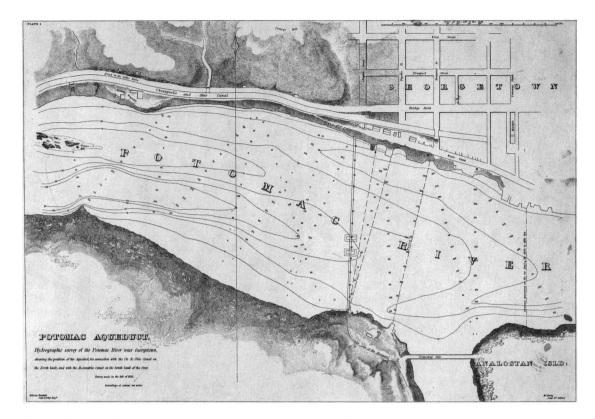

Site location plan of the Potomac Aqueduct. Drawing by Captain William Turnbull and Lieutenant M. C. Ewing, in U.S. House, *Drawings Accompanying the Report of Turnbull*, plate 1. Georgetown and the Chesapeake and Ohio Canal are located in the upper portion of the drawing, Virginia in the lower. The almost vertical, double line (*center right*) is the oblique crossing selected by Benjamin Wright and Nathan Roberts. The dotted lines to the right of this line indicate the alternate locations proposed by Captain William Turnbull but not selected. The aqueduct was built at the original site shown. Author's collection.

peake and Ohio Canal, was directed to supervise the survey of a route of the Alexandria Canal and to take soundings across the Potomac River where the aqueduct was to be built.[16]

SITE SELECTION FOR THE POTOMAC AQUEDUCT

In 1829 engineers Benjamin Wright and his assistant, Nathan Roberts, selected a site for the aqueduct, an oblique river crossing. Army engineer William Turnbull, in his official report on the aqueduct, criticized Wright's selection, pointing out that the oblique siting increased the length and cost of the aqueduct, decreased the section available to pass flood waters and ice, and complicated construction. Turnbull and Alexandria Canal Company engineer W. M. C. Fairfax proposed several alter-

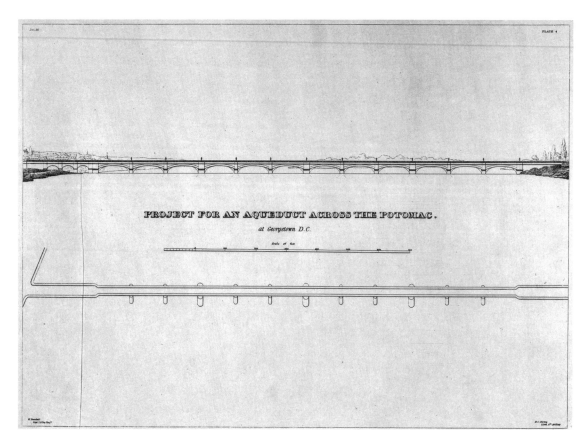

Elevation and plan of the proposed masonry aqueduct. Drawing by Captain William Turnbull and Lieutenant M. C. Ewing, in U.S. House, *Drawings Accompanying the Report of Turnbull*, plate 4. The Potomac Aqueduct was originally designed as a traditional masonry arch structure of twelve spans. Because of a shortage of funds, the directors of the Alexandria Canal Company decided on a wooden superstructure instead of the masonry superstructure shown here. The three spans on the Virginia side (*right*) were deleted in favor of a causeway. Georgetown is shown on the left. Author's collection.

natives that would have been shorter and also would have brought the aqueduct at right angles to the river flow. These alternative sites were not accepted, and the aqueduct was constructed on the original site chosen by Wright and Roberts.[17] The total length of the aqueduct was approximately sixteen hundred feet.

SURVEYING RIVERBED CONDITIONS AT THE CROSSING

Turnbull and Fairfax were also critical of Wright and Roberts's attempts to survey the depth of mud in the Potomac River down to bedrock. Wright and Roberts had surveyed these underwater conditions by using a 50-foot-long iron rod to probe the depth of the river and the thickness of the mud above bedrock. Unfortunately, the iron rod was almost immediately lost.[18] The loss of the iron rod led Wright and Roberts to guess that bedrock would be found at 15 feet. But Turnbull and Fairfax later found that bedrock was much deeper, on average 28 feet below the ordinary high-water mark.[19] This meant that the cofferdams would have to be deeper, much more substantial, and more expensive than originally thought.

TURNBULL'S RIVERBED INVESTIGATIONS

Turnbull, who later took over the responsibility for building the aqueduct, had to undertake his own riverbed investigations. Instead of using iron-rod probes, he used borings done through the use of a square box he had constructed. The box, with interior dimensions of 8 by 8 inches, was constructed of 3-inch-thick heart pine banded by flat iron bars and was 36 feet long. The end of the box that was to be driven into the mud and gravel was shod with steel-tipped flat iron shoes. The box was driven into the bed of the river by a pile driver mounted on a scow, as far down as it would go without damaging the wooden box. The box was then withdrawn and emptied by an auger. At each location Turnbull could thus measure the distance down to bedrock as well as the depth of the mud and gravel above the bedrock.[20] Through these means, Turnbull was able to draw the profile of the riverbed.

DESIGNING THE PIERS AND SUPERSTRUCTURE
OF THE POTOMAC AQUEDUCT

Once the profile of the riverbed had been ascertained, the engineers could begin designing the piers and structure of the aqueduct. Initially they designed a masonry aqueduct of twelve arches supported by eleven piers and two abutments. The masonry arches were to be of 100-foot spans with a 25-foot rise. The piers were to be of two types: every third pier (three total) was to be an abutment pier, 21 feet thick at the springing of the arch; the other support piers (eight total) were to be 12 feet thick.[21]

This initial plan was modified by the Alexandria Canal Company Board of Directors in several ways. First, it was decided that the superstructure would not be constructed of masonry arches. The company, short on money, substituted a less expensive wood-framed superstructure. Second, a 350-foot causeway replaced the three arches on the Virginia side of the aqueduct. In addition, the distance between piers for the masonry arches was increased from 100 feet to 105 feet; the abutment piers were decreased from three to two and the support piers decreased from eight to six. The thicknesses of the piers remained as originally planned. The German railroad and canal observer Von Gerstner described the aqueduct's construction:

> Measured between the two river banks, it is 1,500 feet long. From one abutment to the other—the more southerly is displaced 350 feet out into the stream—the length is 1,065 feet. The aqueduct has nine spans with widths of 109.1 feet in

the clear. Of the eight piers, two have a top width of 16 feet, the others 7-1/8 feet. They are grounded on a solid rock foundation, and their height up to 1 foot beneath the bottom of the canal totals 50 to 62 feet from the foundation. They rise 29 feet above the normal water level at flood stage. The river water varies in depth from 13 to 25 feet, and mud, sand, and gravel from zero to 18 feet. Nonetheless, the catchment dam for each of the first two piers cost only $14,500. The superstructure of the aqueduct is built of wood according to Long's design, and the channel is 18 feet wide and 6 feet deep. Construction of the Potomac Aqueduct was commenced in 1834 [actually 1831]. Its supervision has been entrusted to Turnbull, a major of engineers.[22]

BUILDING COFFERDAMS AND INTERMEDIATE BRIDGE PIERS

The pier construction of the Potomac Aqueduct of the Alexandria Canal had a unique role in the history of American construction. Captain William Turnbull (later promoted to major) wrote, "Experience in founding upon rock, at so great a depth, is very limited in this country, there being but one example, viz. the bridge over the Schuylkill at Philadelphia."[23]

Turnbull was referring to the Market Street Bridge, where building the bridge's piers posed a greater challenge than building the superstructure. Up until 1833, American engineers usually ignored the problem of building cofferdams and in-river pier construction by using large-span wooden structures, such as the Colossus of Philadelphia,[24] or through the use of suspension bridges. The beginning of railroad construction in the 1830s meant that wider rivers would have to be spanned, greater loads carried, and intermediate bridge piers constructed.

Wrote Turnbull on the construction difficulties of building cofferdams for intermediate bridge piers: "No descriptive memoir or drawings of this work (i.e., the Market Street Bridge) ever having been published, nor of the London bridges, (the deepest foundation perhaps in Europe,) the engineers, therefore, had to proceed with the greatest caution. . . . The cofferdams for the construction of bridges of Neuilly and Orleans, designed by the distinguished Peronnet, were selected as models."[25]

There were English-language treatises on cofferdam and pier construction, such as Charles Labelye's account of Westminster Bridge and George Semple's *Treatise*,[26] but these were probably not available to Turnbull. The then newly published American Edition of the *Edinburgh Encyclopedia* (1832) did have plans of deepwater cofferdams, such as the cofferdams for the Neuilly Bridge in Paris, and Turnbull probably was aware of these drawings.[27]

POTOMAC AQUEDUCT CONSTRUCTION BEGINS

By January 1833, the Alexandria Canal Company began to make plans to begin work in the 1833 construction season, on March 1. On January 29, 1833, the company ad-

Schuylkill Bridge, High Street, Philadelphia, by William Russell Birch, 1805. The Permanent Bridge, or Schuylkill Bridge, was built by Timothy Palmer (1751–1821), a well-known wooden bridge builder. It was the first bridge in the United States to have a pier constructed in "deep water," that is, ten feet or more. The deepwater pier (*left*) was designed by English engineer William Weston, originally brought to the United States by financier-speculator Robert Morris to build the Schuylkill and Susquehanna Canal. As shown with the outside sheathing removed, the structure was a variation of the wooden arch–queen-post structure developed as the Burr Truss. Its overall length was 1,300 feet; the center span was 195 feet with a 12-foot rise, and the side spans were 150 feet each. As shown in the smaller inset drawing, it was covered to protect the wooden structural members and extend the bridge's life. Author's collection.

vertised for the construction of the aqueduct. Several offers were received, varying from $99,093 to $247,909. One proposal came from Dr. John Martineau and Stewart. Martineau had a considerable degree of engineering authority, having served as a director of the Chesapeake and Ohio Canal, an engineering position second only to that of the canal's chief engineer, Benjamin Wright.

CIRCULAR COFFERDAMS

Martineau proposed an innovative way to build the piers, by the use of circular cofferdams. The cofferdams would be eighty feet in diameter and would be constructed of two circular rims, one resting on top of the other. These rims were to be of wooden construction, built from 12-by-14-inch boards mounted vertically in 10-foot sections and held together with iron fittings. In essence, Martineau's circular cofferdam was a single row of vertically mounted beams without cross-bracing of any kind. Further, Martineau made no provision to seal this structure with clay puddling or other material.

Turnbull and other engineers objected that Martineau's cofferdam design was inadequate. But because of Martineau's engineering reputation, the company's Board of Directors decided to award him the contract. On June 29, 1833, the company signed a contract with Dr. Martineau to construct the aqueduct's piers and the south (Virginia) abutment. The first pier to be constructed was Pier Number 1, adjacent to the Virginia shoreline.

Martineau delayed proceeding with the contract. It was not until September 2, 1833, that the work for the circular cofferdam was let. By September 26, 1833, the circular frame for the cofferdam was towed into position over the site of the future Pier Number 1. Bad weather intervened. The cofferdam was not weighted down and did not sink until October 2, 1833. Only one pile driver was employed, and not until November 16, 1833, had all the piles been driven, securing the circular cofferdam in place.[28]

The next step was to pump the water out of the cofferdam so that construction, mucking out the mud to bedrock, could proceed. The contractor constructed a platform on piles adjacent to and downstream from the circular cofferdam on which to mount a twenty-horsepower steam engine that would power eight pumps, capable of raising 500 cubic feet of water (approximately 3740 gallons) per minute. On December 13, 1833, the contractor began pumping water. After an hour of work, the water level within the cofferdam had not fallen but had risen eight and a half feet—equal to the rise of the tide outside the cofferdam. It became apparent that without puddling, the circular cofferdam could never be emptied of water.[29]

At this point winter intervened. On December 21, 1833, the circular cofferdam was crushed by a flood accompanied by ice. The contractor showed no desire to repair his work, and on January 4, 1834, the canal company declared the contract abandoned and directed Captain Turnbull to vigorously press forward on the work.

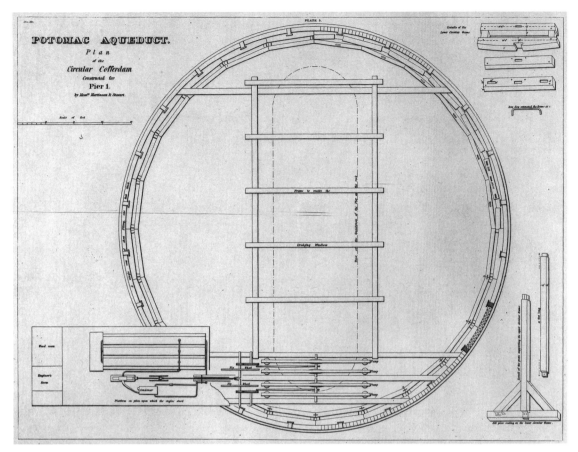

The circular cofferdam. Drawing by Captain William Turnbull and Lieutenant M. C. Ewing, in U.S. House, *Drawings Accompanying the Report of Turnbull*, plate 5. In their bid for the designed Potomac Aqueduct, Martineau and Stewart proposed the use of a circular cofferdam for pier construction. Over the objections of Turnbull and other engineers, the Alexandria Canal Company accepted the proposal, and a circular cofferdam was constructed for Pier Number 1. However, it could not keep the water out. It was destroyed by an ice storm in December 1833. Author's collection.

TURNBULL'S PLAN FOR COFFERDAM CONSTRUCTION

Turnbull's first step was to contract for the necessary equipment to build the Potomac Aqueduct's piers. His first contract was for two steam engines to power the pumps needed to empty the cofferdam.[30] He also contracted for the building of two scows upon which to mount the steam engines.

It was Turnbull's intention to build two piers a year, requiring a total construction time of four years for all eight piers (not including construction time needed for the two abutments). For this purpose, Turnbull needed more pile drivers. He contracted for the construction of three pile drivers: two heavy-duty pile drivers to drive the

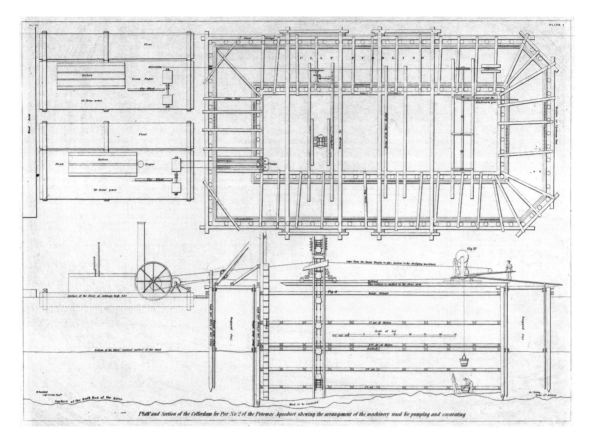

The box-shaped cofferdam for Pier Number 2. Drawing by Captain William Turnbull and Lieutenant M. C. Ewing, in U.S. House, *Drawings Accompanying the Report of Turnbull*, plate 7. In January 1834 Captain William Turnbull took over the construction of the Potomac Aqueduct. He designed a rectilinear cofferdam consisting of a box within a box. Between the outer box and the inner box, Turnbull inserted puddling (i.e., tamped clay) as a water sealant. Shown in this plan and section are the scow-mounted steam engines used to power the pumps and the excavating machine. Also shown in the section drawing are the outer piles, which were not driven to bedrock. The decision not to drive them to bedrock later proved a problem. Author's collection.

oak piles and one light-duty pile driver to drive the sheeting. The heavy-duty pile drivers were to be worked by horses, and the light-duty pile driver by a tread wheel. He also repaired Martineau's and Stewart's pile driver, worked by a crank. So he had a total of four pile drivers.

Turnbull also contracted for an excavating machine to be used to dig the mud out of the bottom of the cofferdam and transport it to the top for transfer to waiting scows.[31] In addition, Turnbull contracted for sixteen wooden pumps, each one 38 feet long. They were made of eight white pine staves, 3 inches thick. Inside, the "barrel" was 18 inches in diameter. These wooden staves were banded together by iron bands and their joints sealed by narrow slips of cotton cloth coated with white lead.

On March 4, 1834, Turnbull's crew had pulled the first pile of what remained of Martineau's and Stewart's circular cofferdam. By March 26 the entire structure had been removed.

Constructing the Cofferdam for Pier Number 2   Turnbull had decided to begin his work at the site of the future Pier Number 2, adjacent to and north of Pier Number 1, where Martineau and his engineer colleague Stewart had worked. At this location bedrock could be found at a depth of 18 feet of water and 17 feet of mud.

Turnbull's plan for a cofferdam was a box within a box. In plan view, this was a parallelogram within a parallelogram. The outer parallelogram was approximately 116 feet long by 61 feet wide, as measured from the outside surfaces. The inner parallelogram was approximately 84 feet long by 29 feet wide. The space between the inner and outer parallelograms was approximately 15 feet on all sides and was where the clay puddling would be tamped into place. It was this clay puddling, long used as a sealer on canal prisms, that Turnbull intended to use to keep water out of the cofferdam.

The inner wall of the cofferdam was formed of pilings of oak that were 40 feet long and 16 inches in diameter. They were spaced at four feet on center around the perimeter of the inner wall. These pilings were iron-shod and were driven to bedrock using a ram weighing 1,700 pounds dropped 40 feet. The pilings were tied together by means of stringers of pine, one foot square, with each piling bolted to the stringer.

The outer wall of the cofferdam was also formed of pilings of oak. These were 36 feet long but not shod with iron and not driven to bedrock. Not driving these outer pilings to bedrock was later found to be an error, since the pressure of the water forced sand and water to enter the cofferdam under these pilings. The outer pilings were secured to stringers in the same way as for the inner wall.[32]

Scaffolding was installed on top of the oak pilings, to hold the pile drivers for the sheeting piles. The sheeting piles were then added. The inner box's sheeting piles were 40 feet long and 6 inches thick, and the outer sheeting piles were 36 feet long. They were formed, using bolts, into 16-foot-long panels, called montants, and driven into the mud with two pile drivers with a 1,300 pound ram dropped 40 feet.

Turnbull provided some data on driving these sheet piles. The tread-wheel pile driver was found to be the most effective. The crank-powered pile driver required eight men and a superintendent and was able to drop the ram from the top of the planes, 40 feet above, every seven and a half minutes. The tread-wheel-powered pile driver required only six men and a superintendent and made a blow from the same height and at the same weight (1,300 pounds) every minute and fifteen seconds. The horse-powered pile drivers were also able to deliver a blow every minute and fifteen seconds.

Next, Turnbull tied the outer parallelogram to the inner one with the use of ties of 11-inch-square pine members spaced every 12 feet. But when the clay puddling was added to the space between the inner and outer boxes, these ties were found insufficient. Turnbull had to add ties for every other oak pile and had to strengthen the lateral stability of the structure through other means.

With the inner and outer structures secure, and with the clay puddling in place in between, the next step was to begin the pumps to empty the cofferdam of water so that excavation could commence. Turnbull began the pumps on September 2, 1834, with less than satisfactory results. Ropes stretched and straps broke. Other things went wrong. Turnbull wrote, "So frequently were accidents happening to the machinery and pump gearing that the time of pumping in each day rarely exceeded . . . fifty minutes."[33]

Slowly Turnbull and his men began to improve the efficiency and reliability of the pumps and the steam engines. The water began to be lowered in the cofferdam. But then it was discovered that the cofferdam structure had to be again strengthened against the outside water pressure. Finally, by mid-September 1834, the cofferdam for Pier Number 2 had been entirely emptied of water.

Trying the Excavating Machine   By October 2, 1834, Turnbull was ready to try the excavating machine that he had constructed. The machine, an endless series of buckets driven by a steam engine in a scow alongside the cofferdam, worked well under no load. But it worked only with great difficulty when subjected to the resistance of the mud to be excavated and the weight of the mud to be carried to the top of the cofferdam. Furthermore, to be used as an excavator, it had to be lifted over the shores installed in the cofferdam, a difficult and lengthy process. Turnbull abandoned the idea of using the machine as an excavator and instead used it as an elevator for mud hand-dug by laborers. A steam-driven windlass, capable of hauling four buckets of 4.6 cubic feet each, was also used to hoist mud to the top of the cofferdam, as was the horse-driven pile driver, now converted into a mud-hoisting machine.

By October 22, 1834, laborers had excavated six feet of mud out of the bottom of the cofferdam. But then it was discovered that several oak piles on the south side of the cofferdam had failed and that that failure had increased pressure on the cross braces at the end, causing them to crack. While Turnbull was moving to shore up this failure of the cofferdam, a major leak was discovered in the northeast corner of the cofferdam. The pressure of water outside the cofferdam forced the mud under the foot of the outside or "montant" piles that had not been driven to bedrock. The clay puddling soon stopped the leak. After this leak, Turnbull had his men operate the pumps to eliminate the water that had entered the cofferdam. The crank wheel of the steam engine almost immediately broke, halting the pumping. It was soon replaced.

More troubles plagued construction of Pier Number 2. On November 3, 1834, a large leak developed, filling the cofferdam with water. Sand accumulated around the pump in the cofferdam, resulting in additional strain on the rope and engine, which caused the crankshaft to break. The engine was repaired, and excavation within the cofferdam continued. On November 15, 1834, in the midst of an early winter storm, the workers refused to go to work; they stayed off work for several days. The storm caused several of the construction scows to break their moorings and be swept to shore. Leakage continued in the cofferdam, and pumping machinery continued to fail, probably because of the extremely cold weather. Despite these setbacks, Turnbull was able to reestablish the stability of the cofferdam and was able to reach bedrock through the mud. Fifteen cubic feet of masonry was set in hydraulic cement by the beginning of January.[34] By that time, January 4, 1835, Turnbull had to close construction operations for the winter and secure the cofferdam and equipment.[35]

Construction in 1835 Instead of being discouraged by the setbacks of 1834, Turnbull was encouraged that he could prevail. Gradually he and his men had been improving the efficiency and reliability of their equipment. They had learned lessons that would allow more efficient construction, sealing, and dewatering of cofferdams for the remaining piers. Through trial and error they had also learned how to construct a cofferdam that would resist the water pressures that they had experienced.

The 1835 construction season started late, in April, because there was a delay in securing adequate funds. The first work needed, beginning on April 22, 1835, was to strip the winter covering off of cofferdam number two. The pumps were then put back in place, as were the steam engines and the windlasses (for hoisting out a small amount of mud in the cofferdam).

Next, a system was needed for unloading stones from the scows: stones that weighed from three to four tons each had to be lifted to the top of the cofferdam and then lowered into place in the pier under construction. A Lieutenant Bartlett devised such a system for Turnbull. Two railways, consisting of 12-inch-square members, were placed parallel to each other and securely fastened to the oak piles. On the inner surface of the rails, flat irons were fastened. Carriages were built on top of these. The end of the carriages extended 10 feet over the edge and was reinforced with braces of iron. On top of the carriages was mounted a derrick, and on the derrick a winch. The derrick was mounted on wheels, allowing travel across the top of the cofferdam. The derricks were operated by four men and a boy. This device allowed the derrick to travel over to the edge of the cofferdam, lower a rope to the waiting scow, have a stone attached, winch the stone up to the top of the derrick, travel back to the appropriate position, and lower the stone to the construction work inside the cofferdam. These two railways permitted the two derricks to travel longitudinally along the length of the cofferdam.[36]

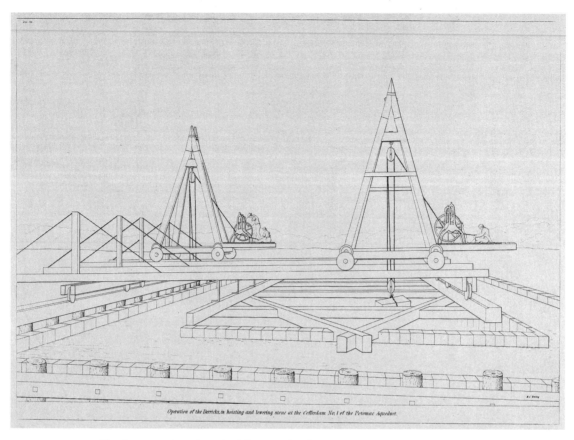

Operation of the Derricks, in hoisting and lowering stone at the Cofferdam No. 1 of the Potomac Aqueduct.

The Potomac Aqueduct derricks. Drawing by Lieutenant M. C. Ewing, in U.S. House, *Drawings Accompanying the Report of Turnbull*, no plate number. At the beginning of 1835, Turnbull had a system designed to off-load three to four tons of stone from scows, lift them vertically to the top of the cofferdam, transport them horizontally across the top of the cofferdam, and then lower the stone to the masons working in the cofferdam below. Author's collection.

Special care was taken with the mortar used in the piers. Hydraulic cement, resistant to water erosion, was used up to the two-foot mark above the high-water mark. Above that, common lime mortar was used.

Unlike the excavation work undertaken in the previous year, the masonry work of Pier Number 2 went quite fast. By June 21, 1835, the masonry of the pier was up to the top of the cofferdam. At this time the derricks and railways were removed and two booms were installed to assist in constructing the pier to 29 feet above the high-water point.

With the masonry at the top of cofferdam number two, it was obvious to all that Turnbull and his engineers had been successful in developing techniques to build piers in deepwater conditions. In celebration of this victory, the president of the

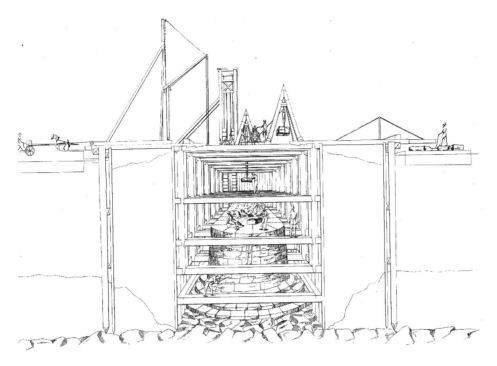

"The Potomac Aqueduct. Perspective View of the Cofferdam for Pier No. 5, Septr. 1838." Drawing by Captain William Turnbull and Lieutenant M. C. Ewing, in U.S. House, *Drawings Accompanying the Report of Turnbull,* no plate number. Shown are the masonry work and the delivery system of derricks, winches, and railway. Note that the outer pilings have been driven to bedrock, unlike those at Pier Number 2. Author's collection.

United States, Andrew Jackson, and his cabinet visited the construction site of Pier Number 2 in June 1835.[37]

BURR TRUSSES FOR THE AQUEDUCT SUPERSTRUCTURE

Once the problems of the cofferdam, pumps, and masonry handling had been worked out on Pier Number 2, construction on the rest of the aqueduct proceeded with few problems. With the completion of the piers, work could commence on the aqueduct's superstructure. It had been decided to span the piers with large wooden "Burr" trusses. They were designed by B. F. Miller, the master carpenter and principal superintendent of the work. White oak was delivered to the aqueduct by the Chesapeake and Ohio Canal. A wood preservative, Kyan's Process, was used on these trusses.[38]

KYAN'S PROCESS FOR WOOD PRESERVATION

Kyan's Process was named for Englishman John H. Kyan, who in 1832 patented a process for preserving wood by saturating it in a solution of corrosive sublimate,

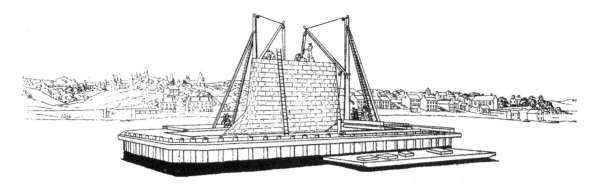

"Potomac Aqueduct. Perspective of the booms. Shewing the manner in which the stone was hoisted upon the Pier after removing the derricks," n.d. Drawing by Captain William Turnbull and Lieutenant M. C. Ewing, in U.S. House, *Drawings Accompanying the Report of Turnbull*, plate 15. The derricks, winches, and railways were dismantled once the masonry had been constructed above the top of the cofferdam. Booms were then erected to continue the work of building the masonry piers. Author's collection.

also called mercuric chloride.[39] The solution was typically one pound of bichloride of mercury in four gallons of water.[40] The timber was saturated for at least one week in soaking pits.

In 1840, at the Georgetown end of the aqueduct, Turnbull set up soaking pits to "kyanize" the timber. Wrote Turnbull: "The white oak timber for the posts of the superstructure was received by the Chesapeake and Ohio canal. I therefore built a carpenter-shed, for dressing the posts, on a level with the towing-path of the canal, on which the posts were landed; and immediately underneath the shed were placed the reservoirs and tanks for saturating the timber, into which it was lowered by means of block and tackle, and by the same means was taken out of the tanks, and placed under a shed adjacent, to dry."[41]

MERCURY POISONING OF THE WORKMEN

In terms of preserving the timber wood superstructure, the process was very effective. Twenty-two years later, the wood superstructure was found to be still serviceable, "where 9 spans of Burr truss bridge stood from 1840 to 1862, and failed even then, not from decay, but from defects in construction."[42] However, the key ingredient in the kyanizing process was the toxin mercuric chloride. Drooling of the workers (hypersalivation), a symptom of mercury poisoning, was extensively reported. Turnbull may have known of these symptoms before kyanizing was used on the Potomac Aqueduct. Two years earlier, in 1838, kyanizing was tried on railroad construction in Maryland. Army engineer General T. J. Cram later reported, on September 9, 1870, "This experiment of Kyanizing timber was the first, I believe, ever practiced in our country. . . . The process, however, was so unhealthy, salivating all the men, it had to be abandoned."[43]

Throughout his lengthy account of the construction of the aqueduct, reported on almost day by day, nowhere does Turnbull mention the health of his workers or whether any workers suffered from mercury poisoning. Nor was Turnbull the only engineer to ignore the health effects of kyanizing. Throughout the remainder of the century, discussion on kyanizing focused on its costs, usually considered excessive, and not its effects on workers' health.

Despite the probable mercury poisoning of his workers, Turnbull was immensely proud of the aqueduct. He wrote of his accomplishment: "When I reflect upon the numerous difficulties which we have overcome in the progress of the work, and recall the disheartening predictions of [the] numerous portions of the community who looked upon the attempt to establish foundations at so great a depth, and in a situation so very exposed and dangerous, and who did not fail to treat it as an absurdity, I cannot but congratulate myself upon having so happily succeeded."[44]

Turnbull had constructed one of the longest bridges in the United States at its time.[45] He had developed a practical approach to pier construction in deepwater conditions. This approach was well illustrated by William Turnbull and M. C. Ewing's drawings, which were widely distributed through US government reports. By the time the reports appeared, the American railroad revolution was in full swing, and these techniques were used by railroad builders to extend their railroads westward.

In the years 1840–1843, a wooden superstructure was added on top of the masonry piers to carry the Potomac Aqueduct from the Chesapeake and Ohio Canal in Georgetown to Alexandria, Virginia. The Chesapeake and Ohio Canal itself reached Cumberland, Maryland, some seven years later. Like its larger cousin, the Chesapeake and Ohio Canal, the Alexandria Canal saw modest traffic but failed to live up to the expectations of its developers. By the time the canal opened in 1843, it was apparent that the future of American transport was with railroads.

With the onset of the Civil War in 1861, the Potomac Aqueduct was drained of water and used as a military vehicular bridge across the Potomac. After the Civil War, in 1868, the wooden superstructure was replaced by another wooden structure. The new structure was two stories high, with a road on top and a canal on the bottom. In 1888 this second wooden structure was replaced by a metal truss structure that served as a highway bridge. In the early 1920s, the present reinforced concrete arch highway bridge, Key Bridge, was constructed just south of the Potomac Aqueduct.[46] Almost immediately after the opening of Key Bridge, by 1926, the army was pressing Congress for funds to tear down the old Potomac Aqueduct, at a cost of $228,000.[47] The army finally, after World War II, received funds to tear down most of the piers of Potomac Aqueduct. Not destroyed were the abutment at the Chesapeake and Ohio Canal, now under the administration of the Chesapeake and Ohio Canal National Historical Park, and a remnant of Pier Number 1 on the Virginia side.

View of the Potomac Aqueduct, the second wooden superstructure in place. Georgetown is behind the aqueduct. Photographed sometime between 1868 and 1877 by an unidentified photographer. Courtesy Georgetown Public Library.

The Chesapeake and Ohio Canal had modest success in bringing the produce of the Potomac River valley to Georgetown and Washington. In 1876, for example, an inclined plane had to be constructed two miles above Georgetown to lower (and raise) boats into the Potomac River in a caisson mounted on rails from the canal to the Potomac, so as to avoid the bottleneck of canal traffic that had developed in Georgetown. The incline operated successfully for only two years. The flood of 1877, the three floods of 1886, and especially the Great Flood of 1889, severely damaged the canal. In the meantime, the Baltimore and Ohio Railroad funneled much of the Potomac River trade to Baltimore. The Port of Baltimore had the advantage of being on the Chesapeake Bay; ocean-bound ships did not have to make the long river transit, as did vessels coming to Georgetown and Alexandria. The problem of sedimentation in the Potomac was never solved, and that sedimentation limited the draft of vessels at Georgetown and Washington. Lighters had to be used to transfer cargoes from canal boats, thereby increasing shipping costs. In the end the commercial aspirations of the founders of the new federal city were never achieved.

THE EARLY VISION FOR A NEW FEDERAL CITY

In 1790, developing a capital city for a new country in a sparsely settled region was an extraordinarily ambitious undertaking. Unlike European capitals, such as London and Paris, the American capital was not established in a city built several thousand years previously. President Washington, assisted by Jefferson, attempted to short-circuit the extremely long development time of a major city into ten short years. If the new capital had been integrated as a federal district within an existing city such as Philadelphia or New York, Washington and his team would have needed only to construct public buildings; they could have benefited from an already established city infrastructure of roads, bridges, docks, sources of building supplies, worker housing, hospitals, and so forth. Because that infrastructure was lacking, and because the new capital was initially planned without funding from Congress, it was an audacious enterprise.

Still, it was George Washington who had the vision to establish the new capital on the banks of the Potomac. As the first president of the new republic, he held the ideal position to bring about the needed legislation. And as the country's iconic Revolutionary War hero, he had the respect of the American people and the moral authority to initiate such a plan. Even his death in 1799, one year before the government intended to move from Philadelphia to Washington, reinforced the consensus among Americans that Washington's vision for the federal city should be realized. Years after his death, as construction and development continued, the memory of Washington was evoked in decisions large and small regarding the federal city. Changes in the design of the Capitol, for example, were fought for and against based on what Washington had approved years earlier. The memory of Washington was also involved in the decision in 1815 to keep the capital on the Potomac River in the aftermath of the British destruction. Developing a new federal city from scratch without funds may have been a crazy idea, but it was Washington's crazy idea, and this was enough to ensure the support of Americans long after he had died. Despite the rough start, it was Washington's vision that triumphed.

As Washington's Secretary of State—it was a far more comprehensive cabinet position than it is today—Jefferson was the key adviser in developing specifics for raising the new city from bare farm fields in ten short years. His integral role in the development of Washington, DC, continued after the government relocated to the Potomac region in 1800. As the third president of the United States, Jefferson exercised hands-on supervision of work on the incomplete Capitol building. He hired the most competent architect-engineer in the United States at the time, Benjamin Latrobe, and with his and Latrobe's joint effort, the House of Representatives building was designed and constructed to complement the Senate building. Never before or again has a president of the United States personally supervised an engineering and building project.

OVERCOMING FINANCIAL PROBLEMS AND ADAPTING TO CHANGING BUILDING PRACTICES

Washington and Jefferson launched the federal city building project without any clear vision of how much it might cost or how long it might take. Had they reviewed such estimates, perhaps the principal participants in the project would have been discouraged and the city never built. Financing the building project was a constant burden. Notwithstanding the lack of an initial congressional appropriation and the woefully inadequate grants from the Maryland and Virginia legislatures, Washington and Jefferson's successors gradually overcame those financial obstacles. In the end, it was Pierre L'Enfant's initially rejected recommendation to fund construction by borrowing money and using the unsold lots as collateral that helped get the project off the ground. After 1800, Jefferson and Latrobe used congressional appropriations to build the House of Representatives building and other structures. Slowly, the financial problems of the new federal city were brought under control as the original ideas for funding development and construction evolved into systematic and more easily accountable business practices—the funding of construction through borrowed money, the control of cost overruns, and periodic audits of accounts — that are still used to this day.

When planning for the city of Washington first began, the method of organizing the workforce still followed the Old World master-apprentice model. Washington and Jefferson initially looked to the building practices of their former mother country as their guide. There, as in the United States, a man of the landed gentry would produce an elevation and floor plan of the house he desired built. Then a master builder, using the knowledge of building that he had "under his cap" would fill in the details needed for construction. Journeymen and apprentices, under the supervision of the master builder, would then do the work. But in England in the late eighteenth century, building practices and the management of engineering and construction were beginning to change. With immigrant forces arriving in America and looking for work, and with many of the old methods for construction becoming obsolete, the architecture-engineering leaders were presented with many challenges. One source of this change was the Industrial Revolution. Innovative materials and techniques appeared, and larger, more intricate building projects were newly possible to undertake. The new federal city became an example of those extensive, complex projects. The Capitol, for example, was the largest single building erected in the country at that time, the President's House was the largest single house in the country, and the bridges across the Potomac were built with new materials and using new techniques. These larger, more complicated building projects put the new disciplines in architecture and in canal, civil, and military engineering in demand.

THE EMERGENCE OF PROFESSIONAL ARCHITECT-ENGINEERS

The US Capitol design competition brought the older gentleman-patron and master-builder model for building organization into collision with the newer professional architect-engineer model. It became apparent that the construction of large complex structures could not be entrusted to gentleman amateurs. Architect Stephen Hallet's voluminous criticisms of William Thornton's design for the Capitol demonstrated that the gentlemen of the gentry could not design the public buildings needed in the new capital. These buildings required more comprehensive drawings, prepared by men trained and experienced in the art of building. The simple plan and elevation drawings of the landed gentry gave way to large collections of drawings prepared by professional architect-engineers, showing the workmen what needed to be built and, in many cases, how it was to be built.

The intense conflict between William Thornton and Benjamin Latrobe further demonstrated the disparity between gentleman-patrons and trained, professional architect-engineers. Thornton believed that the master-builder should take his drawings of the Capitol and make them work. Latrobe (and his predecessors Hallet and Hadfield) thought that amateurs had no place in the design of complex buildings. Although both the gentleman-patron and master-builder model and the professional architect-engineer model managed to coexist during the first fifty years of Washington's construction, the advent of new technologies — as with the iron suspension bridges built at Little Falls — made it apparent how necessary it was to hire a trained architect-engineer for the job. Both models, however, left their legacy and have provided solutions to aspects of building the new federal city that have stood the test of time. Smaller, less complicated structures continued to be built in the older manner. But larger, more complicated structures were designed and built in the newer mode of architect-engineer. And as the Industrial Revolution progressed, society became more dependent on professionally trained architect-engineers.

Other factors contributed to the emergence of the newer mode of design by professionals trained in the building craft. It is clear that Washington, and perhaps Jefferson, intended to make the new federal city the physical embodiment of the new republic. It was Washington's intent that the it should be comparable to the great European capitals of London and Paris.

Washington was greatly attracted to Thornton's drawing of the Capitol, because it was big and bold and worthy, in Washington's view, of the newly emerging republic. This view, that the new public buildings of Washington should reflect the uniqueness and greatness of the American democratic ideal, was reflected in the early decision that the public buildings were to be stone-clad at a time when very few buildings in the United States were so built.

Master planning was another field where the intention of Washington to create a world-class city worthy of the new republic can be seen. At the time most new towns in America were planned by surveyors who laid out streets and lots on a rectilinear basis. This was quick and easy. For the new federal city, Washington chose an experienced architect-planner, Pierre L'Enfant, to lay out a different kind of city, one that would inspire the populace of the new country. L'Enfant's grand plan with radial avenues and expansive spaces was like nothing previously seen in America. At the time his plan was criticized as being exceptionally grandiose. Nevertheless, L'Enfant's plan has stood for more than two hundred years and has created a beautiful capital city. Even during times of great enlargement of the city, such as during the Civil War, World War I, and World War II, the plan has guided the successful and beautiful evolution of the city.

THE NEW FEDERAL CITY AS A COMMERCIAL CENTER

Not everything went as George Washington originally intended. The new federal city did not ultimately become the commercial center he envisioned. The Potomac Canal was never an efficient conduit for shipping agricultural goods to Washington, and the increasing sedimentation limited oceangoing ships' ability to load at

Georgetown and Washington. Warmer winters ensured that the ice flood of 1784 would never recur, thus dashing hopes that such an event would again gouge out a shipping channel through the developing shoals opposite Georgetown. Competition from other cities further limited the commercial growth of Washington. The hugely successful Erie Canal, opened in 1825, guaranteed that New York City would become the commercial capital of the new country, as it largely monopolized agricultural shipments from the Northwest Territory. Ironically, it was the growth of the federal government into a vast enterprise that eventually brought Washington its greatest commercial success. Washington did eventually become a great commercial city, but its commerce was of government.

Despite the conflict between the landed gentry and the professionally trained architect-engineers, an aesthetically pleasing city of classical proportions, wide avenues, and splendid vistas emerged. The city design is one that allows for expansion; Washington, DC, has successfully integrated Victorian excess, lush Beaux-Arts structures, and the twentieth-century aesthetic without losing the grace of Washington and Jefferson's original vision, which continues to dominate. L'Enfant's plan for the federal city was resurrected by the Commission of Fine Arts in the early twentieth century and continues in use today. The adaptability of this plan to ever-changing urban needs has been amply demonstrated by its continued use throughout the twentieth century and into the twenty-first. Similarly, the neoclassical architecture adopted for the first public buildings provided an architectural model that has been followed for the succeeding centuries. From an audacious and precarious beginning, the engineering and construction of Washington has allowed it to evolve into a first-class capital.

ABBREVIATIONS

CL1   John C. Van Horne and Lee W. Formwalt, eds., *The Correspondence and Miscellaneous Papers of Benjamin Henry Latrobe*, vol. 1, *1784–1804* (New Haven, CT: Yale University Press for the Maryland Historical Society, 1984).

CL2   John C. Van Horne, Jeffrey A. Cohen, Darwin H. Stapleton, Lee W. Formwalt, William B. Forbush III, and Tina H. Sheller, eds., *The Correspondence and Miscellaneous Papers of Benjamin Henry Latrobe*, vol. 2, *1805–1810* (New Haven, CT: Yale University Press for the Maryland Historical Society, 1986).

CL3   John C. Van Horne, Jeffrey A. Cohen, Darwin H. Stapleton, William B. Forbush III, and Tina H. Sheller, eds., *The Correspondence and Miscellaneous Papers of Benjamin Henry Latrobe*, vol. 3, *1811–1820* (New Haven, CT: Yale University Press for the Maryland Historical Society, 1988).

DH   U.S. House of Representatives, *Documentary History of the Construction and Development of the United States Capitol Building and Grounds,* Report No. 646, 58th Cong. 2nd sess. (1904).

NARA   National Archives and Records Administration, Washington, DC

RDCC   *Records of the District of Columbia Commissioners and of the Offices Concerned with Public Proceedings, 1797–1867*, National Archives Microfilm Publications Microcopy M-371 (Washington, DC: National Archives and Records Service, 1964).

INTRODUCTION. GEORGE WASHINGTON, THOMAS JEFFERSON, AND
THE VISION FOR A NEW FEDERAL CITY ON THE POTOMAC
Epigraph. As quoted by Tindal, *Standard History*, 67–68.
Epigraph. Rochefoucauld-Liancourt, "Voyage to Federal City in 1797," 35–60, 58.
1. Thomas Jefferson, Collection of Notes Labeled "Ana," as quoted in Tindall, *Standard History*, 29–30.
2. Jefferson note, November 29, 1790, "Proceedings to Be Had under the Residence Act," in Padover, *Jefferson and the Capital*, 31.
3. Jefferson's note, dated November 29, 1790, is quoted in full in Padover, *Jefferson and the Capital*, 33; and in Tindall, *Standard History*, 40.
4. Padover, *Jefferson and the Capital*, 67–68.
5. Commissioners to George Brent, December 23, 1791, in District of Columbia Commis-

sioners, *Records of the District of Columbia Commissioners* (hereafter cited as *RDCC*), Letters Sent, roll 3, 48.

6. Then in Stafford County, now King George County, Virginia. Located east by northeast of Richmond.

7. "Archives of Maryland Online," in *The Laws of Maryland*, 3181:625, http://aomol.msa .maryland.gov/html/index.html.

8. Ibid., 3185:1333.

9. Herrick, *Ambitious Failure*, 47.

10. Georgetown Potomac Bridge Company, "Memorial."

11. "An Act Authorizing the Erection of a Bridge over the River Potomac, within the District of Columbia," February 5, 1808, in U.S. Office of Chief of Engineers, *Laws Relating to Construction of Bridges*, 6–13.

12. The Washington Bridge remained the country's longest bridge only for a short time. In 1812–1814, the Columbia-Wrightsville Bridge, with a total length of 5,690 feet, was built across the Susquehanna River in Pennsylvania.

13. Van Horne et al., *Correspondence of Latrobe*, vol. 2, *1805–1810* (hereafter cited as *CL2*), 209.

1. PIERRE L'ENFANT'S TWO PLANS FOR EXECUTING THE PRESIDENT'S VISION

1. Pierre L'Enfant to President George Washington, report on renewing the work at the federal city in 1792, January 17, 1792, as quoted by Kite, *L'Enfant and Washington*, 118–132, 120.

2. Commissioners to City Surveyor, March 14, 1792, in *RDCC*, Letters Sent, roll 3, 71.

3. L'Enfant to Washington, January 17, 1792, as quoted by Kite, *L'Enfant and Washington*, 110–132, 110. This letter report was forwarded to the commissioners by Thomas Jefferson on March 6, 1792.

4. Ibid.

5. Ibid., 111.

6. Ibid.

7. Ibid.

8. Ibid.

9. Ibid., 120.

10. Ibid., 117.

11. Proceedings, September 24, 1791, in *RDCC*, roll 1, 30. Although L'Enfant included excavation in his plan submitted to Washington, Isaac Roberdeau, who worked for L'Enfant, reported to the commissioners on January 10, 1792, that L'Enfant had issued no instructions for this work to proceed. January 10, 1792, Letters Received, ibid., roll 8, 72.

12. Notes on Commissioners Meeting, September 8, 1791, in Padover, *Jefferson and the Capital*, 70–74, 72.

13. Thomas Jefferson to the commissioners, March 6, 1792, in Padover, *Jefferson and the Capital*, 103–106, 106. The strikeovers are as in the original.

14. Ibid., 106n.

15. L'Enfant, report on renewing the work, in Kite, *L'Enfant and Washington*, 110–132, 117.

16. Ibid.

17. Tindall, *Standard History*, 120.

18. L'Enfant, report on renewing the work, in Kite, *L'Enfant and Washington*, 110–132, 118.

19. Ibid.

20. *RDCC*, roll 1, 89.

21. L'Enfant, report on renewing the work, in Kite, *L'Enfant and Washington*, 110–132, 118–119.

22. Ibid., 119.

23. Ibid.

24. Ibid.

25. Ibid.

26. The commissioners reported: "Mr. Hoban is desirous to write to South Carolina to get information of the experience of a Wind Saw Mill and to enquire if a good workman can be got to do the work." Proceedings, November 18–21, 1793, in *RDCC*, roll 1, 204. The commissioners also had investigated a water-powered stonecutting machine. Neither device was built.

27. L'Enfant, report on renewing the work, in Kite, *L'Enfant and Washington*, 110–132, 119.

28. Ibid., 120.

29. Ibid.

30. Ibid.

31. Ibid.

32. L'Enfant had estimated 1,070 men. The difference may have been the 117 mechanics.

33. Ibid., 130–131.

2. FINANCING THE FEDERAL CITY

Epigraph. Thomas Jefferson note, November 29, 1790, "Proceedings to Be Had under the Residence Act," in Padover, *Jefferson and the Capital*, 31.

1. Ibid., 30–36, 33; also in Tindall, *Standard History*, 40.

2. Ibid.

3. Padover indicates that this meeting occurred on September 8 or 9, 1791. However, it actually occurred a year earlier; Jefferson refers to the meeting in his note of November 29, 1790.

4. Explained Jefferson to Washington on this meeting (in part): "To obtain this sum [needed for the construction of the new federal city]; this expedient was suggested to them. To procure a declaration from the proprietors of those spots of land most likely to be fixed for the town, that if the President's location of the town should comprehend their lands, they would give them up for the use of the U.S. on condition they should receive the double of their value, estimated as they would have been had there been no thought of bringing the federal seat into their neighborhood." Thomas Jefferson, "Notes on the meeting between Wm. Deakins, Benjamin Stoddert, Charles Carroll, Mr. Madison and himself on Sept. 8, 1791," September 8 or 9, 1791, in Padover, *Jefferson and the Capital*, 69–70.

5. Each of the states issued its own currency at the time, usually in pounds. The value of the different state currencies varied. The books of the commissioners usually were listed in US dollars and converted to Maryland pounds. Each pound of Virginia currency equaled $3.33. Later that currency may have changed value, as the audit of 1793 treats Virginia pounds as equal to Maryland pounds, at $2.66.

6. Jefferson, "Notes on the meeting between Wm. Deakins, Benjamin Stoddert, Charles Carroll, Mr. Madison and himself on Sept. 8, 1791," September 8 or 9, 1791, in Padover, *Jefferson and the Capital*, 69–70.

7. U.S. House, *Report of Committee on Expenditure*, 1:246.

8. The commissioners sent numerous letters to the state of Virginia to secure the promised grant, such as one dated September 23, 1793, which read (in part): "It is with regret we feel ourselves under the necessity of applying to you again on the subject of the Virginia

Donation. Our situation makes it proper for us to tell you frankly that we shall not be able to carry on the public buildings unless we can soon have what is behind of the last sum, the receipt of which we made ourselves certain of long before this time." Commissioners to Governor Henry Lee, September 23, 1793, as quoted by Tindall, *Standard History*, 188–189.

9. Jefferson note, November 29, 1790, "Proceedings under Residence Act," in Padover, *Jefferson and the Capital*, 30–36, 34–35; and in Tindall, *Standard History*, 41–42.

10. U.S. House, *Report of Committee on Authorizing Loan*, 1:143.

11. Clark, *Greenleaf and Law*, 67.

12. District of Columbia Commissioners, C. Copy of Morris and Greenleaf's contract with the Commissioners, dated 24 December, 1793, in "Report of the Commissioners," January 30, 1801, 1:223–224.

13. In 1801 the commissioners, in their report to Congress, estimated that the average price of lots sold north and east of Massachusetts Avenue since the passage of the Guarantee Act on May 1, 1796, was $105, as compared to an average price of $343 for lots sold south and west of Massachusetts Avenue. Ibid., 1:221.

14. George Washington to Daniel Carroll, January 7, 1795, as quoted by Clark, *Greenleaf and Law*, 92.

15. Deakins's financial report to the commissioners of April 22, 1794, in Tindall, *Standard History*, 189–190.

16. Historian Allen C. Clark has estimated the extent to which Greenleaf and his partners had bought lots in the new city: "A computation has been figured by which it appears that the triumvirate owned 7,234 lots. That Greenleaf individually owned 1,341. So that, the aggregate of lots in which he was interested was 8,575. *Washington in Embryo* states that at this time the public lots numbered 10,136; and, of course, the proprietors' numbered the same. Of the public lots Greenleaf had 6,000, of proprietors' 2,575 or expressed in percentage 60% of the public, 25% of the proprietors' and 42% of all." Clark, *Greenleaf and Law*, 72.

17. Washington to Carroll, January 7, 1795, as quoted ibid., 92.

18. Ibid.

19. Ibid.

20. District of Columbia Commissioners, "Report of the Commissioners," January 30, 1801, 1:219–231.

21. Historian Allen C. Clark reported that the lots sold by the commissioners were only a portion of the lots sold to the Greenleaf-Morris-Nicholson partnership. Clark "James Greenleaf," 5:236.

22. Not all historians have agreed with this conclusion, that the Greenleaf agreement had been a disaster for the commissioners. Greenleaf biographer Allen C. Clark wrote: "Until Greenleaf appeared the Commissioners' sales were slight; he purchased six thousand lots and concerned two capitalists in the enterprise; he undertook to procure for the Commissioners a loan in Holland; they expected from the sale and loan ample funds; from the sale they did receive two hundred and eighty-five thousand dollars. . . . But for Greenleaf, individually and instrumentally, in probability, would the transfer of the government seat have been made?" Ibid., 5:237.

This amount cited by Clark, $285,000, is derived by Clark from an incorrect reading of the commissioners' report of January 30, 1801. It is clear that the commissioners received only a small part of that money, the payment of May 1, 1794. See ibid., 5:236–237; and Clark, *Greenleaf and Law*, 175–176.

23. Morris wrote Washington (in part): "You will readily believe that I have suffered severe mortification at being in arrears with my Payments to the Commissioners of the Federal City,

but my feelings are still more deeply wounded at the idea of an application from them to you upon this subject. The only apology I can make for being in that situation, is the Impossibility of obtaining money for the sale of property or upon Loan." Robert Morris to President Washington, September 21, 1795, as quoted by Clark, *Greenleaf and Law*, 114–115, 114.

24. Ibid.

25. Ibid.

26. Morris wrote to Washington: "Our embarrassments have arisen from another source, Mr. Greenleaf is under Contract with his hand & seal, to provide us with money to carry through the operations which at his instance we were tempted to undertake, but the French invasion of Holland put it out of his power to fulfill his engagement." Ibid.

27. Morris wrote to Washington: "I am of opinion that under existing circumstances the Commrs. would not stand justified were they to advertise our property for sale to discharge the present debt due to them by Mr. Nicholson & myself, and at any rate I hope they will not do it, for you and they may rely that our exertions shall possess them of the money much sooner than they could obtain it by such sales." Ibid., 114–115, 115.

28. The Griswold committee reported: "The committee have likewise understood that, in consequence of the contracts entered into with Morris, Greenleaf, and Nicholson, for the sale of a large number of the building lots belonging to the public, and the failure of payment on their part, and the subsequent transactions which have grown out of that contract, some doubts have been entertained respecting the titles to a considerable part of these lots. *These doubts*, whether well or ill founded, must necessarily embarrass the sale of those lots, and ought, in the opinion of the committee, to be removed by a law of the National Legislature to establish titles both in law and chancery, where lands shall be resold, under proper regulations, to raise the purchase money which shall fall due from the first purchaser." U.S. House, *Report of Committee on Expenditure*, 1:243–244.

29. The commissioners reported to Congress: "That, on the 1st day of December, 1795, the wall of the President's house, in the city of Washington, was raised within six feet of the eave; that the sashes, sash-frames, doors, window shutters, and trimmings for doors and windows for the basement story, were made and ready to be put together; and the sashes and sash-frames for the principal story were expected to be finished by March [1796], with the labor of two hands. The work of all the doors of the principal story was cut out, and in a state of forwardness. There was mahogany laid in sufficient to complete the House, as sashes, doors, hand-rails, stairs, balusters, etc. for the two principal stories; flooring plank nearly sufficient to finish the two principal stories, was procured and preparing. The principal timbers were prepared ready to be put up together, and the roof considered as half done." U.S. House, *Report of Committee on Authorizing Loan*, 1:142.

30. The commissioners reported: "The foundation of the Capitol is laid; the foundation wall under ground and above is of different thicknesses, and is computed to average fourteen feet high and nine feet thick. The freestone work is commenced on the north wing; it is of different heights, but may average three feet and a half; the interior walls are carried up the same height." Ibid.

31. Their memorial specified the administration buildings required: "Two buildings [that] may be erected on the President's square, at the expense of $100,000, sufficient to accommodate, in a handsome manner, the Departments of State, Treasury, and War, and the General Post Office; and $100,000 he [Commissioner Alexander White] conceives sufficient to erect a Judiciary." Ibid., 1:143.

32. District of Columbia Commissioners, "Memorial of the Commissioners," 1:133–134, 133.

33. Ibid., 1:134. Also U.S. House, *Documentary History* (hereafter cited as *DH*), 38–40, 40.

34. U.S. House, *Report of Committee on Memorial*, 1:137.

35. Commissioner Alexander White to commissioners Gustavus Scott and William Thornton, February 26, 1796, in Tindall, *Standard History*, 201.

36. White to Scott and Thornton, February 25, 1796, in Tindall, *Standard History*, 201–202.

37. Tindall, *Standard History*, 202–204.

38. Ibid., 204.

39. Ibid., 205.

40. Ibid., 205.

41. District of Columbia Commissioners, "Report of the Commissioners," January 30, 1801, 1:221.

42. Tindall, *Standard History*, 205.

43. The commissioners' letter read (in part): "The present situation of our funds renders it impossible to pay the time-Role due yesterday, or to discharge the arrears of last month. The quarterly salaries which became due on the first of this month, are still unpaid. We wish those in public employment to be fully acquainted with these facts that no cause of complaint may exist. In the present State of things it is almost impossible to say when we shall receive a sufficient sum to discharge the claims against it; but we request you to assure all those in public employment under you, that the time Role shall be paid out of the first funds competent to the object, and shall be preferred to quarterly salaries or any other demands, except small debts due to Labourers and people residing at a distance." In *RDCC*, Letters Sent, October 3, 1797, roll 4, 144–145.

44. Tindall, *Standard History*, 206.

45. Ibid., 206.

46. Ibid., 207.

47. Ibid., 207. The commissioners later reported the circumstances and detail of this $100,000 loan:

> The extent of these loans being thus ascertained, and the purchasers of lots still failing to pay the amount of their purchases, it became obvious that the views of Government could not be accomplished without further aid.
>
> The commissioners, therefore, prepared a second memorial to Congress, stating the situation of the federal seat and the resources which remained in their hands, which memorial was transmitted by the President to Congress on the 23d February, 1798, and in consequence, an act was passed authorizing the Treasurer of the United States to advance the sum of $100,000, at the times in the said act mentioned, which was declared to be in full of the sums previously guarantied. District of Columbia Commissioners, "Report of the Commissioners," January 30, 1801, 1:220.

48. The commissioners explained the need for this additional money: "From the difficulty of collecting outstanding debts, as well from purchasers themselves, it became evident that the several objects considered as necessary previous to the removal of Government could not be accomplished with the means at the disposal of the commissioners; application was therefore made to the Assembly of Maryland for a loan, and a resolution of that Legislature was obtained on the 23rd December 1799, directing the trustee of the State to transfer to the commissioners of the federal buildings in the city of Washington the sum of $50,000 of the stock of the United States, bearing a present interest of six per cent. Per annum, on their giving such real and personal security as the Governor and Council should approve, for the payment of the principal sum, on or before the 1st day of November, 1802, and the punctual payment of the interest quarter-yearly." Ibid.

49. Tindall, *Standard History*, 208.

50. The commissioners explained: "The trustee of the State [of Maryland] transferred the stock accordingly, which has been sold at different times, including the interest received thereon, for the sum of $42,738.36, and interest has been paid thereon up to the 30th September last, inclusive, to the amount of $2,250, leaving the nett sum of $40,488.36 to be applied to public use." District of Columbia Commissioners, "Report of the Commissioners," January 30, 1801, 1:220.

51. The first day that a quorum of both houses was present was November 21, 1800. Tindall, *Standard History*, 231.

52. Not all of the administration of the federal government moved to the new federal city in November 1800. Not until June 16, 1801, were all the administrative personnel and associated documents relocated to the new city. Ibid.

53. U.S. House, *Report of Committee on Expenditure*, 1:243.

54. Ibid.

55. Ibid.

56. The commissioners were then being severely criticized for mismanagement and therefore interested in applying the highest value to the assets held by the U.S. Government. They argued: "The land which has been accepted, or purchased by the commissioners for the use of the United States, and yet remains unsold (exclusive of lots forfeited for non-payment of the purchase money, and which for that cause are liable to be resold,) consists of 24,655,735 square feet of ground in the city of Washington, equal to 4,682 lots of 5,265 square feet each, exclusive of lots which bind on navigable water; these form fronts in the extent of 2,043 feet, and are generally sold by the foot front. It is impossible to ascertain with precision the value of this property; some idea may be formed of it by taking the average price at which similar property has heretofore been sold; in this case, a reference must be had to different situations; 3,178½ of the city lots lie northeast of Massachusetts Avenue, in which situation only five standard lots (except 1,500 part of Morris & Greenleaf's selection of 6,000 lots) have been sold by the commissioners; but many lots (of private property) on the same side of that avenue have been sold. We have been able to ascertain the price of 355 of these, which, united with the price of the five lots sold by the commissioners, made an average of upwards of $105 per standard lot, which rate would produce, by the sale of the whole number on that side of the avenue, the sum of $333,747. The remaining 1,504 lots are situated to the southwest of Massachusetts Avenue; the average price of lots sold in that division of the city, since passing the guaranty law in May, 1796, is $343, at which rate the above-mentioned 1,504 lots would produce $515,873. The average price of lots binding on navigable water, sold during the same period, is $12.71 per foot front. The property of this description, remaining to be sold at the same rate, would produce $25,979.24." District of Columbia Commissioners, "Report of the Commissioners," January 30, 1801, 1:220–221.

57. U.S. House, *Report of Committee on Expenditure*, 1:244.

58. Ibid.

59. Commissioner Scott died December 25, 1800, and was replaced by William Cranch. Cranch resigned after six weeks and was replaced by Tristram Dalton.

60. Plus an additional $7,613 for payments for lots already sold and paid to the commissioners to prevent them from being resold for nonpayment. Commissioners William Thornton, Alexander White, and Tristram Dalton to President Jefferson, December 19, 1801, in District of Columbia Commissioners, "Recapitulations of Facts," 1:256.

61. President Jefferson wrote on the need to avoid lot sales to pay the interest due: "The moneys now due, and soon to become due, to the State of Maryland, on the loan guaran-

teed by the United States, call for an early attention. The lots on the city, which are charge-able with the payment of these moneys, are deemed not only equal to the indemnification of the public, but to ensure a considerable surplus to the city, to be employed for its improvement, provided they are offered for sale only in such numbers to meet the existing demand: but the act of 1796 requires that they shall be positively sold in such numbers as shall be necessary for the punctual payment of the loans; 9,000 dollars of interest are lately become due; 3,000 dollars quarterly-yearly will continue to become due; and 50,000 dollars, an additional loan, are reimbursable on the 1st day of November next. These sums would require sales so far beyond the actual demand of the market, that it is apprehended that the whole property may be thereby sacrificed, the public security destroyed, and the residuary interest of the city entirely lost." President Jefferson to Congress, January 11, 1802, in District of Columbia Commissioners, "Recapitulations of Facts," 1:254.

62. Wrote President Jefferson: "Under these circumstances, I have thought it my duty, before I proceed to direct a rigorous execution of the law, to submit the subject to the consideration of the Legislature. Whether the public interest will be better secured in the end, and that of the city saved, by offering sales commensurate only to the demand at market, and advancing from the Treasury, in the first instance, what these may prove deficient, to be replaced by subsequent sales, rests for the determination of the Legislature." President Jefferson to Congress, January 11, 1802, in District of Columbia Commissioners, "Recapitulations of Facts," 1:254.

63. U.S. House, *Report Recommending Offices of Commissioners Be Abolished*, 1:260.

64. Ibid.

65. U.S. Congress, "An Act to Abolish the Board of Commissioners in the City of Washington, and for Other Purposes," May 1, 1802, in U.S. Congress, *Laws of the United States*.

3. CONSTRUCTING THE FEDERAL CITY

Epigraph. Pierre L'Enfant to President George Washington, report on renewing the work at the federal city in 1792, January 17, 1792, quoted by Kite, *L'Enfant and Washington*, 111.

1. Proceedings, September 23, 1793, in DCC, *RDCC*, roll 1, 197.

2. *George-Town (DC) Weekly Ledger*, supplement, June 18 and June 25, 1791, 1.

3. *Georgetown (DC) Columbian Chronicle*, March 25, 1794, 4; Proceedings, January 1, 1793, in *RDCC*, roll 1, 156.

4. Hamburg was also called Funkstown. Jacob Funk had divided this area of the federal city into building lots in 1768. Bryan locates Hamburg between the present Nineteenth and Twenty-Third Streets, NW, and from H Street, NW, to the Potomac River. See Bryan, *History of the National Capital*, 1:59. Payroll for February and March 1795, in RG 42, National Archives and Records Administration, Washington, DC (hereafter, NARA), box 1, nos. 37 and 101.

5. Proceedings, October 21, 1791, in *RDCC*, roll 1, 52.

6. Proceedings, June 6, 1792, in *RDCC*, roll 1, 115–116.

7. Thomas Jefferson to Thomas Johnson, March 8, 1792, in Padover, *Jefferson and the Capital*, 109–112, 111–112.

8. Harley J. McKee, "Brick and Stone: Handicraft to Machine," in *Building Early America: Proceedings of the Symposium held at Philadelphia to Celebrate the 250th Birthday of the Carpenters' Company of the City and County of Philadelphia*, ed. Charles E. Peterson (Radnor, PA: Chilton, 1976), 75. McKee adds that American colonists complained about the lack of freestone in the United States, not realizing that it was plentiful.

9. The standard histories of Washington, DC, have referred to the island as Hissington Island. The commissioners' records, written by hand, can be interpreted as Wigginton, Hiss-

ington or Higginton Island. The name was subsequently misspelled by substituting the "W" for an "H" and the double "g" for a double "s."

10. Owen, "Report on Aquia Creek," 109–110.

11. Proceedings, November 18, 1791, in *RDCC*, roll 1, 150.

12. On November 18, 1791, the commissioners authorized the expenditure of £1,800 Virginia currency ($6,000) for the purchase of the island, which became known as Government Island. Proceedings, November 18, 1791, in *RDCC*, roll 1, 58–59.

13. Proceedings, April 10, 1792, in *RDCC*, roll 1, 93.

14. Owen, "Report on Aquia Creek," 110.

15. See Carter et al., *Latrobe's View of America, 1795–1820*, 270.

16. Benjamin Latrobe, "An Account of the Freestone Quarries on the Potomac and Rappahannoc Rivers, February 19, 1807," in *CL2*, 380–389, 386; and in *Transactions of the American Philosophical Society* 6 (1809): 283–293. Present-day visitors to Government Island can see the undercutting and the channels described by Latrobe.

17. Ibid. Four tons, at, say, fifteen cubic feet to the ton, would be a block of sixty cubic feet, perhaps 6 ft. × 4 ft. × 2.5 ft. Writing later, Owen reported, "The columns of the east portico of the Capitol, each of a single piece weighing 18 tons, were obtained at these United States quarries." Owen, "Report on Aquia Creek," 110.

18. Proceedings, May 1, 1792, in *RDCC*, roll 1, 103. Smith was paid one dollar per ton. RG42, NARA, box 1, no. 109.

19. RG42, NARA, box 1, no. 109.

20. Owen, "Report on Aquia Creek," 113.

21. In a Congressional hearing, the stone cutters' cosmetic technique of using gum shellac to repair defective masonry units was discussed:

Q. You speak about the use of gum shellac, is it not a fact that stone cutters have been always in the habit, if they happen to break off a piece of stone they are cutting, to put it on again if they can do it without detection?

A. Yes, it has always been the case so long as I have known the trade.

Testimony of True Putney, construction superintendent, master mason, U.S. House of Representatives, Seneca Stone Investigation, March 18, 1872, NARA, CIS No. 42 HPub-T.3 (unpublished hearing), 31.

22. Samuel Milliken to the commissioners, October 27, 1792, in *RDCC*, Letters Received, roll 9, 145.

23. Commissioners to Jefferson, November 5, 1792, in Padover, *Jefferson and the Capital*, 156–157, 157.

24. Jefferson to the commissioners, November 13, 1792, in Padover, *Jefferson and the Capital*, 157–158, 158.

25. Commissioners to Jefferson, December 5, 1792, in Padover, *Jefferson and the Capital*, 161–162, 162.

26. Jefferson to the commissioners, December 13, 1792, in Padover, *Jefferson and the Capital*, 162–163, 163.

27. Latrobe, "Report on the Public Buildings," December 1, 1804, in Van Horne and Formwalt, *Correspondence of Latrobe*, vol. 1, *1784–1804* (hereafter cited as *CL1*), 582; and in *DH*, 111–114 (omits report on President's House); Latrobe, *Private Letter to Congress*, 296–316, 299–300.

28. Latrobe, *Private Letter to Congress*, 299–300.

29. Owen, "Report on Sandstones," 37.

30. Testimony of Adolph Cluss, architect, in U.S. House of Representatives, Committee

on Public Buildings and Grounds, Seneca Stone Investigation, April 10, 1872, NARA, CIS No. 42 HPub-T.8 (2) (unpublished hearing), 15–17.

31. Ibid., 11.

32. Proceedings, May 1–4, 1792, in *RDCC*, roll 1, 101–102.

33. Proceedings, June 5, 1792, in *RDCC*, roll 1, 112; Proceedings, June 6, 1792, in *RDCC*, roll 1, 114.

34. Proceedings, October 22, 1791, in *RDCC*, roll 1, 55.

35. Francis Cabot to the commissioners, December 11, 1791, in *RDCC*, Letters Received, roll 8, no. 62.

36. Commissioners to Jefferson, October 22, 1791, in Padover, *Jefferson and the Capital*, 127–129, 128–129.

37. Advertisement of William Prentiss, from the *Washington Gazette*, June 21, 1797.

38. Proceedings, December 10 to 24, 1793, in *RDCC*, roll 1, 208.

39. James Hoban to the commissioners, January 1, 1793, in *RDCC*, Letters Received, roll 8, 169; Proceedings, March 6, 1793, in *RDCC*, roll 1, 167.

40. Proceedings, December 16 to 20, 1793, in *RDCC*, roll 1, 209–210.

41. Both New York and Philadelphia were constructing residences for the president in anticipation that the capital would be located in their city.

42. Jefferson to commissioners, March 6, 1792, in Padover, *Jefferson and the Capital*, 103–106, 105.

43. Washington to David Stuart, March 8, 1792, in Padover, *Jefferson and the Capital*, 112.

44. Commissioners to Jefferson, March 14, 1792, in Padover, *Jefferson and the Capital*, 116–118, 118.

45. Jefferson to the commissioners, March 21, 1792, in Padover, *Jefferson and the Capital*, 124–125.

46. Jefferson to the commissioners, April 20, 1792, in Padover, *Jefferson and the Capital*, 137–138, 138.

47. Commissioners to Jefferson, June 2, 1792, in Padover, *Jefferson and the Capital*, 142–143.

48. Proceedings, July 5, 1792, in *RDCC*, roll 1, 120.

49. The month of August was an exception. The existing time rolls indicate that every day in that month was worked.

50. Commissioners to William Prout, August 3, 1792, in *RDCC*, Letters Sent, roll 3, 108.

51. Proceedings, September 4, 1792, in *RDCC*, roll 1, 136.

52. Proceedings, February 4–8, 1793, in *RDCC*, roll 1, 165.

53. Commissioners to Traquair, January 2, 1793, in *RDCC*, Letters Sent, roll 3, 133.

54. Joseph Fenwick to commissioners, April 4, 1793, in *RDCC*, Letters Received, roll 8, 241.

55. Jefferson to commissioners, December 23, 1792, in Padover, *Jefferson and the Capital*, 164–165.

56. Commissioners to Jefferson, January 5, 1793, in Padover, *Jefferson and the Capital*, 165–167, 166.

57. Commissioners to Jefferson, April 11, 1792, in Padover, *Jefferson and the Capital*, 134–136, 135.

58. Proceedings, January 3, 1793, in *RDCC*, roll 1, 160–161.

59. In 1790, as Letitia Woods Brown reported, there were 8,043 free blacks in Maryland and 12,866 free blacks in Virginia. Some of the skilled workers may have been free blacks from eastern urban areas. Brown, *Free Negroes in District of Columbia*, 12.

60. Charles Firrer to commissioners, February 26, 1793, in *RDCC*, Letters Received, roll 9, 214.

61. In the first month of 1795, Dermot earned £28 ($74.66, including £1 for working Sundays). Of this, £27 1s was applied to the second payment on Lot Number 3 in Square 725. Miscellaneous Treasury Records, RG 217, NARA, box 1, no. 8.

62. Commissioners to Jefferson, January 5, 1793, in Padover, *Jefferson and the Capital*, 165–167, 166.

63. Captain Williams was asked to obtain as many as one hundred Negro men on these terms. Proceedings, November 3, 1794, in *RDCC*, roll 1, 292.

64. Miscellaneous Treasury Records, RG 217, NARA.

65. Ibid.

66. Ibid.

67. Ibid.

68. Thomas Jefferson, "Memorandum Relative to Commissioners for Laying Off the Federal City," March 11, 1793, in Padover, *Jefferson and the Capital*, 177–178.

69. Proceedings, February 18, 1795, in *RDCC*, roll 1, 361–365.

4. DEVELOPING A COMMERCIAL CENTER

Epigraph. Lear, *Observations on the River Potomack*, 14.

1. See Hains, "Improvement of Harbors," 770–771. Also, a recent article in the *Washington Post* quoted Washington, DC, historian-geographer Don Hawkins: "I feel pretty strongly that the idea that this was Braddock's rock grew up much later." Hawkins argues that it does not make sense that Braddock would have disembarked himself and his troops south of Rock Creek, necessitating crossing the then bridgeless creek, when wharves existed just upstream at Georgetown. See John Kelly, "Once-Grand Braddock's Rock Has Nearly Eroded from Memory," *Washington Post*, February 16, 2014, C4.

2. Moore, *Address to Citizens of Georgetown*, 9–10.

3. As quoted by Tindall, *Standard History*, 42–43.

4. See Kapsch, *Potomac Canal*.

5. The winter of 1783–1784 was extremely harsh and set record cold temperatures across North America and northern Europe. It was considered the harshest winter in five hundred years. During that winter, ice in Charleston Harbor permitted the citizens to ice skate, and the Mississippi River froze at New Orleans. The harsh weather that year has usually been attributed to the volcanic ash thrown into the atmosphere by the eruption of the Iceland volcano Laki in June 1783. But this explanation has been recently challenged by scientists who attribute the unusually cold winter of 1783–1784 to the oscillation of the negative phase of the North Atlantic Oscillation, combined with a warm event of the El Niño Southern Oscillation, an event similar to what caused the unusually severe winter of 2009–2010. See Roseann D'Arrigo, Richard Seager, Jason E. Smerton, Allegra N. LeGrange, and Edward R. Cook, "The Anomalous Winter of 1783–1784: Was the Laki Eruption or an Analog of the 2009–2010 Winter to Blame?," *Geophysical Research Letters* 38, no. 5 (March 16, 2011), doi: 10, 1029/2011GL046696, 2011 (accessed September 2, 2013).

6. B. [Benjamin] H. [Henry] Latrobe, "Navigation of the Potomac," April 28, 1812, in Latrobe, *Opinion on a Project*, vii–viii. Latrobe did not arrive in America until 1796, twelve years after the 1784 ice flood, so he was relating the effects of the ice flood as observed by others.

7. Ibid.

8. Hains, "Improvement of Harbors," 771. Hains, usually a reliable source of historical information, incorrectly dates this event to the winter of 1779–1780.

9. Latrobe, "Navigation of the Potomac," in Latrobe, *Opinion on a Project*, viii.

10. Benjamin Latrobe to Thomas Fitzsimmons, March 27, 1811, in Van Horne et al., *Correspondence of Latrobe*, vol. 3, *1811–1820* (hereafter cited as *CL3*), 44–48, 46.

11. Hains, "Improvement of Harbors," 771.

12. Ibid., 770. Soundings for this map were made by Ignatius Fenwick and Richard Johns.

13. Located on the south side of the mouth of the Anacostia, or Eastern Branch, River.

14. Located at the eastern end of the present-day Theodore Roosevelt Bridge.

15. Hains, "Reclamation of the Potomac Flats," 55–56.

16. Latrobe to Fitzsimmons, March 27, 1811, in *CL3*, 46.

17. Michler, "Report of Michler," in *Report of the Chief of Engineers to the Secretary of War, for the Year 1869*, 519–525, 519; in the separate report of the same title (Washington, DC: US Government Printing Office, 1869), 28.

18. Jackson, *Chronicles of Georgetown*, 72.

19. *RDCC*, roll 1, 112.

20. Hains, "Improvement of Harbors," 772–773.

21. Moore, *Address to Citizens of Georgetown*, 2.

22. Ibid., 6, 4.

23. The hearing was on January 22, 1811. *CL3*, 19n5.

24. Ibid., 47–48n5.

25. Latrobe, "Navigation of the Potomac," in Latrobe, *Opinion on a Project*, 1–2.

26. Ibid., 7.

27. Ibid., 8.

28. Ibid.

29. *CL3*, 47n5–48. For details of the dispute between Thomas Moore and Benjamin Latrobe on this subject, see Moore, *Ship Navigation to Georgetown*; and Latrobe, "Navigation of the Potomac," in Latrobe, *Opinion on a Project*.

30. Benjamin Latrobe, as quoted by *CL3*, 48n5.

31. *Raleigh North-Carolina Standard*, June 14, 1837, 4.

32. Proceedings, October 22, 1791, in *RDCC*, roll 1, 55.

33. Proceedings, November 25, 1791, in *RDCC*, roll 1, 60–65.

34. Proceedings, June 4, 1792, in *RDCC*, roll 1, 111.

35. Latrobe to Thomas Tingey, May 18, 1805, in *CL2*, 77–82, 78; and Hibben, *Navy-Yard, Washington*, 22; *CL2*, 81n4, based on Edwin W. Small, *Early Wharf Building*, rev. ed. (N.p.: Eastern National Park & Monument Association, 1970).

5. EARLY INFRASTRUCTURE AND TRANSPORT IMPROVEMENTS

1. Tindall, *Standard History*, 83.

2. Ibid., 120.

3. Rochefoucauld-Liancourt, "Voyage to Federal City in 1797," 59.

4. Proceedings of the Commissioners, March 15, 1792, in *RDCC*, roll 1, 84.

5. Proceedings of the Commissioners, March 29, 1792, in *RDCC*, roll 1, 89.

6. Leonard Harbaugh to the commissioners, "Informs the Comrs. of the Ceremonies to be observed in laying Corner Stone of Bridge on July 4th," July 3, 1792, in *RDCC*, Index to Letters Received, roll 8. The account of the ceremonial cornerstone laying was printed in the *Maryland Journal and Baltimore Advertiser*, July 10, 1792. See Bryan, *History of the National Capital*, 1:190–191n3. It was also published in the *Georgetown (DC) Weekly Ledger*, July 7, 1792. See Kimball and Bennett, "Competition for the Federal Buildings," no. 5, 209.

7. Kimball and Bennett, "Competition for the Federal Buildings," no. 5, 209. Contracts

at that time were rudimentary compared to the lengthy contracts executed today. The contract is recorded in Proceedings of the Commissioners, September 1, 1792, in *RDCC*, roll 1, 133–134.

8. Annotation made by Chancellor James Kent in his copy of Tobias Lear's *Observations on the River Potomack* (New York, 1793), now in the Library of Congress, as quoted by Spratt, "Rock Creek's Bridges," 133.

9. Spratt, "Rock Creek's Bridges," 133.

10. Ibid., 133–134; and Crew, *Centennial History of Washington, D.C.*, 333.

11. Spratt, "Rock Creek's Bridges," 134.

12. Benjamin Stoddert to President Washington, October 24, 1792, Founders Online, National Archives, http://founders.archives.gov/documents/Washington/05-1-2-0142, ver. 2014-05-09, from *The Papers of George Washington: The Presidential Series*, vol. 11, *16 August 1792–15 January 1793*, ed. Christine Sternberg Patrick (Charlottesville: University of Virginia Press, 2007), 261–263.

13. Tobias Lear to Washington, June 17, 1793, Founders Online, National Archives (http://founders.archives.gov/documents/Washington/05-13-0069, ver. 2014-05-09), from: Patrick, *The Papers of George Washington: Presidential Series*, vol. 13, *1 June–31 August 1793*, edited by Christine Sternberg Patrick (Charlottesville, VA: University of Virginia Press), 96–99.

14. Commissioners to Harbaugh, October 16, 1794, in *RDCC*, Letters Sent, roll 10.

15. James Hoban to the Commissioners, December 15, 1794, in *RDCC*, Letters Received, roll 10.

16. Spratt, "Rock Creek's Bridges," 134.

17. D. B. Warden, *Description of the District of Columbia* (Paris, 1816), 100, as quoted by Myer, *Bridges and City of Washington*, 1983 and 1992 reprints, 55.

18. Viator, *The Washington Sketch Book* (New York: Mohun, Ebbs and Hough, 1864), 255.

19. *Washington Gazette*, November 23, 1796, 3.

20. Ibid., November 26, 1796, 4.

21. *Georgetown (DC) Columbian Chronicle*, July 15, 1794, 3.

22. U.S. House, *Report of Committee on Expenditure*, 245–246, tables.

23. *Washington National Intelligencer*, November 11, 1807.

24. A Stockholder, "To the Stockholders of the Potomac Company," *Washington National Intelligencer*, April 24, 1807, 1; Commissioners of the States of Maryland and Virginia, *Letter from Governor and Council of Maryland*. The Senate version was transmitted on January 27, 1823 (Senate Document No. 535, 17th Cong., 2nd sess.). The State of Maryland version is Commissioners of the States of Maryland and Virginia, *Message of the Governor of Maryland Transmitting a Report of the Commissioners Appointed to Survey the River Potomac*, January 1, 1823 (Annapolis: J. Hughes, 1822).

25. A Stockholder, "To Stockholders of Potomac Company," 1; *Washington National Intelligencer*, July 17, 1807, 2.

26. A Stockholder, "To the Stock Holders of the Potomac Company," *Washington National Intelligencer*, April 17, 1807, 1.

27. Tindall, *Standard History*, 83.

28. See District of Columbia Inhabitants, "Memorial for Erecting a Bridge," 422; and Mason, "Memorial of John Mason."

29. Engineer Charles Ellet proposed a suspension bridge for this location: Ellet, *Report of a Suspension Bridge*, 1st ed.

30. See Henry Grattan Tyrrell, *Bridge Engineering: A Brief History of This Constructive Art*

from the Earliest Times to the Present Day (Evanston, IL: Self-published, 1911), 135; and Nelson, *Colossus of 1812.*

31. U.S. House, *Letter from Secretary of War,* April 21, 1828, 12.

32. Kilty, *Laws of Maryland;* "Archives of Maryland Online," in *The Laws of Maryland,* http ://aomol.msa.maryland.gov/html/search1.html, 3181:625.

33. Ibid., 3185:1333.

34. *Georgetown (DC) Columbia Mirror and Alexandria Gazette,* June 9, 1795, as quoted by Herrick, *Ambitious Failure,* 46.

35. In 1804, after the first bridge at Little Falls had failed, the company published a summation of these dealings: "It is known to the generality of the Stockholders, that Subscriptions were originally opened for 400 Shares, at 200 each, amounting to 80,000 dollars—a sum which it was supposed would be required to complete the Bridge in a manner to put it out of danger of injury by floods. As soon as one half the shares were subscribed, the work was commenced, not doubting that the rest would also be subscribed but in this the directors were disappointed, never being able to obtain subscriptions for more than 245 shares . . . and they finished the Bridge, by loans of money, little short of 30,000 dollars." *Washington Federalist,* May 14, 1804, as quoted by Herrick, *Ambitious Failure,* 47.

36. *Georgetown (DC) Columbian Chronicle,* June 12, 1795, 2.

37. Carter et al., *Latrobe's View of America, 1795–1820,* 134, illustration on 135.

38. *Centinel of Liberty and George-town (DC) Advertiser,* August 26, 1796, 3.

39. Griggs, "Timothy Palmer," 34–35; and F. E. Griggs Jr., "*Timothy Palmer*: The Nestor of American Bridge Builders," www.ce.memphis.edu/3121/stuff/general/timothy_palmer.html (accessed July 30, 2017).

40. Georgetown Potomac Bridge Company, "Memorial."

41. In some sources described as 36 or 37 feet high.

42. *Centinel of Liberty and George-town (DC) Advertiser,* August 26, 1796, 3. Foundation stone, known as blue stone, was also abundant on the Maryland side of the Little Falls Bridge and was later extensively quarried and used in the construction of buildings in Washington, DC. See U.S. House of Representatives, Committee on Public Buildings and Grounds, *Seneca Stone Investigation,* unpublished hearings, March 11, 18, 20, 22, 25; April 3, 10, 12, 22, 24, 1872.

43. Herrick, *Ambitious Failure,* 47.

44. Timoshenko, *History of Strength of Materials,* 181–182.

45. John Shippen to Colonel Joseph Shippen, December 13, 1801, in Balch, *Letters and Papers,* 309–312, 311.

46. In 1826 the company reported to Congress a somewhat different date of its destruction: "The wooden arch of this bridge, from a defect in its construction, fell in about the year 1805." Georgetown Potomac Bridge Company, "Memorial."

47. *Washington Federalist,* May 14, 1804, in Herrick, *Ambitious Failure,* 58–59.

48. *Washington National Intelligencer,* October 20, 1802, in Herrick, *Ambitious Failure,* 58.

49. *Washington Federalist,* May 14, 1804, in Herrick, *Ambitious Failure,* 58–59.

50. *Washington Federalist,* May 29, 1805, in Herrick, *Ambitious Failure,* 62–63.

51. *Alexandria (VA) Daily Advertiser,* November 25, 1806, in Herrick, *Ambitious Failure,* 63.

52. The company reported to Congress in 1826, "In 1808 [*sic*—1807], the bridge last mentioned was carried off by a flood." Georgetown Potomac Bridge Company, "Memorial."

53. Finley, "Description of Patent Chain Bridge," 441–442.

54. Finley described the chains for the suspension bridge at Cumberland as being fabricated from inch-and-a-quarter bar iron, and it is possible that this was what was used here.

55. Finley here is referring to the wood-frame Market Street or "Permanent Bridge" over the Schuylkill River in Philadelphia, completed by Timothy Palmer in 1805. It was 550 feet long on two intermediate piers.

56. Finley, "Description of Patent Chain Bridge," 444.

57. *Washington Federalist*, February 17, 1808, as quoted by Herrick, *Ambitious Failure*, 76–77.

58. *Alexandria (VA) Gazette and Daily Advertiser*, November 21, 1810, in Herrick, *Ambitious Failure*, 78–79.

59. *Washington National Intelligencer*, April 23, 1811, 2.

60. Finley, "Description of Patent Chain Bridge," 442–443.

61. Ibid., 441–453.

62. "Chain Bridge, [signed by James Findley]," *Washington National Intelligencer*, January 24, 1811, 4.

63. See Kemp, "Finley and the Origins," 164–165; and *Centinel of Liberty and George-town (DC) Advertiser*, February 2, 1798, 3.

64. Finley, "Description of Patent Chain Bridge," 453.

65. "Cast Iron Bridges," *Washington National Intelligencer*, June 18, 1811, 2.

66. Crew, *Centennial History of Washington, D.C.*, 335.

67. A. B. McLean, letter (estimate to repair Washington Bridge), June 6, 1836, in Woodbury, "Letter from Secretary of Treasury," 2.

68. Crew, *Centennial History of Washington, D.C.*, 335.

69. For a summary of those later bridges at Little Falls, see Myer, *Bridges and City of Washington*.

70. John Mason, president of the Falls Bridge Turnpike Company, in Virginia House of Delegates, Report, December 6, 1820.

71. "An Act Authorizing the Erection of a Bridge over the River Potomac, within the District of Columbia," February 5, 1808, in U.S. Office of Chief of Engineers, *Laws Relating to Construction of Bridges*, 6–13.

72. *CL2*, 209n6.

73. Crew, *Centennial History of Washington, D.C.*, 334.

74. Julius Caesar used a similar design to cross the Rhine River.

75. Benjamin Latrobe to Elias B. Caldwell, January 17, 1810, in *CL2*, 824–829, 825.

76. Latrobe to Caldwell, January 17, 1810, in *CL2*, 824–829, 827.

77. Latrobe, *Opinion on a Project*, 20.

78. Hains, "Improvement of Harbors," 783.

79. Ibid.

80. Michler, "Report of Michler," in *Report of the Chief of Engineers to the Secretary of War, for the Year 1869*, 519–525, 519; and in the separate report of the same title (Washington, DC: US Government Printing Office, 1869), 28; and "An Act Authorizing the Erection of a Bridge over the River Potomac to Alexander's Island," approved February 5, 1808.

81. *Raleigh (NC) Minerva*, June 8, 1809, 2.

82. Royall, *Sketches of History, Life and Manners*, 145.

83. *Raleigh (NC) Minerva*, June 8, 1809, 2.

84. Royall, *Sketches of History, Life and Manners*, 145.

85. "An Act for the Relief of the President and Directors of the Washington Bridge Company," approved April 26, 1818, in U.S. Office of Chief of Engineers, *Laws Relating to Construction of Bridges*, 15.

86. Crew, *Centennial History of Washington, D.C.*, 335.

87. Ibid., 337.

88. Loammi Baldwin to Congressman C. F. Mercer, April 5, 1834, in Baldwin, "Letters from Loammi Baldwin," 3–4.

89. Ibid., 1.

90. A. B. McLean to Levi Woodbury, June 6, 1836, enclosure "A," in Woodbury, "Letter from Secretary of Treasury," 2.

91. Michler, "Report of Michler," in *Report of the Chief of Engineers to the Secretary of War, for the Year 1869*, 519–525, 525; and in the separate report of the same title (Washington, DC: US Government Printing Office, 1869), 34.

92. "An Act for Erecting a Bridge over the Eastern Branch of the Patowmack River," in Kilty, *Laws of Maryland*.

93. *Washington National Intelligencer*, January 11, 1804, in Bryan, *History of the National Capital*, 1:491n5.

94. "An Act for Erecting a Bridge over the Eastern Branch, or Anacostia River," November 1797, in Kilty, *Laws of Maryland*.

95. *Washington National Intelligencer*, December 20, 1822; February 4, 1825, in Bryan, *History of the National Capital*, 1:99n1.

96. Congress responded with "An Act for the Relief of the Eastern Branch Bridge Company" and "An Act for the Relief of the Anacostia Bridge Company," both approved March 3, 1815. Michler, "Report of Michler," in *Report of the Chief of Engineers to the Secretary of War, for the Year 1869*, 519–525, 519; and in the separate report of the same title (Washington, DC: US Government Printing Office, 1869), 28.

97. "An Act to Incorporate a Company to Build a Bridge over the Eastern Branch of [the] Potomac, between Eleventh and Twelfth Streets East, in the City of Washington," approved February 24, 1819, in U.S. Office of Chief of Engineers, *Laws Relating to Construction of Bridges*, 15–17.

98. Brief of "An Act to Incorporate a Company to Build a Bridge over the Eastern Branch between Eleventh and Twelfth Streets East," February 24, 1819, as reported in Michler, "Report of Michler," in *Report of the Chief of Engineers to the Secretary of War, for the Year 1869*, 519–525, 519; and in the separate report of the same title (Washington, DC: US Government Printing Office, 1869), 28.

99. *City of Washington Gazette*, October 19, 1819.

100. Tindall, *Standard History*, 119.

101. Formwalt, *Latrobe and Internal Improvements*, 211.

102. Clussman and Kammerhueber, *Report on the Washington Canal*, 5–6.

103. Tindall, *Standard History*, 120.

104. *CL1*, 284n10. Actual amounts spent on the canal were $2,133.33 in 1793, $1,600.00 in 1794, and $1,241.42 in 1795. U.S. House, *Report of Committee on Expenditure*, tables on 245–246.

105. *RDCC*, roll 1.

106. Tindall, *Standard History*, 186. See *RDCC*, roll 1, 131–132.

107. U.S. House, *Report of Committee on Petition*, February 11, 1802, 259.

108. Washington Canal Company, *Report of Washington Canal Company*, January 31, 1817, 1.

109. Tindall, *Standard History*, 214–215; Heine, "Washington City Canal," 3.

110. *CL1*, 388n2.

111. Law, *Observations on Intended Canal*, 159–168.

112. *CL1*, 453n4.

113. Ibid., 388n3.

114. Van Horne points out that Latrobe's original report on the Washington City Canal has not been found but that parts of it were printed in the *Washington National Intelligencer* on May 26, 1809, and in one of the versions of Law, *Observations on Intended Canal*. See *CL2*, 822n2.

115. *CL2*, 823.

116. Ibid.; Heine, "Washington City Canal," 5–6; Formwalt, *Latrobe and Internal Improvements*, 213.

117. *CL2*, 824n8; and Formwalt, *Latrobe and Internal Improvements*, 213.

118. Latrobe to Caldwell, president, Washington Canal Company, January 19, 1810, in *CL2*, 830–833, 831.

119. Latrobe to the president and directors, Washington Canal Company, February 6, 1810, in *CL2*, 838–842, 838.

120. The wooden locks originally constructed at Little Falls for the Potomac Canal were always a problem. Very early, as early as 1802, the company realized that they would deteriorate and would have to be replaced by masonry locks. In that year, the company began stockpiling stone at Little Falls for the purpose. See Kapsch, *Potomac Canal*.

121. Latrobe to the president and directors, Washington Canal Company, February 6, 1810, in *CL2*, 838–842, 840.

122. Ibid., 838.

123. *CL2*, 848n4; Formwalt, *Latrobe and Internal Improvements*, 215.

124. Latrobe to the president and directors, Washington Canal Company, February 6, 1810, in *CL2*, 838–842, 839.

125. Ibid., 842n5.

126. Ibid., 841, 842n8, 848n4.

127. Ibid., 859n6. Charles Randle was later given a contract to survey and prepare an estimate for the National Road from Cumberland to Wheeling. See Latrobe to Charles Randle, July 31, 1810, in *CL2*, 889–890, 889.

128. Historian Heine wrote that Sanderlin in his book *Great National Project*, Bryan in his *History of the National Capital*, vol. 1, and perhaps others, were incorrect in stating that the groundbreaking date for the canal was May 9, 1810. Heine claimed that it was May 2, 1810. See Heine, "Washington City Canal," 6n17.

129. *CL2*, 859n6.

130. Heine, "Washington City Canal," 6.

131. Latrobe to Isaac Hazlehurst, June 11, 1810, in *CL2*, 871–873, 872.

132. The "old cut" of the canal was the section linking St. James Creek to Tiber Creek. This work was undertaken by the commissioners between 1793 and 1795. A total of approximately five thousand dollars was expended at that time on the connection.

133. See *CL2*, 877n2.

134. Latrobe to Thomas Law, July 17, 1810, in *CL2*, 875–879, 875.

135. Ibid.

136. *CL2*, 877n6.

137. Latrobe to Hazlehurst, September 1, 1810, in *CL2*, 892–895, 892.

138. Latrobe to canal company president Caldwell, December 31, 1810, in *CL2*, 951–952, 951.

139. Latrobe to Law, June 27, 1811, in *CL3*, 100–103, 101.

140. Latrobe to James Cochran, July 8, 1811, in *CL3*, 110–112, 110–111.

141. Latrobe to Law, June 27, 1811, in *CL3*, 100–103, 100.

142. Latrobe to Cochran, July 8, 1811, in *CL3*, 110–112, 111.

143. Heine, "Washington City Canal," 7; William Elliot, *The Washington Guide* (Washington, DC: Franck Taylor, 1837), 277–278.

144. The Clubfoot Canal linked New Bern, North Carolina, on the Neuse River, with Beaufort, North Carolina, on Bogue Sound. Latrobe to Randle, August 19, 1815, in *CL3*, 688–691, 689, 690. Also see 691n5.

145. Clussman and Kammerhueber, *Report on the Washington Canal*, 7.

146. Latrobe to Randle, August 19, 1815, in *CL3*, 688–691, 690.

147. Washington Canal Company, *Report of Washington Canal Company*, January 31, 1817, 1.

148. Washington Canal Company, (Annual Report) *To the Senate and House*, April 21, 1824, 2; (Annual Report) *To the Senate and House*, February 24, 1825, 2; and (Annual Report) *To the Senate and House*, March 10, 1826, 2.

149. Chesapeake and Ohio Canal Company, *Proceedings of the President and Directors*, 477, as cited by Bearrs, *Bridges*, 47.

150. Swift and Hale, *Report on Chesapeake and Ohio Canal*, 9.

151. Heine, "Washington City Canal," 13.

152. Ibid., 11–12; U.S. Senate, *Report on Senate bill 310*, 2.

153. U.S. Senate, *Report on Senate bill 310*, 3.

154. Swift and Hale, *Report on Chesapeake and Ohio Canal*, 9.

155. U.S. Senate, *Report on Senate bill 310*, 3.

156. Ibid., 4.

157. Ibid.

158. Abert and Kearney, *Report upon Chesapeake and Ohio Canal*; and Bearrs, *Bridges*, 7–8.

6. BUILDING MILITARY DEFENSES FOR THE CAPITAL

Epigraph. Lear, *Observations on the River Potomack*, 26.

Epigraph. Royall, *Sketches of History, Life and Manners*, 144.

1. H. Knox, *Instructions to John Vermonnet*, May 12, 1794, regarding fortifications at Annapolis, Maryland, and Alexandria, Virginia, in *American State Papers*, 1:93.

2. Harris and Preston, *Papers of William Thornton*, xlvi, 126n.

3. Knox, *Instructions to John Vermonnet*, 1:93.

4. Ibid.

5. Ibid.

6. Ibid. Martello towers, small forts, had been used from the time of the French Revolution. American blockhouses modeled on British Martello towers were frequently constructed of wood instead of masonry.

7. Ibid.

8. John Vermonnet to the Secretary of War, June 17, 1794, in *A Copy of Letters from John Vermonnet to the Secretary of War, respecting Fortifications*, in *American State Papers*, 1:94.

9. Vermonnet to the Secretary of War, July 5, 1794, in *A Copy of Letters from John Vermonnet to the Secretary of War, respecting Fortifications*, in *American State Papers*, 1:94.

10. Vermonnet to the Secretary of War, November 5, 1794, in *A Copy of Letters from John Vermonnet to the Secretary of War, respecting Fortifications*, in *American State Papers*, 1:95.

11. "Fortifications," communicated to the Senate by the Secretary of War (Timothy Pickering), January 18, 1796, in State of the Fortifications of the United States, in *American State Papers*, 1:111.

12. "Fortifications," communicated to Congress (by President Thomas Jefferson), Febru-

ary 18, 1806, No. 60; January 6, 1809, No. 84, in *American State Papers*, 1:192–197, 194, 223–224, 224.

13. Jonathan Williams to Secretary of War, February 13, 1808, War Department, Chief of Engineers, Buell's Collection, 58510/134, NARA, as quoted in *Pen Portraits of Alexandria, Virginia, 1739–1900*, ed. T. Michael Miller (Bowie, MD: Heritage Books, 1987), 63–64.

14. "Fortifications and Gunboats," communicated to the Senate on the Third of December, 1807 (by Secretary of War H. Dearborn), November 20, 1807, No. 74, in *American State Papers*, 1:219–222, 221.

15. "Fort Washington," National Park Service, www.nps.gov/fowa/historyculture/warburton.htm (accessed February 16, 2015).

16. "Fortifications," communicated to the House of Representatives (by H. Dearborn, Secretary of War), December 8, 1807, No. 76, in *American State Papers*, 1:236–239, 237.

17. "Fortifications," communicated to the House of Representatives (by W. Eustis, Secretary of War), December 21, 1809, No. 89, in *American State Papers*, 1:245–247, 246.

18. "Fortifications," communicated to the House of Representatives (by W. Eustis, Secretary of War), December 17, 1811, No. 106, in *American State Papers*, 1:307–311, 310.

19. "Fort Washington," National Park Service.

20. cehp, *Historic Resources Study*, 6.

21. "Fort Washington," National Park Service.

22. Report of Lieutenant James L. Edwards on Fort Washington, July 25, 1814, in U.S. House, "Capture of City of Washington," 1:545.

23. Brigadier General Winder to Secretary of War John Armstrong, August 19, 1814 in U.S. House, "Capture of City of Washington," 1:547.

24. U.S. House, "Capture of City of Washington," 1:566.

25. The ships were the *Seahorse* (38-gun frigate); the *Euryalus* (36-gun frigate); the bomb ships *Aetna*, *Devastation* and *Meteor* (each with 2 mortars and from 8 to 10 guns); and the rocket ship *Erebus* (16 guns on upper deck and 32-pound Congreve rockets stored below). Snow, *When Britain Burned White House*, 151–152.

26. Captain Samuel T. Dyson to Armstrong, August 29, 1814, in U.S. House, "Capture of City of Washington," 1:588; Francis S. Belton, Assistant Adjutant General Tenth Military District, "General Orders," November 17, 1815, in U.S. House, "Capture of City of Washington," Communicated to the House of Representatives on the 29th of November, 1814, 13th Cong., 3d sess. No. 137, in *American State Papers*, 1:588–589.

27. L'Enfant did submit a report on the reconstruction of the fort to Secretary of War A. J. Dallas on June 23, 1815. See Pierre L'Enfant to Secretary of War A. J. Dallas, June 23, 1815, reprinted as Appendix H in Caemmerer, *Life of L'Enfant*, 451–462.

28. National Register Nomination, "Fort Washington," https://npgallery.nps.gov/GetAsset/2d04aefc-d0dd-4c75-88af-5c17febc5ac4 (accessed July 9, 2017), item 8.3.

29. Ibid., items 8.4, 8.7.

30. Ibid., item 8.10.

31. Uncited reference in McClellan, *Silent Sentinel on the Potomac*, 8.

32. John Norton, Deputy Commissary United States' Ordnance, to Colonel R. M. Johnson, October 21, 1814, in U.S. House, "Capture of City of Washington," 1:586–587.

33. Snow, *When Britain Burned White House*, 143–144; Pitch, *Burning of Washington*, 138–139; McClellan, *Silent Sentinel on the Potomac*, 12.

34. Snow, *When Britain Burned White House*, 109–110; Pitch, *Burning of Washington*, 101–103.

35. "Dry Docks" (Report of Benjamin Latrobe on the proposed dry dock, locks, and canal to be constructed in and to the Navy Yard), December 4, 1802, and transmitted to Congress by President Thomas Jefferson on December 28, 1802, No. 30, in *American State Papers*, 1:104–108.

36. Benjamin Latrobe to Thomas Tingey, May 18, 1805, in *CL2*, 77–82, 78; Hibben, *Navy-Yard, Washington*, 22.

7. THE FIRST PUBLIC BUILDING CAMPAIGN (1791–1802)

Epigraph Tindall, *Standard History*, 223.

1. Proceedings, November 26, 1791, in *RDCC*, roll 1, 64.

2. Francis Cabot to the commissioners, December 11, 1791, in *RDCC*, Letters Received, roll 8, 62.

3. Proceedings, September 24, 1791, in *RDCC*, roll 1, 30.

4. Isaac Roberdeau, working for Pierre L'Enfant, reported to the commissioners, "With respect to any Clay you may order to be turned up at the President House; I can say nothing, but that I have received no instructions from the Major [L'Enfant] to proceed in that part of the City." Roberdeau to the commissioners, January 10, 1792, in *RDCC*, Letters Received, roll 8, 72.

5. Proceedings, January 7, 1792, in *RDCC*, roll 1, 72.

6. President Washington to Thomas Jefferson, August 29, 1791, in Padover, *Jefferson and the Capital*, 67–68, 68.

7. Notes on Commissioners Meeting, September 8, 1791, in Padover, *Jefferson and the Capital*, 70–74, 72.

8. The final advertisement for the President's House design competition, which was very similar to Jefferson's draft, was approved March 14, 1792. Proceedings, March 14, 1792, in *RDCC*, roll 1, 81–82.

9. Commissioners, "Draft of Competition for Plan of a Capitol," March 14, 1792, in Padover, *Jefferson and the Capital*, 119–120; and in *DH*, 14.

10. Proceedings, July 18, 1792, in *RDCC*, roll 1, 124.

11. Tindall, *Standard History*, 173–174.

12. Ryan and Guinness, *White House*, 88. This was probably the *Charleston (SC) City Gazette*, November 15, 1792.

13. Proceedings, October 13, 1792, in *RDCC*, roll 1, 145.

14. Commissioners to Samuel Blodget, January 5, 1793, in *DH*, 22.

15. Butler, "Competition 1792," 75.

16. Ibid., 74.

17. The story of the design competition of the United States Capitol has been told many times: Howard, "Architects of the Capitol"; Brown, *History of the Capitol*; Brown, *Glenn Brown's History*; Hazleton, *National Capitol*; Kimball and Bennett, "Competition for the Federal Buildings"; Frary, *They Built the Capitol*; Butler, "Competition 1792"; Scott, *Temple of Liberty*, 27–43; Allen, *History of the Capitol*.

18. President Washington to the commissioners, July 23, 1792, in *DH*, 18.

19. Commissioners to Jefferson, August 29, 1792, in *DH*, 19.

20. "Fellow Citizen" [William Thornton] to the commissioners, in *RDCC*, Index to Letters Received, roll 8, 112.

21. Harris and Preston, *Papers of William Thornton*, xlvii.

22. Commissioners to Thornton, December 4, 1792, in *DH*, 19.

23. Harris and Preston, *Papers of William Thornton*, xlvi, 239.

24. Ibid., 239. These drawings have not been found and are presumed lost.

25. Washington to the commissioners, January 31, 1793, in Fitzpatrick, *Writings of Washington*, 32:325.

26. Jefferson to Daniel Carroll, February 1, 1793, in Padover, *Jefferson and the Capitol*, 171; and in *DH*, 23.

27. Commissioners to Jefferson, February 7, 1793, in Padover, *Jefferson and the Capital*, 172–173, 173; and in *DH*, 23.

28. Commissioners to Stephen Hallet, March 13, 1793, in *DH*, 25.

29. Commissioners to Thornton, April 5, 1793, in Harris and Preston, *Papers of William Thornton*, 238; and in *DH*, 25.

30. *CL1*, 448n3; Kimball and Bennett, "Thornton and the Design of the United States Capitol," 89–90.

31. Thornton to the commissioners, ca. April 10, 1793, Harris and Preston, *Papers of William Thornton*, 242–249.

32. Thornton to Jefferson, July 8–12, 1793, Harris and Preston, *Papers of William Thornton*, 262–268, 262, 266.

33. Jefferson to Washington, in Padover, *Jefferson and the Capital*, 184–186, 184; and in *DH*, 26–27.

34. Washington to Jefferson, June 30, 1793, in Padover, *Jefferson and the Capital*, 181–183, 182.

35. Ibid., 181–183, 183.

36. Jefferson to Thornton, July 8, 1793, in Harris and Preston, *Papers of William Thornton*, 259.

37. Harris and Preston, *Papers of William Thornton*, 262–268.

38. Jefferson to Washington, July 17, 1793, in Padover, *Jefferson and the Capital*, 184–186, 184–185; and in *DH*, 26–27.

39. Ibid.

40. Washington to the commissioners, July 25, 1793, in *DH*, 27–28, 27.

41. *DH*, 27–28, 28.

42. Ibid.

43. Ibid.

44. *Columbian Mirror and Alexandria (VA) Gazette*, September 25, 1793, as quoted by Allen, *History of the Capitol*, 24; and in *DH*, 29.

45. Miscellaneous Treasury Records, RG 217, NARA.

46. Ibid.

47. Ibid.

48. Joseph Dove to the commissioners, March 31, 1800, in *RDCC*, Letters Received, roll 8, 1843.

49. Miscellaneous Treasury Records, RG 217, NARA.

50. Proceedings, April 12, 1792, in *RDCC*, roll 1, 95.

51. Proceedings, March 14, 1793, in *RDCC*, roll 1, 173.

52. Proceedings, September 23, 1793, in *RDCC*, roll 1, 197; and in *DH*, 29.

53. Ibid.

54. "1793 Yellow Fever Epidemic," Wikipedia, https://en.wikipedia.org/wiki/1793_Philadelphia_yellow_fever_epidemic (accessed September 20, 2014); and Carey, *Short Account of Malignant Fever*.

55. Collen Williamson to the commissioners, March 11, 1796, in *RDCC*, Letters Received, roll 12, 765.

56. Proceedings, January 1, 1793, in *RDCC*, roll 1, 156.

57. C. M. Harris, "Editorial Note: Thornton's Capitol Modified: The Philadelphia Conference," in Harris and Preston, *Papers of William Thornton*, 255–259, 257.

58. Commissioners to Hallet, June 24, 1794, in *DH*, 30.

59. Commissioners to Hallet, June 26, 1794, in *DH*, 31.

60. Commissioners to Hallet, June 27, 1794, in *DH*, 32.

61. Hallet to the commissioners, June 28, 1794, in RG 42, NARA.

62. Daniel D. Reiff, "Hallet, Etienne Sulpice," in Placzek, *Macmillan Encyclopedia of Architects*, 2:298.

63. Proceedings, October 16th to 22nd, 1793, in *RDCC*, roll 1, 201.

64. Cornelius McDermot Roe to the commissioners, January 30, 1794, in *RDCC*, Letters Received, roll 1, 235–236. A *perch* of *masonry* is the volume of a stone wall one *perch* (16½ feet) long, 18 inches high, and 12 inches wide.

65. Commissioners to Collen Williamson, May 19, 1794, in *RDCC*, Letters Sent, roll 3, 61.

66. Commissioners to Hicks, Maitlands, Ore, and Hakesly, May 17, 1794, in *RDCC*, Letters Sent, roll 3, 62.

67. Tindall, *Standard History*, 186.

68. R. Brown and others to the commissioners, June 6, 1794, in *RDCC*, Letters Received, roll 10, 384.

69. Proceedings, June 6–8, 1794, in *RDCC*, roll 1, 219.

70. Proceedings, June 22 to 28, 1794, in *RDCC*, roll 1, 256.

71. Commissioners to Williamson, June 7, 1794, in *RDCC*, Letters Sent, roll 3, 65.

72. *Washington Gazette*, June 22, 1796, 3.

73. Williamson to the commissioners, June 5, 1794, in *RDCC*, Letters Received, roll 10, 383.

74. Proceedings, June 22 to 28, 1794, in *RDCC*, roll 1, 256.

75. Proceedings, February 18, 1795, in RDCC, roll 1, 361–365.

76. Commissioners to Secretary of State Edmund Randolph, June 26, 1795, in *DH*, 35.

77. Commissioners to Randolph, July 13, 1795, in *DH*, 36.

78. Roe to commissioners, August 4, 1795, in RG 42, NARA; Allen, *History of the Capitol*, 30.

79. Commissioners to Randolph, July 13, 1795, in *DH*, 36.

80. Proceedings, May 28, 1795, in *RDCC*, roll 1, 391.

81. Williamson to the commissioners, March 11, 1796, in *RDCC*, Letters Received, roll 12, 765.

82. Allen, *History of the Capitol*, 25.

83. Padover, *Jefferson and the Capital*, Appendix, 507; and *CL1*, 284n11. On February 20, 1804, Benjamin Latrobe wrote of George Blagden: "After the departure of Mr. Hatfield, the public became indebted to Mr. Geo. Blagden, of whose integrity and abilities, as the principal Stone Mason, his work bears honorable testimony, for the excellent execution of the Freestone work of the North wing." Benjamin Latrobe, "The Report of the Surveyor of the Public Buildings," February 20, 1804, in *CL1*, 443–449, 445.

84. Allen, *History of the Capitol*, 29.

85. George Hadfield to the commissioners, ca. October 15, 1795, RG 42, NARA.

86. Hadfield to the commissioners, October 28, 1795, RG 42, NARA.

87. Thornton to Washington, November 2, 1795, in Thornton Papers, Library of Congress, as quoted in Allen, *History of the Capitol*, 31–32.

88. Washington to the commissioners, November 9, 1795, in *DH*, 36–37, 37.

89. Ibid.

90. Ibid.

91. Allen, *History of the Capitol*, 32.

92. Commissioners to Washington, June 29, 1796, RG 42, NARA.

93. U.S. House, *Report of Committee on Authorizing Loan*, 1:142.

94. James Hoban makes reference to the original "raising" when he requests funds for a "raising" in rebuilding the President's House in 1816. James Hoban to the commissioners, October 10, 1816, in *RDCC*, Letters Received, roll 8, 2379.

95. U.S. House, *Report of Committee on Authorizing Loan*, 1:142.

96. Commissioners to Washington, January 29, 1795, in RG 42, NARA; Allen, *History of the Capitol*, 28.

97. Farm and Others to the commissioners, June 6, 1796, in *RDCC*, Letters Received, roll 12, 829.

98. Francis Cochran and others to the commissioners, March 6, 1797, in *RDCC*, Letters Received, roll 13, 1081.

99. Washington to the commissioners, January 29, 1797, in RG 42. NARA; and in *DH*, 76–77.

100. Commissioners to Hoban and Hadfield, October 3, 1797, in *RDCC*, Letters Sent, roll 4, 144–145.

101. Proceedings, November 14, 1797, in *RDCC*, roll 1, 35.

102. Proceedings, November 15, 1797, in *RDCC*, roll 1, 37.

103. Proceedings, December 5, 1797, in *RDCC*, roll 1, 45.

104. Pierce Purcell to the commissioners, September 13, 1798, in *RDCC*, Letters Received, roll 15, 1440.

105. Purcell to the commissioners, September 20, 1798, in *RDCC*, Letters Received, roll 15, 1449.

106. Proceedings, November 15, 1797, in *RDCC*, roll 1, 37.

107. Proceedings, December 5, 1797, in *RDCC*, roll 1, 46.

108. Hoban to the commissioners, December 4, 1797, in *RDCC*, roll 14, 1255.

109. Proceedings, December 5, 1797, in *RDCC*, roll 1, 44–45.

110. Emmett C. Davison, ed., "History of Organized Labor in Washington, D.C.," in Proctor, *Washington*, 2:867.

111. Commissioners to Alexander White, March 27, 1798, in *RDCC*, Letters Sent, roll 4, 967.

112. White to John Miller, April 18, 1798, in *RDCC*, Letters Sent, roll 4, 988.

113. Ibid.

114. Proceedings, April 17, 1798, in *RDCC*, roll 1, 109.

115. Archie Burns and others to the commissioners, April 16, 1798, in *RDCC*, Letters Received, roll 13, 1332.

116. John McDouagh and others to the commissioners, April 16, 1798, in *RDCC*, Letters Received, roll 13, 1330.

117. Alexander Reid and others to the commissioners, April 17, 1798, in *RDCC*, Letters Received, roll 13, 1334.

118. Commissioners to the stone cutters at the Presidents House, April 16, 1798, in *RDCC*, Letters Sent, roll 4, 979.

119. Commissioners to Hadfield, April 18, 1798, in *RDCC*, Letters Sent, roll 4, 985.

120. This is the same Robert Stewart who owned a one-acre quarry on Government Island in Aquia Creek, Stafford County, "containing the best stone on the island." *CL1*, 427n4.

121. Commissioner Scott to Robert Stewart, April 18, 1798, in *RDCC*, Letters Sent, roll 4, 983.

122. Commissioner White to Miller, April 18, 1798, in *RDCC*, Letters Sent, roll 4, 988.

123. Miller to the commissioners, May 1, 1798, in *RDCC*, Letters Received, roll 13, 1349.

124. From George Blagdin [Blagden], April 21, 1798, in *RDCC*, Letters Received, roll 14, 1339.

125. Hadfield to the commissioners, April 30, 1798, in *RDCC*, Letters Received, roll 13, 1347.

126. Proceedings, April 23, 1798, in *RDCC*, roll 1, 114–115.

127. Emmett C. Davison, ed., "History of Organized Labor in Washington, D.C.," in Proctor, *Washington*, 2:867.

128. Commissioner White to Miller, May 7, 1798, in *RDCC*, Letters Sent, roll 4, 1000.

129. Peter Lennox to the commissioners, April 25, 1798, in *RDCC*, Letters Received, roll 13, 1344. The records also spell Lennox with a single *n*.

130. Munroe to Hadfield, April 16, 1798, in *RDCC*, Letters Sent, roll 4, 978.

131. Commissioners to Hadfield and Hoban, May 4, 1798, in *RDCC*, Letters Sent, roll 4, 995.

132. Commissioners to Hoban, May 12, 1798, in *RDCC*, Letters Sent, roll 4, 1013.

133. Commissioners to Hadfield, May 23, 1798, in *RDCC*, Letters Sent, roll 4, 1028.

134. Thomas Munroe to Joseph Huddleston, May 29, 1798, in *RDCC*, Letters Sent, roll 4, 1041.

135. Commissioners to Hadfield, May 18, 1798, in *RDCC*, Letters Sent, RG 42, NARA.

136. Commissioners to John Adams, June 25, 1798, in *RDCC*, Letters Sent, RG 42, NARA.

137. Edward C. Carter II, John C. Van Horne, and L. W. Formwalt, eds., *The Journals of Benjamin Henry Latrobe, 1799–1820: From Philadelphia to New Orleans*, vol. 3 of Carter et al., *Virginia Journals of Latrobe*, 3:72.

138. Commissioners to Hoban, June 23, 1798, in *RDCC*, Letters Sent, roll 4, 1060.

139. Commissioners to Hoban, July 2, 1798, in *RDCC*, Letters Sent, roll 4, 1076.

140. Commissioners to Redmund Purcell, May 25, 1798, in *RDCC*, Letters Sent, roll 4, 1034.

141. Samuel Smallwood to the commissioners, June 8, 1798, in *RDCC*, Letters Received, roll 14, 1375.

142. Proceedings, April 10, 1799, in *RDCC*, roll 2, 103.

143. Commissioners to Robert Auld, April 11, 1799, in *RDCC*, Letters Sent, roll 4, 231.

144. Commissioners to Richard Gridley, June 13, 1798, in *RDCC*, Letters Sent, roll 4, 1048.

145. Proceedings, August 20, 1797, in *RDCC*, roll 1, 206.

146. Gridley to the commissioners, June 11, 1798, in *RDCC*, Letters Received, roll 14, 1376.

147. Gridley to the commissioners, June 27, 1798, in *RDCC*, Letters Received, roll 14, 1390.

148. Commissioners to Gridley, June 27, 1798, in *RDCC*, Letters Sent, roll 4, 669.

149. Gridley to the commissioners, June 28, 1798, in *RDCC*, Letters Received, roll 14, 1392.

150. *Centinel of Liberty and Georgetown (DC) Advertiser,* January 4, 1799, 1.

151. Proceedings, January 9, 1799, in *RDCC*, roll 2, 61.

152. Commissioners to Hoban, April 11, 1799, in *RDCC*, Letters Sent, roll 4, 231.

153. Proceedings, April 4, 1799, in *RDCC*, roll 2, 101.

154. James Tompkins to the commissioners, April 9, 1799, in *RDCC*, Letters Received, roll 16, 1609.

155. John Dickey to the commissioners, April 11, 1799, in *RDCC*, Letters Received, roll 16, 1611.

156. Redmund Purcell to the commissioners, April 3, 1799, in *RDCC*, Letters Received, roll 16, 1606.

157. Commissioners to Hoban, April 2, 1800, in *RDCC*, Letters Sent, roll 4, 395–396.

158. Proceedings, October 20, 1800, in *RDCC*, roll 2, 11.

159. Carpenters at the President's house to the commissioners, October 21, 1800, in *RDCC*, Letters Received, roll 17, 1968.

160. Commissioners to Secretary of State and Secretary of the Navy, October 24, 1800, in *RDCC*, Letters Sent, roll 4, 94–95.

161. Proceedings, October 22, 1800, in *RDCC*, roll 2, 14.

162. Ibid.

163. Proceedings, October 24, 1800, in *RDCC*, roll 2, 17.

164. Hoban to the commissioners, October 29, 1800, in *RDCC*, Letters Received, roll 17, 1972.

165. Commissioners to Hoban, October 30, 1800, in *RDCC*, Letters Sent, roll 4, 28–29.

166. Public Buildings & Grounds, in *RDCC*, Index to Letters Received, roll 8, 1357.

167. Bryan, "Removal of Government to Washington," 253–262.

168. *Centinel of Liberty and Georgetown (DC) Advertiser*, June 15, 1798, 4. There was some discussion whether these dimensions were sufficient for a building of such size. See George Blagden (who maintained that the walls were not sufficient) to the commissioners, September 17, 1798, in *RDCC*, Letters Received, roll 8, 1444; Charles Cook and others (who maintained that the walls were adequate and the work good), September 11, 1798, in *RDCC*, roll 8, 1442; and Leonard Harbaugh (who maintained the walls were sufficiently strong), September 16, 1798, in *RDCC*, roll 8, 1444.

169. Tindall, *Standard History*, 213–214.

170. Commissioners to Hoban, July 23, 1799, in *RDCC*, Letters Sent, roll 3, 241; also RG 42, NARA.

171. Tindall, *Standard History*, 213–214. This building came to be known as the War Department building, but initially it housed other departments as well, such as the State Department. Tindall gives the overall dimensions of both buildings as 148 feet by 57 feet 6 inches.

172. Bryan, "Removal of the Government to Washington," 252–262.

173. *Washington National Intelligencer and Washington Advertiser*, November 3, 1800, 3; and *Washington Universal Gazette*, November 6, 1800, 3.

174. Tindall, *Standard History*, 223.

175. Index to Letters Received, June 1, 1802, in *RDCC*, roll 8, and RG 42, NARA.

8. THE SECOND PUBLIC BUILDING CAMPAIGN (1803–1811)

Epigraph. Benjamin Latrobe to President Jefferson, February 28, 1804, on his meeting with William Thornton to discuss needed changes in the design of the Capitol, in *CL1*, 441–443, 442.

Epigraph. William Thornton to Jonathan Smith Findlay, editor of the *Washington Federalist*, May 1, 1808, in *CL2*, 614–619.

1. Commissioners to John Emory, July 9, 1800, in *RDCC*, Letters Sent, roll 2; and in *DH*, 91.

2. Allen, *History of the Capitol*, 45.

3. Commissioners to President Thomas Jefferson, June 1, 1801, in Padover, *Jefferson and the Capital*, 209.

4. Jefferson to commissioners, June 2, 1801, in Padover, *Jefferson and the Capital*, 210–211; and in *DH*, 96–97.

5. Commissioners' Proceedings, June 10, 1801, in *RDCC*.

6. Allen, *History of the Capitol*, 46.

7. Commissioners to Jefferson, August 24, 1801, in Padover, *Jefferson and the Capital*, 222–223, 223.

8. James Hoban, "Report of James Hoban, Superintendent of the Capitol, of the Work Done at That Building from the 18th of May, 1801, to the 14th December, 1801," undated, in *DH*, 99.

9. Allen, *History of the Capitol*, 46.

10. District of Columbia Commissioners, "Report of the Commissioners," January 30, 1801.

11. U.S. House, *Report of Committee on Expenditure*, 243; and *DH*, 95–96, 95.

12. Ibid.

13. U.S. House, *Report Recommending Offices of Commissioners Be Abolished*, 1:260; and in *DH*, 100–101, 100.

14. U.S. Congress, "An Act to Abolish the Board of Commissioners in the City of Washington and for Other Purposes," in U.S. Congress, *Laws of the United States*, 1:498–500; and in *DH*, 101–102.

15. C. M. Harris and Daniel Preston, "Biographical Sketch of William Thornton, from 1759 to 1802," in Harris and Preston, *Papers of William Thornton*, xxxi–lxxv, li.

16. Jefferson to Benjamin Latrobe, March 6, 1803, in *CL1*, 260–261, 260; and in Padover, *Jefferson and the Capital*, 296.

17. Jefferson to Latrobe, March 6, 1803, in *CL1*, 262; and in Padover, *Jefferson and the Capital*, 297.

18. Latrobe to Jefferson, March 13, 1803, in *CL1*, 263.

19. Benjamin Latrobe, "Report on the U.S. Capitol," April 4, 1803, in letter to Jefferson, April 4, 1803, in *CL1*, 268–284.

20. Ibid., 278.

21. Ibid.

22. Ibid., 279.

23. Ibid., 280.

24. Ibid., 269.

25. Ibid., 270.

26. Ibid., 270–271.

27. Ibid., 272.

28. Thornton to Jefferson, July 8–12, 1793, in Harris and Preston, *Papers of William Thornton*, 262–268, 264.

29. Latrobe, "Report on U.S. Capitol," April 4, 1803, 276.

30. Allen, *History of the Capitol*, 53.

31. Benjamin Latrobe, "Report on the Public Buildings," February 20, 1804, in letter to the President of the United States (Thomas Jefferson), in *CL1*, 429–433, 429, 430; and in *DH*, 104–106, 105.

32. Latrobe, "Report on Public Buildings," February 20, 1804, in letter to Jefferson, in *CL1*, 433n5, 431.

33. Jefferson to Latrobe, February 28, 1804, in *CL1*, 439–440.

34. Latrobe, memorandum, February 27, 1804, in *CL1*, 438.

35. Latrobe to Jefferson, February 27, 1804, in *CL1*, 436–439, 437.

36. Benjamin Latrobe, "Letter from B. Henry Latrobe to the Chairman of the Committee of the House of Representatives in Congress, [Philip R. Thompson], to whom was referred the, Message of the President of the United States of the 22d of February 1804, transmitting a report of the Surveyor of the public buildings of the 20th of February 1804," February 28, 1804, 443–449, 444–445; in *DH*, 107–110, 108.

37. Latrobe to Jefferson, February 28, 1804, in *CL1*, 441–443, 442.

38. Ibid.; and in *DH*, 106–107, 107.

39. Latrobe, "Letter to [Thompson]," February 28, 1804, in *CL1*, 444, 445; and in *DH*, 108.

40. *DH*, 110–111.

41. Latrobe to Jefferson, March 29, 1804, in *CL1*, 466–473; and in Padover, *Jefferson and the Capital*, 343–344 (includes first two paragraphs only).

42. Latrobe to Jefferson, March 29, 1804, in *CL1*, 467; and in Padover, *Jefferson and the Capital*, 343–344 (includes first two paragraphs only).

43. Ibid.

44. "Corps de logis" is an architectural term referring to the principal block of a building, usually a classical building. Here Latrobe is referring to the domed central section of the Capitol.

45. Latrobe to Jefferson, March 29, 1804, in *CL1*, 468; and in Padover, *Jefferson and the Capital*, 343–344 (includes first two paragraphs only).

46. Jefferson to Latrobe, March 31, 1804, in *CL1*, 473–475, 473, 474.

47. Jefferson to Latrobe, April 9, 1804, in *CL1*, 475–476, 475; and in Padover, *Jefferson and the Capital*, 344–345, 344.

48. Thornton to Latrobe, April 23, 1804, in *CL1*, 479–480.

49. Latrobe to Thornton, April 28, 1804, in *CL1*, 481–483, 481–482.

50. Thornton to Latrobe, June 27, 1804, in *CL1*, 518–519, 518.

51. Ibid., 519.

52. Latrobe to Thornton, July 21, 1804, in *CL1*, 523–524; and Columbia Historical Society, *Records of the Columbia Historical Society, Washington, D.C.* (1915), 18:150, 175–176, as quoted by *CL1*, 524n3.

53. Latrobe, "Report on the Public Buildings," December 1, 1804, in *CL1*, 578.

54. Ibid., 579.

55. Ibid.; and in *DH*, 112.

56. Latrobe, "Report on the Public Buildings," December 1, 1804, in *CL1*, 579.

57. Latrobe, "Report on the Public Buildings," December 1, 1804, in *CL1*, 581; and in *DH*, 112.

58. Ibid.

59. Ibid.

60. Latrobe, "Report on the Public Buildings," December 1, 1804, in *CL1*, 581; and in *DH*, 112–113.

61. Latrobe, "Report on the Public Buildings," December 1, 1804, in *CL1*, 581; and in *DH*, 113.

62. Latrobe to Philip R. Thompson, December 30, 1804, in *CL1*, 586–588, 586; and in *DH*, 114.

63. Latrobe to Thompson, December 30, 1804, in *CL1*, 588n1; and "An Act making an appropriation for completing the south wing of the Capitol, at the city of Washington, and for other purposes," approved January 25, 1805, in *DH*, 115.

64. Latrobe to Jefferson, May 5, 1805, in *CL2*, 62–67.

65. Latrobe to Jefferson, June 28, 1805, in *CL2*, 95–97, 96.

66. Latrobe to Jefferson, August 31, 1805, in *CL2*, 130–138, 132.

67. Ibid., 130–131.

68. Wiebenson, "Domes of the Halle au Blé," 266n33; Hahmann, "How Stiff Is Curved-Plank Structure," 1501.

69. John Robison, *A System of Mechanical Philosophy*, 4 vols. (Edinburgh: Printed for John Murray, London, 1822), 1:611–622.

70. Wiebenson, "Domes of the Halle au Blé," 264; Robison, *System of Mechanical Philosophy*, 1:611–612.

71. The wooden dome had burned on October 16, 1803, and was replaced by a smaller dome with iron ribs and copper sheathing. See Wiebenson, "Domes of the Halle au Blé," 266.

72. Latrobe to Jefferson, August 31, 1805, in *CL2*, 130–138, 134–136.

73. Wiebenson, "Domes of the Halle au Blé," 266.

74. Jefferson to Latrobe, September 8, 1805, in *CL2*, 139–140, 140.

75. Latrobe, "Report on the Public Buildings," December 22, 1805, in *CL2*, 168, 172n1; and in *DH*, 115.

76. Latrobe, "Report on the Public Buildings," December 22, 1805, in *CL2*, 171; and in *DH*, 117.

77. Latrobe, "Report on the Public Buildings," December 22, 1805, in *CL2*, 171–172; and in *DH*, 117.

78. *Annals of Congress*, 9th Cong., 1st sess. (1806), 1284; and in *DH*, 119.

79. Jefferson to Latrobe, July 1, 1806, in *CL2*, 236–237, 236.

80. Ibid.

81. Latrobe to John Lenthall, July 3, 1806, in *CL2*, 237–239, 237.

82. Latrobe to George Blagden, July 13, 1806, in *CL2*, 246. Hunt was listed in Philadelphia directories as late as 1817. Barr was a stone cutter in Philadelphia from 1802 to 1810.

83. Jefferson to Latrobe, July 17, 1806, in *CL2*, 247.

84. Ibid.

85. Latrobe to Jefferson, August 15, 1806, in *CL2*, 262–266, 263.

86. Ibid.

87. Latrobe to Jefferson, August 27, 1806, in *CL2*, 269–270, 269.

88. Ibid., 270.

89. Latrobe to Jefferson, August 31, 1805, in *CL2*, 130–138, 134–137.

90. The tension rings would have been "purlins and iron straps, which made so many hoops to the whole," described by Robison in his description of the wooden dome of the Halles au Blé in *System of Mechanical Philosophy*, 1:611–612.

91. Latrobe to Jefferson, October 29, 1806, in *CL2*, 277–282, 279.

92. Ibid.

93. Ibid.

94. Jefferson to Lenthall, October 21, 1806, as quoted in *CL2*, 281n2; and in Padover, *Jefferson and the Capital*, 371–372.

95. Jefferson to Latrobe, October 31, 1806, in *CL2*, 282–284, 282–283.

96. Latrobe, "Report of the Surveyor," November 25, 1806.

97. Latrobe, *Private Letter to Congress*, 296–321.

98. *CL2*, 316n1.

99. Jefferson, introductory paragraph to Latrobe, "Report of the Surveyor," November 25, 1806, in Padover, *Jefferson and the Capital*, 372; and in *DH*, 120.

100. Latrobe, introductory paragraph to "Report of the Surveyor," November 25, 1806, in Padover, *Jefferson and the Capital*, 373; and in *DH*, 120.

101. Latrobe, "Report of the Surveyor," November 25, 1806, in Padover, *Jefferson and the Capital*, 373; and in *DH*, 120.

102. Latrobe, "Report of the Surveyor," November 25, 1806, in Padover, *Jefferson and the Capital*, 373–374; and in *DH*, 120.

103. Latrobe, "Report of the Surveyor," November 25, 1806, in Padover, *Jefferson and the Capital*, 375–376; and in *DH*, 121.

104. Latrobe, "Report of the Surveyor," November 25, 1806, in Padover, *Jefferson and the Capital*, 378–379; and in *DH*, 122–123.

105. *DH*, 123.

106. House proceedings of December 15, 1806, in *Annals of Congress*, 9th Cong., 2nd sess. (1806), 159; and in *DH*, 124.

107. Latrobe, *Private Letter to Congress*, 299–300.

108. Ibid., 300–302.

109. Ibid., 302–303.

110. Ibid., 304.

111. Ibid., 305–306.

112. Ibid., 309.

113. House Proceedings of February 13, 1807, in *DH*, 124.

114. Latrobe, *Private Letter to Congress*, 317n2. Van Horne et al. provide a summary of Thornton's comments on Latrobe's *Private Letter* in *CL2*, 317–321.

115. Latrobe to Jefferson, January 6, 1807, in *CL2*, 356–358, 356–357.

116. Latrobe to Jefferson, April 14, 1807, in *CL2*, 408–409, 409.

117. Jefferson to Latrobe, April 22, 1807, in *CL2*, 410–411, 410; and in Padover, *Jefferson and the Capital*, 386–387, 386.

118. Latrobe to Jefferson, August 13–15, 1807, in *CL2*, 461–467, 463; and in Padover, *Jefferson and the Capital*, 394–396, 394.

119. Latrobe to Jefferson, August 13–15, 1807, in *CL2*, 464, 465.

120. Jefferson to Latrobe, August 18, 1807, in *CL2*, 469–470, 469.

121. Latrobe to Jefferson, August 21, 1807, in *CL2*, 472–474, 473, 474.

122. Latrobe to Jefferson, September 1 1807, in *CL2*, 475–476, 475; and in Padover, *Jefferson and the Capital*, 396–397, 396.

123. Latrobe to Jefferson, September 5, 1807, in *CL2*, 477–478, 477, 478n1.

124. Latrobe to Jefferson, September 17 1807, in *CL2*, 482–486, 484.

125. Ibid., 482, 484.

126. *CL2*, 489n1.

127. Latrobe to Lenthall, October 18, 1807, in *CL2*, 488–490, 489n.

128. Benjamin Latrobe, *Letter and Report from Benjamin H. Latrobe, Surveyor of the Public Buildings at the City of Washington, on the Subject of the Said Buildings*, October 27, 1807 (Washington, DC, 1807), in *CL2*, 491–494; in *Annals of Congress*, 10th Cong., 1st sess. (1807), 790, 800; in *DH*, 129–131; and later published in *Washington National Intelligencer*, November 11, 1807.

129. Latrobe, *Letter and Report*, in *CL2*, 493, 494; in *Annals of Congress*, 10th Cong., 1st sess. (1807), 790, 800; in *DH*, 129–131; and later published in *Washington National Intelligencer*, November 11, 1807.

130. *Washington National Intelligencer*, October 27, 1807, in *CL2*, 505n1.

131. Latrobe to Samuel Harrison Smith, editor of the *Washington National Intelligencer*, November 22, 1807, in *CL2*, 499–506; and published in *Washington National Intelligencer*, November 30, 1807.

132. Ibid., 499.

133. Ibid., 500–501.

134. Ibid.

135. Ibid., 502–503.

136. Ibid., 503.

137. Ibid., 504.

138. Ibid., 505.

139. Latrobe, "Report of the Surveyor," March 23, 1808, in *CL2*, 565; and in *DH*, 131.

140. Latrobe, "Report of the Surveyor," March 23, 1808, in *CL2*, 566; and in *DH*, 131–132.

141. Latrobe, "Report of the Surveyor," March 23, 1808, in *CL2*, 567; and in *DH*, 132.

142. Latrobe, "Report of the Surveyor," March 23, 1808, in *CL2*, 569, 577n8; in *DH*, 134; in No. 249, *American State Papers*, 1:719–724; and in Padover, *Jefferson and the Capital*, 403–404.

143. Latrobe, "Report of the Surveyor," March 23, 1808, in *CL2*, 570; in Padover, *Jefferson and the Capital*, 405–406; and in *DH*, 134.

144. Latrobe, "Report of the Surveyor," March 23, 1808, in *CL2*, 570–571; in Padover, *Jefferson and the Capital*, 406–407; and in *DH*, 134.

145. Latrobe, "Report of the Surveyor," March 23, 1808, in *CL2*, 572; in Padover, *Jefferson and the Capital*, 408–409; and in *DH*, 131–137.

146. Latrobe, "Report of the Surveyor," March 23, 1808, in *CL2*, 572; in Padover, *Jefferson and the Capital*, 408; and in *DH*, 131–137.

147. Latrobe, "Report of the Surveyor," March 23, 1808, in *CL2*, 573–575; in Padover, *Jefferson and the Capital*, 410–412; and in *DH*, 131–137.

148. Latrobe, "Report of the Surveyor," March 23, 1808, in *CL2*, 575; in Padover, *Jefferson and the Capital*, 412; and in *DH*, 131–137.

149. *DH*, 138.

150. Latrobe to Richard Stanford, April 8, 1808, in *CL2*, 584–591, 585.

151. Ibid.

152. Ibid., 585, 586.

153. Latrobe to Thompson, December 30, 1804, in *CL1*, 586–588, 586, 588n1; and "An Act making an appropriation for completing the south wing of the Capitol, at the city of Washington, and for other purposes," approved January 25, 1805, in *DH*, 114, 115.

154. Latrobe to Stanford, April 8, 1808, in *CL2*, 586.

155. Ibid.

156. Ibid., 587, 590n7.

157. Ibid., 587.

158. Ibid., 587, 590n8.

159. U.S. House, *Report of Committee on Deficit*, in *American State Papers*, 1:955–956; and in *DH*, 140–141.

160. House proceedings of April 25, 1808, in *Annals of Congress*, 10th Cong., 1st sess. (1807), 2276; and in *DH*, 143.

161. Jefferson to Latrobe, April 25, 1808, in Padover, *Jefferson and the Capital*, 414–416, 414; and in *DH*, 145.

162. Thornton to Smith, editor of the *Washington National Intelligencer*, April 20, 1808, in *CL2*, 600–607; published in *Washington Federalist*, April 26, 1808.

163. Ibid., 600.

164. Ibid., 602–605.

165. Latrobe to Findlay, editor of *Washington Federalist*, April 26, 1808, in *CL2*, 607–610, 607; also in *Washington Federalist*, April 30, 1808.

166. Ibid., 607–608.

167. Ibid., 609.

168. Thornton to Findlay, May 1, 1808, in *CL2*, 614–619; and in *Washington Federalist*, May 7, 1808.

169. Latrobe to Findlay, May 9, 1808, in *CL2*, 619–620, 619; and in *Washington Federalist*, May 11, 1808.

170. Latrobe to Walter Jones and John Law, memorandum, June 26, 1808; "Memoranda of facts relating to the causes of the difference between Doctor William Thornton and B. Henry Latrobe," in *CL2*, 637–641.

171. Allen, *History of the Capitol*, 94–95.

172. Jefferson to Latrobe, April 26, 1808, in *CL2*, 611–613, 612.

173. Jefferson to Latrobe, April 25, 1808, in Padover, *Jefferson and the Capital*, 414–416, 415; and in *DH*, 145.

174. Latrobe to Jefferson, May 23, 1808, in *CL2*, 621–628, 621, 622; and in Padover, *Jefferson and the Capital*, 420–427, 421–422.

175. Latrobe to Jefferson, May 23, 1808, in *CL2*, 624; in Padover, *Jefferson and the Capital*, 424; and in *DH*, 131–137.

176. Ibid.

177. Jefferson to Latrobe, June 2, 1808, in *CL2*, 631–633, 632.

178. "Bilibous fever" (bilious fever) is an obsolete medical term used in the eighteenth and nineteenth centuries for a disorder marked by vomiting and nausea. "Bilious Fever," Wikipedia, https://en.wikipedia.org/wiki/Bilious_fever (accessed November 21, 2014).

179. Latrobe to George Clymer, June 18, 1808, in *CL2*, 634–635, 634.

180. Latrobe to Isaac Hazlehurst, September 3, 1808, in *CL2*, 656–657, 656, 657n1.

181. Latrobe to Jefferson, September 11, 1808, in *CL2*, 657–660, 657.

182. Ibid., 658.

183. Latrobe to John B. Colvin, editor of the *Washington Monitor*, September 19, 1808, in *CL2*, 661; and in the *Washington Monitor* of September 20, 1808.

184. Latrobe to Samuel Harrison Smith, editor of the *Washington National Intelligencer*, September 20, 1808, in *CL2*, 662–664, 662; and in the *Washington National Intelligencer* of September 23, 1808.

185. Latrobe to Smith, September 20, 1808, in *CL2*, 663–664; and in *Washington National Intelligencer*, September 23, 1808.

186. Ibid., 663.

187. Latrobe to the editor of the *Washington Monitor*, September 21, 1808, in the *Washington Monitor* of September 22, 1808.

188. Ibid.

189. Latrobe gave several conflicting numbers for the dimensions of this dome. The editors of *Correspondence of Latrobe* note that he may have rounded his numbers or inconsistently included and excluded the thickness of the vaults, which would result in different values. See *CL2*, 767n2.

190. Latrobe to Jefferson, September 23, 1808, in *CL2*, 665–669, 666.

191. Ibid.

192. Ibid., 666–667.

193. "A Plain Man" to the editor, *Washington Federalist*, October 8, 1808. A summary of "A Plain Man" newspaper attacks on Latrobe and Latrobe's published defense is contained in *CL2*, 668n4.

194. "A Plain Man" to the editor, *Washington Federalist*, October 8, 1808.

195. Ibid.

196. John Lenthall, entry for May 24, 1808, for the north wing of the U.S. Capitol, *Day Book, U.S. Public Buildings*, as published in Latrobe's letter to the editor, *Washington Federalist*, October 20, 1808. In the same letter, Alexander McIntire and Sam Burch attested that this was a true copy of the *Day Book* and that it was written in the hand of John Lenthall.

197. Latrobe to the editor, *Washington Federalist*, October 14, 1808, as published in Latrobe to the editor, *Washington Federalist*, October 20, 1808.

198. Latrobe to the editor, *Washington Federalist*, October 17, 1808, as published in Latrobe to the editor, *Washington Federalist*, October 20, 1808.

199. James Hoban to the editor, *Washington Federalist*, October 27, 1808.

200. Hoban to the editor, *Washington Federalist*, November 15, 1808.

201. Benjamin Latrobe, "The Report of the Surveyor of the Public Buildings of the United States in the City of Washington," November 18, 1808, in *CL2*, 670–676; and in *DH*, 146–149.

202. Benjamin Latrobe, Report to Senator Stephen Row Bradley, "To the honorable general Bradley, chairman of the committee of the Senate of the United States, appointed on the 12th of December, 1808, on the subject of the public buildings of the United States," December, ca. 13–21, 1808, in *CL2*, 685–692.

203. Latrobe, "Report of the Surveyor," November 18, 1808, in *CL2*, 670; and in *DH*, 146–149.

204. Latrobe, "Report of the Surveyor," November 18, 1808, in *CL2*, 670–671; and in *DH*, 146–149.

205. Latrobe, "Report of the Surveyor," November 18, 1808, in *CL2*, 671, 675n4; and in *DH*, 146–147.

206. Latrobe, "Report of the Surveyor," November 18, 1808, in *CL2*, 671; and in *DH*, 146–149, 147.

207. Latrobe, "Report of the Surveyor," November 18, 1808, in *CL2*, 672–673, 674; and in *DH*, 147, 148.

208. Latrobe, "Report of the Surveyor," November 18, 1808, in *CL2*, 674.

209. Latrobe, "Report of the Surveyor," November 18, 1808, in CL2, 674; and *in DH, 149 (omits President's House)*.

210. Latrobe, Report to Bradley, December 13–21, 1808, in *CL2*, 685–692, 686.

211. *DH*, 155. For a summary of the history of this appropriation, see *CL2*, 693–694n1.

212. *DH*, 156–157; also see *CL2*, 752n3.

213. Latrobe to President James Madison, September 8, 1809, in *CL2*, 763–767, 764.

214. See *CL2*, 767n2.

215. Latrobe to Madison, "The Report of the Surveyor of the public buildings of the U. States," December 11, 1809, in *CL2*, 794–801, 795.

216. Ibid.

217. Ibid., 796.

218. Ibid., 797.

219. Ibid., 795–796.

220. Ibid., 797–798.

221. Ibid., 798.

222. Ibid.

223. Ibid., 798–799.

224. Ibid., 798.

225. *CL2*, 837n5.

226. *CL2*, 861n4.

227. Latrobe to Madison, "Report of the Surveyor," December 28, 1810, in *CL2*, 945–949; and in *DH*, 163–165.

228. Latrobe to Madison, "Report of the Surveyor," December 28, 1810, in *CL2*, 945, 947; and in *DH*, 163–165, 163, 164.

229. Benjamin Latrobe, "The Report of the Surveyor of the Public Buildings of the United States," December 28, 1810, in *CL2*, 945–948, 947–948; and in *DH*, 163–165, 164–165.

9. THE THIRD PUBLIC BUILDING CAMPAIGN (1815–1824)

Epigraph. Mrs. Samuel Harrison Smith, *First Forty Years*, 111.

1. *DH*, 185.

2. Benjamin Latrobe to Thomas Jefferson, July 12, 1815, in *CL3*, 667–674, 668.

3. Samuel Lane to Lewis Condict, January 4, 1817, in *RDCC*, Letters Sent, roll 19, 182–183.

4. Latrobe recommended that the limestone belt in Frederick County, MD, and Loudon

County, VA, be developed for the lime needed in the third public building campaign. Latrobe to Commissioners of the Public Buildings, August 8, 1815, in *CL3*, 681–685, 682–683.

5. Latrobe to the Commissioners of Public Buildings, August 8, 1815, in *RDCC*, Letters Received; and in *CL3*, 681–685, 683.

6. See Attachment, "Report on the Seneca Sandstone," to Owen H. Ramsburg, Memorandum to Architect of the Capitol Mario E. Campioli, March 15, 1965, reporting on a tour of Lee's Quarry conducted by Frank X. Kuhn and Ramsburg on March 12, 1965. Records of the Architect of the Capitol, Office of the Architect of the Capitol.

7. Latrobe to Joseph Gales and William Seaton, editors of the *Washington National Intelligencer*, January 18, 1817, in *CL3*, 851–856, 852.

8. Latrobe to the Commissioners of Public Buildings, August 8, 1815, in *CL3*, 681–685, 681–682. Following Latrobe's recommendation, a quarry was opened on the banks of the Potomac River. Today it is located at the hiker-biker stop at mile 38.2 on the Chesapeake and Ohio Canal.

9. Ibid.

10. Latrobe to William Lee, August 13, 1816, in *CL3*, 798–802, 799.

11. *Washington Daily National Intelligencer*, May 21, 1816, 3.

12. Ibid., April 4, 1816.

13. Commissioners to New York Stone Cutters, September 20, 1815, in *RDCC*, Letters Sent, roll 5, 87.

14. James McKeon to Lane, August 17, 1816, in *RDCC*, Letters Received, roll 19, 2359.

15. Commissioners to John Brannan, August 28, 1815, in *RDCC*, Letters Sent, roll 5, 80–81.

16. Commissioners to Colonel Bomford, July 25, 1815, in *RDCC*, Letters Sent, roll 5, 60; Captain Morton to the commissioners, July 26, 1815, in *RDCC*, Letters Received, roll 19, 2214.

17. *Washington City Weekly Gazette*, January 16, 1816, 4.

18. Bricklayers to the commissioners, July 21, 1815, in *RDCC*, Letters Received, roll 19, 2214.

19. Commissioners to Messrs. Nathan U. Hays, Jos. Smith & others, Carpenters employed at the Presidents' House, October 3, 1815, in *RDCC*, Letters Sent, roll 5, 89.

20. Commissioners to Stone Cutters at the Capitol, October 27, 1815, in *RDCC*, Letters Sent, roll 5, 102–104.

21. Commissioners to Peter Lenox, April 4, 1816, in *RDCC*, Letters Sent, roll 3, 147.

22. Laborers at Capitol to the commissioners, April 15, 1816, in *RDCC*, Letters Received, roll 19.

23. James Hoban to the commissioners, June 10, 1816, in *RDCC*, Letters Received, roll 19, 2329.

24. Commissioners of Public Buildings to Latrobe and Shadrach Davis, March 8, 1816, in *RDCC*, Letters Sent, roll 5, 137.

25. *CL3*, 743n6.

26. Commissioners of Public Buildings to Latrobe, in *RDCC*, Letters Sent, March 30, 1816, roll 5, 147.

27. "Labourers will work the whole day, without regard to the above regulation." Commissioners to Hoban and Latrobe, April 4, 1816, in *RDCC*, Letters Sent, roll 5, 147.

28. Latrobe to Lane, May 1, 1816, in *RDCC*, Letters Received, roll 19, 2309.

29. Hoban to Lane, October 10, 1816, in *RDCC*, Letters Received, roll 20, 2379.

30. Presidents' House Carpenters to commissioners, April 3, 1817; Carpenters at the Capitol to the commissioners, April 1, 1817, both in *RDCC*, Letters Received, roll 20, 2432.

31. Lane to Lewis Condict, January 4, 1817, in *RDCC*, Letters Sent, roll 5, 183–184; and (in part) in *DH*, 193–194.

32. Commissioners to George Blagden, March 4, 1816, in *RDCC*, Letters Sent, roll 5, 131–132.

33. Index to Letters Sent, in *RDCC*, May 21, 1818, roll 3, 268.

34. Charles Bulfinch, report to Lane, November 21, 1818, in *DH*, 207–209, 208.

35. George Hadfield to the commissioners, October 13, 1814, in *RDCC*, Letters Received, roll 18, 2162.

36. U.S. House, "Capture of City of Washington," 596.

37. "An Act making appropriations for repairing or rebuilding the public buildings within the city of Washington," approved February 13, 1815, in *DH*, 185.

38. Hoban to the commissioners, April 25, 1815, in *RDCC*, Letters Received, roll 18, 2188.

39. Commissioners John P. Van Ness and T. Ringgold to Latrobe, March 14, 1815, in *CL3*, 634–635, 635n3.

40. Latrobe to Commissioners of the Public Buildings, April 19, 1815, in *CL3*, 647–653, 649.

41. Latrobe to Jefferson, July 12, 1815, in *CL3*, 667–674, 671.

42. Benjamin Latrobe, "Report on the U.S. Capitol," November 28, 1816, in *CL3*, 831–835, 833; and in *DH*, 190–193, 191–192.

43. Latrobe to Jefferson, July 12, 1815, in *CL3*, 667–674, 670.

44. Latrobe to Abner Lacock, U.S. senator from Pennsylvania, September 24, 1814, in *CL3*, 575–577, 575.

45. Latrobe to Commissioners of the Public Buildings, April 19, 1815, in *CL3*, 647–653, 648.

46. Ibid., 648–649.

47. Latrobe to Jefferson, July 12, 1815, in *CL3*, 667–674, 672.

48. Latrobe to Commissioners of the Public Buildings, March 7, 1816, in *CL3*, 736–740, 737.

49. Latrobe to Commissioners of the Public Buildings, April 27, 1815, in *CL3*, 654–660, 654.

50. Ibid., 654, 655. The changes proposed by Latrobe would have infuriated Latrobe's old nemesis William Thornton. Van Horne cites an undated letter from Thornton to Thomas Jefferson on this subject, believed to be from this time, which begins, "Mr. Latrobe is arrived and I heard this morning he is preparing to alter the Representatives Chamber entirely." *CL3*, 674n9.

51. Latrobe to Lane, January 23, 1817, in *CL3*, 857–859, 857–858.

52. Ibid., 858.

53. Lane to Condict, chairman of the Congressional Committee on Public Buildings, February 15, 1817, in *DH*, 196.

54. "Act making further provision for repairing the public buildings, and improving the public square," approved March 3, 1817, in *DH*, 198.

55. *DH*, 198–199, 198. See *CL3*, 877–879n2, n3, n4.

56. President James Monroe to Lane, April 4, 1817, in *DH*, 198–199.

57. Lane to Latrobe, October 31, 1817, in *CL3*, 962–964, 963.

58. Latrobe to Jacob Small, November 5, 1817, as quoted in *CL3*, 964n2.

59. *CL3*, 107n4.

60. Lee to Bulfinch, as quoted in *CL3*, 954n4.

61. Latrobe to Monroe, November 20, 1817, in *CL3*, 968–969.

62. Submitted to Congress on February 10, 1818, by commissioner Lane, in *DH*, 200–202.

63. Charles Bulfinch, "Report in part of the Committee of Public Buildings," April 4, 1818, in *DH*, 205–206, based on Bulfinch's estimate of March 31, 1818.

64. "Act making appropriations for public buildings, and for furnishing the Capitol and the President's house," approved April 20, 1818, in *DH*, 206.

10. LATER TRANSPORTATION IMPROVEMENTS

Epigraph. Von Gerstner, *Die innern Communicationen*, 639.

1. Chesapeake and Ohio Canal Company, *President and Directors of the Chesapeake and Ohio Canal Company to the Stockholders in General Meeting*, June 1, 1829, first annual report (Washington, DC: Gales & Seaton, 1829), 4; and in manuscript, RG 79, NARA, entry 180, vol. A, 32–56.

2. Swift and Hale, *Report on Chesapeake and Ohio Canal*, 33.

3. John J. Abert and James Kearney, "Letter Report on the Chesapeake and Ohio Canal to Brig. Gen. Gratiot, June 13, 1831," in Abert and Kearney, *Report upon Chesapeake and Ohio Canal*, 88–89. The ruins of these structures may be seen immediately downriver of the Thompson Boat Center.

4. McNeil, "Report on Chesapeake and Ohio Canal," 142.

5. This location was above the stop lock and levee built in 1852 to prevent flood waters from the Potomac from entering the old channel. The levee, between the canal and river, is 15 feet high and 500 feet long.

6. Abert and Kearney *Report upon Chesapeake and Ohio Canal*, 95.

7. Ibid., 98–99.

8. McNeil, "Report on Chesapeake and Ohio Canal," 144. Abert and Kearney give a somewhat longer length: 438 feet between abutments, plus 96-foot ends, for a total length of 534 feet. Abert and Kearney, *Report upon Chesapeake and Ohio Canal*, 101.

9. Abert and Kearney, *Report upon Chesapeake and Ohio Canal*, 101–102. For a history of the aqueduct and its subsequent rehabilitation, see Kapsch and Kapsch, *Monocacy Aqueduct*.

10. This portion of the chapter is partially adapted from two previously published works: Robert J. Kapsch, "The Construction of the Potomac Aqueduct (1833–1841): Solving the Pier Construction Problem of Deep Water," in *Canal History and Technology Proceedings*, vol. 28, March 14, 2009, ed. Lance E. Metz (Easton, PA: Canal History and Technology Press, 2009), 63–90; and Kapsch, "Construction of the Potomac Aqueduct."

11. U.S. House, *Canal Company—Alexandria, Report*.

12. U.S. House, *Letter from Secretary of War*, 12.

13. U.S. House, *Canal Company—Alexandria, Report*.

14. Hahn and Kemp, *Alexandria Canal*, 19.

15. *Washington Niles Register*, July 9, 1831, 328.

16. Chesapeake and Ohio Canal Company, *Papers*, RG79, NARA, entry 182.

17. U.S. House, *Potomac Aqueduct, 1838, Turnbull's Report*, 3.

18. Ibid., 2. The loss of the 50-foot iron probe was greeted with great amusement by the other workers.

19. Ibid., 4.

20. Ibid., 3.

21. Ibid., 4.

22. Von Gerstner, *Die innern Communicationen*, 639.

23. U.S. House, *Potomac Aqueduct, 1838, Turnbull's Report*, 7.

24. Nelson, *Colossus of Philadelphia*, 93.

25. U.S. House, *Potomac Aqueduct, 1838, Turnbull's Report*, 2.

26. Labelye, *Short Account of Westminster Bridge*; Semple, *Building in Water*.

27. The Pont de Neuilly, across the Seine at Paris, between Courbevoie on the right bank and Puteaux on the left, was constructed in 1774 by Jean-Rodolphe Perronet, the founder of Ecole Nationale des Ponts et Chaussées. A multiple-arch segmental masonry arch structure, it was demolished in the mid-twentieth century. It was depicted in "Plan Showing the

Method of Emptying the Foundations of Neuilly Bridge," in *Edinburgh Encyclopedia,* ed. David Brewster, American ed. (Philadelphia: J. and E. Parker, 1832), plate 95 (see plate 36).

28. U.S. House, *Potomac Aqueduct, 1838, Turnbull's Report,* 6.

29. Ibid.

30. The steam engines were manufactured by Mr. Smith of the Alexandria Foundry and Watchman & Bratt of Baltimore. See William Turnbull and W. M. C. Fairfax, engineers, Report to the President and Directors, May 4, 1834, in Alexandria Canal Company, *Annual Report to the Stockholders,* May 6, 1833, 6–7, 6.

31. This mud machine was supplied by Mr. Smith of the Alexandria Foundry, along with the necessary gearing and pumps. See ibid.

32. U.S. House, *Potomac Aqueduct, 1838, Turnbull's Report,* 8.

33. Ibid., 11.

34. Hydraulic cement has the ability to set in water. It was supplied by Boteler and Reynolds, of Shepherdstown, VA (now WV), and the Washington Lime and Cement Company. This cement "has been proved of a quality superior to any specimens from the North that we have tried." See Turnbull and Fairfax, Report to the President and Directors, May 4, 1834, 6–7, 7. The remains of the Botelor and Reynolds cement mill and kilns are still in existence south of Shepherdstown. For a history of this operation, see Hahn and Kemp, *Cement Mills along the Potomac,* 31–72.

35. U.S. House, *Potomac Aqueduct, 1838, Turnbull's Report,* 15–18.

36. Ibid., 19.

37. Ibid., 20.

38. [Turnbull], *Report from Secretary of War,* 37.

39. See Blanc, *Everyday Products Make People Sick,* 329; James Grant Wilson and John Fiske, eds., *Appleton's Cyclopædia of American Biography* (New York: D. Appleton, 1887), 1:579.

40. Thurston, *Materials of Engineering,* 137.

41. [Turnbull], *Report from Secretary of War,* 37.

42. Octave Chanute, "The Preservation of Timber: Report of the Committee on the Preservation of Timber, Presented and Accepted at the Annual Convention, June 25th, 1885," *Transactions of the American Society of Civil Engineers* 14 (1885): 255.

43. Ibid., 254.

44. [Turnbull], *Report from Secretary of War,* 35.

45. Mahan, *Elementary Course on Civil Engineering,* 225.

46. Myer, *Building the Potomac Aqueduct 1830.*

47. U.S. House, *Aqueduct Bridge.*

Abbott, W. W., and Dorothy Twohig, eds. *The Papers of George Washington: Confederation Series 1: January–July 1784.* Charlottesville: University of Virginia Press, 1992.

Abbott, W. W., and Dorothy Twohig, eds. *The Papers of George Washington:* vol. 3, *June–September 1789.* Charlottesville: University of Virginia Press, 1989.

Abert, J. J. *Approximate Estimate of That Portion of the "Brookville Route," of the Maryland Canal.* Letter Report to the Maryland Senate and House of Delegates, February 15, 1839. Maryland State Archives, Annapolis, MD.

———. *Report from J. J. Abert in Reference to the Canal to Connect the Chesapeake and Ohio Canal with the City of Baltimore.* 1838. Reprint, Washington, DC: US Government Printing Office, 1874.

Abert, John J., and James Kearney. "Letter Report on the Chesapeake and Ohio Canal to Brig. Gen. Gratiot, June 13, 1831." In *Report of Col. John J. Abert and Col. James Kearney of the United States Topological Engineers upon an Examination of the Chesapeake and Ohio Canal From Washington City to the "Point of Rocks."* In U.S. House, *Chesapeake and Ohio Canal,* April 17, 1834, 88–105.

Achenbach, Joel. *The Grand Idea: George Washington's Potomac, and the Race to the West.* New York: Simon & Schuster, 2004.

An Account of the Rise, Progress and Termination of the Malignant Fever Lately Prevalent in Philadelphia. Anonymous pamphlet. Philadelphia: Benjamin Johnson, 1793.

Adams, Charles Francis, ed. *The Works of John Adams.* Vol. 1. Boston: Little, Brown, 1850–1856.

Adams, Herbert B. "Washington's Interest in the Potomac Company." In *Maryland, Virginia and Washington,* edited by George William Catt, 79–91. Baltimore: Johns Hopkins University, 1885.

Adler, Bill, ed. *Washington: A Reader: The National Capital as Seen through the Eyes of Thomas Jefferson . . . and Others.* New York: Meredith Press, 1967.

Alexander, Sally Kennedy. "A Sketch of the Life of Major Andrew Ellicott." In *Records of the Columbia Historical Society, Washington, D. C.,* 2:158–202. Washington, DC: Columbia Historical Society, 1899.

Alexandria Canal Company. *Annual Report of the President and Directors of the Alexandria Canal Company, to the Stockholders* (for 1832), May 6, 1833. Alexandria, DC: Alexandria Gazette Office, 1833.

———. *Annual Report of the President and Directors of the Alexandria Canal Company, to the Stockholders* (for 1834), May 4, 1835. Alexandria, DC: Alexandria Gazette Office, 1835.

———. *Annual Report of the President and Directors of the Alexandria Canal Company to the Stockholders* (for 1836). Alexandria, DC: Alexandria Gazette Office, 1836.

———. *Annual Report of the President and Directors of the Alexandria Canal Company* (for 1839). Alexandria, DC: Alexandria Gazette Office, 1839.

Allard, Dean C. "When Arlington Was Part of the District of Columbia." *Arlington Historical Magazine* 6, no. 2 (1978): 36–47.

Allen, William C. *The Dome of the United States Capitol: An Architectural History.* Washington, DC: US Government Printing Office, 1992.

———. "History of Slave Laborers in the Construction of the United States Capitol." Typescript, June 1, 2005. Office of the Architect of the Capitol, Washington, DC.

———. *History of the United States Capitol: A Chronicle of Design, Construction and Politics.* Senate Doc. 106-29, 106th Cong., 2nd sess. (2001).

———. *In the Greatest Solemn Dignity: The Capitol's Four Cornerstones.* Senate Doc. 103-28, 103rd Cong., 2nd sess. (1995).

———. "Remembering Paris: The Jefferson-Latrobe Collaboration at the Capitol." In *Paris on the Potomac: The French Influence on the Architecture and Art of Washington, D.C.,* compiled and edited by Cynthia R. Field, Isabelle Gournay, and Thomas P. Somma, 36–55. Athens: Ohio University Press for the U.S. Capitol Historical Society, 2007.

———. *The United States Capitol: A Brief Architectural History.* House Doc. 101-144, 101st Cong., 1st sess. (1990).

Allgot, Catherine. "'Queen Dolley' Saves Washington City." *Washington History* 12, no. 1 (2000): 54–69.

American Society of Civil Engineers, Committee on History and Heritage of American Civil Engineering. *A Biographical Dictionary of American Civil Engineers.* Vol. 1. New York: American Society of Civil Engineers, 1972.

American State Papers: Documents, Legislative and Executive of the Congress of the United States, from the First Session of the First to the Second Session of the Tenth Congress, Inclusive. March 3, 1789, to March 3, 1809. Vol. 1, *Miscellaneous.* Washington, DC: Gales & Seaton, 1834.

Annals of the Congress of the United States: Fourth Congress. Washington, DC: Gales & Seaton, 1849.

Annals of the Congress of the United States: Thirteenth Congress. Washington, DC: Gales & Seaton, 1854.

Arbuckle, Robert D. *Pennsylvania Speculator and Patriot: The Entrepreneurial John Nicholson, 1757–1800.* University Park: Pennsylvania State University Press, 1975.

Arnebeck, Bob. *Through a Fiery Trial: Building Washington, 1790–1800.* Lanham, MD.: Madison Books, 1991.

———. "The Use of Slaves to Build the Capitol and White House." Online at www.geocities.com/bobarnebeck/slaves.html.

Bacon-Foster, Corra. *Early Chapters in the Development of the Patomac Route to the West.* New York: Burt Franklin, 1971. Reprinted from *Records of the Columbia Historical Society, Washington, D.C.,* 15:96–322. Washington, DC: Columbia Historical Society, 1912.

Baily, Francis. *Journal of a Tour in Unsettled Parts of North America, in 1796 and 1797.* London: Baily Brothers, 1856.

Baker, Mrs. Abby Gunn. "The Erection of the White House." In *Records of the Columbia Historical Society, Washington, D.C.,* 16:120–149. Washington, DC: Columbia Historical Society, 1913.

Baker, O. K., Bernard Geehan, and Alice A. Royaltey, comps. "Index of Persons Known to Have Been Employed on the Original Construction of the President's House and/or the Capitol." Typescript, undated. National Archives Volunteers, Washington, DC.

Baker, William Spohn. *Washington after the Revolution, 1784–1799.* Philadelphia: J. B. Lippincott, 1898.

Balch, Thomas, ed. *Letters and Papers Relating Chiefly to the Provincial History of Pennsylvania with Some Notices of the Writers.* Philadelphia: Crissy & Marley, 1855.

Baldwin, Loammi. "Letters from Loammi Baldwin: Potomac Bridge—Washington." In U.S. House of Representatives, Doc. 356, 23rd Cong., 1st sess., April 28, 1834.

Baltimore City Council. *Report of the Joint Special Committee of the City Council to Whom Was Referred the Subject of the Construction of a Cross-Cut Canal to Connect with the Chesapeake & Ohio Canal, at Georgetown.* Baltimore: James Lucas, printer, 1851.

Bearrs, Edwin C. *The Bridges: Chesapeake and Ohio Canal National Monument, Historic Structures Report—Part II, Historical Data Section.* Washington, DC: National Park Service, Division of History, 1968.

Bedini, Silvio A. *The Life of Benjamin Banneker.* New York: Scribner, 1972.

———. *With Compass and Chain: Early American Surveyors and Their Instruments.* Frederick, MD: Professional Surveyors, 2001.

Bednar, Michael. *L'Enfant's Legacy: Public Open Spaces in Washington, D.C.* Baltimore: Johns Hopkins University Press, 2006.

Belt, Norman B. "The History and Construction of Highway Bridge(s) across the Potomac River at Washington, D.C." Tau Beta Pi thesis, University of Maryland, 1933.

Benjamin, Charles F. *A History of Federal Lodge No. 1.* Washington, DC: Gibson, 1901.

Bennett, Wells. *Stephen Hallet and His Designs for the National Capitol.* Harrisburg, PA, 1916.

———. "Stephen Hallet and His Designs for the National Capitol, 1791–1794." Parts 1–4. *Journal of the American Institute of Architects* 4 (1916): 290–295, 324–330, 376–383, 411–418.

Berg, Scott W. *Grand Avenues: The Story of the French Visionary Who Designed Washington, D.C.* New York: Pantheon, 2007.

Bernstein, Lawrence R. *Minerals of the Washington, D.C. Area.* Educational Series no. 5. Baltimore: Maryland Geological Survey, 1980.

Bethel, Elizabeth. "Material in the National Archives Relating to the Early History of the District of Columbia." In *Records of the District of Columbia Historical Society*, 42–43:168–87. Washington, DC: Columbia Historical Society, 1942.

Bickford, Charlene Bangs, Kenneth R. Bowling, William Charles diGiacomantonio, and Helen E. Veit, eds. *Debates in the House of Representatives.* 5 vols. Documentary History of the First Federal Congress of the United States of America, nos. 10–14. Baltimore: Johns Hopkins University Press, 1992–1996.

Birkbeck, George. *A Lecture on the Preservation of Timber by Kyan's Patent for Preventing Dry Rot.* Delivered at the Society of Arts, December 9, 1834. London: John Weale, 1835.

Blanc, Paul D. *How Everyday Products Make People Sick: Toxins at Home and in the Workplace.* Berkeley: University of California Press, 2007.

Bodey, Hugh. *Nailmaking.* Oxford, UK: Shire, 2008.

Bordewich, Fergus M. "Capitol Fellow" (William Thornton). *Smithsonian Magazine*, December 2008, 78–85.

———. *Washington: The Making of the American Capital.* New York: Amistad/HarperCollins, 2008.

Borneman, Walter R. *1812: The War That Forged a Nation.* New York: HarperCollins, 2004.

Bowling, Kenneth R. *Creating the Federal City, 1774–1800: Potomac Fever.* Washington, DC: American Institute of Architects Press, 1991.

———. *The Creation of Washington, D.C.: The Idea and Location of the American Capital.* Washington, DC: American Institute of Architects Press, 1988.

———. *The Creation of Washington, D.C. The Idea and Location of the American Capital.* Fairfax, VA: George Mason University Press, 1991.

———. "A Foreboding Shadow: Newspaper Celebration of the Federal Government's Arrival." *Washington History* 12, no. 1 (2000): 4–7.

————. "From 'Federal Town' to 'National Capital': Ulysses S. Grant and the Reconstruction of Washington, DC." *Washington History* 14, no. 1 (2002): 8–25.

————. "The Other G. W.: George Walker and the Creation of the National Capital." *Washington History* 3, no. 2 (Fall–Winter 1991–1992): 4–21.

————. *Peter Charles L'Enfant: Vision, Honor, and Male Friendship in the Early American Republic.* Washington, DC: Friends of the George Washington University Libraries, 2002.

Bowling, Kenneth R., and Donald R. Kennon, eds. *Neither Separate or Equal: Congress in 1790.* Athens: Ohio University Press, 2000.

Brissot de Warville, Jacques-Pierre. *New Travels in the United States of America, 1788.* Edited by Durand Echeverria. Translated by Mara Siceanu Vamos and Durand Echeverria. Cambridge, MA: Harvard University Press, 1964.

Brockett, F. I. *The Lodge of Washington: A History of the Alexandria Washington Lodge, No. 22.* Westminster, MD: Willow Bend Books, 2001.

Brown, Alexander Crosby. *The Patowmack Canal: America's Greatest Eighteenth Century Engineering Achievement.* Alexandria, VA: Virginia Canals & Navigations Society, ca. 1992. Reprinted from *Virginia Cavalcade* 12 (1963): 40–47.

Brown, Glenn. *Glenn Brown's History of the United States Capitol.* With an introduction and annotations by William B. Bushong. House Doc. 108-240, 108th Cong. 2nd sess. (1998).

————. *History of the United States Capitol.* 2 vols. Washington, DC: US Government Printing Office, 1900, 1902.

————, comp. *Papers Relating to the Improvement of the City of Washington, District of Columbia.* Senate Doc. 94, 56th Cong., 2nd sess. (1901).

Brown, Gordon S. *Incidental Architect: William Thornton and the Cultural Life of Early Washington, D.C., 1794–1828.* Athens: Ohio University Press for the U.S. Capitol Historical Society, 2009.

Brown, Letitia Woods. *Free Negroes in the District of Columbia, 1790–1846.* New York: Oxford University Press, 1972.

Bryan, Wilhelmus B. *Bibliography of the District of Columbia . . . to 1898.* Senate Doc. 61, 56th Cong., 1st sess. (1900).

————. *A History of the National Capital.* 2 vols. New York: Macmillan, 1914–1916.

————. "Removal of the Seat of Government to Washington." In Cox, *Celebration of Anniversary,* 253–262.

————. "Something about L'Enfant and His Personal Affairs." In *Records of the Columbia Historical Society, Washington, D. C.,* 2:111–117. Washington, DC: Columbia Historical Society, 1899.

Bryan, Wilhelmus B., and Samuel C. Busey. *The Removal of the Seat of Government to the District of Columbia: Two Papers Read before the District of Columbia Historical Society.* U.S. Senate Doc. 62, 56th Cong., 1st sess., December 20, 1899.

Bullock, Steven C. *Revolutionary Brotherhood: Freemasonry and the Transformation of the American Social Order, 1730–1840.* Chapel Hill: University of North Carolina Press, 1996.

Burch, Gary A., and Steven M. Pennington, the Committee on History and Heritage of the National Capital Section, and the American Society of Civil Engineers. *Civil Engineering Landmarks of the Nation's Capital.* Washington, DC: American Society of Civil Engineers, 1982.

Busey, Samuel C. *Pictures of the City of Washington in the Past.* Washington, DC: W. M. Ballantine, 1898.

Bushong, William B. *Uncle Sam's Architects: Builders of the Capitol.* Washington, DC: United States Capitol Historical Society, 1994.

Butler, Jeanne F. "Competition 1792: Designing a Nation's Capitol." In *Capitol Studies,* vol. 4, no. 1. Washington, DC: United States Capitol Historical Society, 1976.

Caemmerer, H. Paul. "Architects of the United States Capitol." In *Records of the Columbia His-*

torical Society of Washington, D. C., edited by H. Paul Caemmerer, 48–49:1–28. Washington, DC: Columbia Historical Society, 1949.

———. *The Life of Pierre Charles L'Enfant: Planner of the City Beautiful, the City of Washington.* Washington, DC: National Republic, 1950.

———. *A Manual on the Origin and Development of Washington.* Senate Doc. 178, 75th Cong., 3rd sess. (1939).

———. *Washington: The National Capital.* Senate Doc. 332, 71st Cong., 3rd sess. (1932).

Carey, Mathew. *A Short Account of the Malignant Fever, Lately Prevalent in Philadelphia.* Philadelphia: Mathew Carey, 1793.

Carne, William F. "Life and Times of William Cranch, Judge of the District Circuit Court, 1801–1855." In *Records of the Columbia Historical Society of Washington, D.C.*, 5:294–310. Washington, DC: Columbia Historical Society, 1902.

Carter, Edward C., II. "Benjamin Henry Latrobe and the Growth and Development of Washington, 1798–1818." In *Records of the Columbia Historical Society of Washington, D. C., 1971–1972*, edited by Francis Coleman Rosenberger, 48:128–149. Charlottesville: University Press of Virginia for the Columbia Historical Society, 1973.

Carter, Edward C., II, and Thomas E. Jeffrey, eds. *The Guide and Index to the Microfiche Edition of the Papers of Benjamin Henry Latrobe.* Clifton, NJ: James T. White for the Maryland Historical Society, 1976.

Carter, Edward C., II, Angeline Polites, Lee W. Formwalt, and John C. Van Horne, eds. *The Virginia Journals of Benjamin Henry Latrobe 1795–1820.* 3 vols. New Haven, CT: Yale University Press, 1977–1980.

Carter, Edward C., II, John C. Van Horne, Charles E. Brownell, and Tina H. Sheller, eds. *Latrobe's View of America, 1795–1820: Selections from the Watercolors and Sketches.* New Haven, CT: Yale University Press for the Maryland Historical Society, 1985.

CEHP, Incorporated. *A Historic Resources Study: The Civil War Defenses of Washington.* Part I. Washington, DC: US Government Printing Office, n.d.

Cerami, Charles A. *Benjamin Banneker: Surveyor, Astronomer, Publisher, Patriot.* New York: John Wiley, 2002.

Chappell, Gordon. *Historic Resource Study: East and West Potomac Parks: A History.* Denver: Denver Service Center, National Park Service, 1973.

Chernow, Barbara Ann. *Robert Morris: Land Speculator, 1790–1801.* New York: Arno Press, 1978.

Chesapeake and Ohio Canal Company. *Proceedings of the President and Directors of the Chesapeake and Ohio Canal Company and of the Corporations of Washington, Georgetown & Alexandria in Relation to the Eastern Termination of the Chesapeake and Ohio Canal.* Washington, DC: Gales & Seaton, 1828.

Chesapeake and Ohio Canal Convention. *Proceedings at a General Convention of Delegates Representing Counties in Virginia, Maryland, Ohio, Pennsylvania and the District of Columbia.* Washington, DC, 1823.

Chesapeake and Ohio Canal Convention. *Report of the Central Committee of the Chesapeake and Ohio Canal Convention Which Met in the City of Washington. . . .* Washington, DC, 1826.

Chevalier, Michel. *Histoire et description des vois de commiunication aux Etat Unis et des travaux d'art qui en dependent.* Paris: C. Gosselin, 1840–1841.

Chisolm, William, and Fred Hardin. *The First 150 Years of Federal Lodge No. 1, F.A.A.M., District of Columbia.* Washington, DC: US Government Printing Office, 1943.

Christianson, Justine, and Christopher Marston, eds. *Covered Bridges and the Birth of American Engineering.* Washington, DC: Historic American Engineering Record, National Park Service, 2015.

Church, Randolph W. "John Ballendine: Unsuccessful Entrepreneur of the Eighteenth Century." *Virginia Cavalcade* 3 (1959): 39–46.

Clark, Allen C. "Daniel Carroll of Duddington." In *Records of the Columbia Historical Society of Washington, D.C.*, 39:1–48. Washington, DC: Columbia Historical Society, 1938.

———. *Greenleaf and Law in the Federal City.* Washington, DC: Press of W. F. Roberts, 1901.

———. "James Greenleaf." In *Records of the Columbia Historical Society of Washington, D.C.*, 5:212–237. Washington, DC: Columbia Historical Society, 1902.

———. "Origin of the Federal City." In *Records of the Columbia Historical Society of Washington, D.C.*, 35–36:1–97. Washington, DC: Columbia Historical Society, 1935.

———. *Thomas Law: A Biographical Sketch.* Washington, DC: Press of W. F. Roberts, 1900.

Cleary, Richard L. *Bridges.* New York: W. W. Norton, 2007.

Clinton, Amy Cheney. "Historic Fort Washington." *Maryland Historical Magazine* 32, no. 3 (September 1937): 228–247.

Clussman and Kammerhueber. *Report on the Present State, and the Improvement of the Washington City Canal*, May 12, 1865. Transmitted by Mayor Richard Wallach to the Board of Alderman and Board of Common Council, May 15, 1865. Washington, DC: R. A. Waters, 1865.

Cohen, Jeffrey A., and Charles E. Brownell. *The Architectural Drawings of Benjamin Henry Latrobe.* Vol. 2, parts 1 and 2. New Haven, CT: Published for the Maryland Historical Society and the American Philosophical Society by Yale University Press, 1994.

Columbia Historical Society Committee on Publication and the Recording Secretary, comps. "The Writings of George Washington Relating to the National Capital." In *Records of the Columbia Historical Society, Washington, D.C.*, 17:3–232. Washington, DC: Columbia Historical Society, 1914.

Commission on the Bicentenary of the U.S. House of Representatives, Commission on the Bicentennial of the U.S. Senate, and Office of the Architect of the Capitol. *The United States Capitol: A Brief Architectural History.* House Doc. 101-144, 101st Cong., 1st sess. (1990).

Commission on the Renovation of the Executive Mansion. *Report of the Commission on the Renovation of the Executive Mansion.* Compiled by Edwin Bateman Morris. Washington, DC: US Government Printing Office, 1952.

Commissioners of the States of Maryland and Virginia. *Letter from the Governor and Council of Maryland Transmitting a Report of the Commissioners Appointed to Survey the River Potomac.* January 28, 1823. House Doc. 46, 17th Cong., 2nd sess. Washington, DC: Gales & Seaton, 1823. An extract of the report was published in *Washington Quarterly Magazine* 1, no. 1 (July 1823): 7–29.

Committee of the Corporation of Georgetown. *Memorial to the Committee of the Senate, for the District of Columbia* (for the authority to build a Free Bridge over the Potomac River, at or near a place called "The Three Sisters"), n.d., enclosed in Georgetown Potomac Bridge Company, "Memorial" (opposing the bill "authorizing the Corporation of Georgetown to erect a Bridge over the river Potomac, within the District of Columbia," at Three Sisters), May 1, 1826, in Senate Report 86, 19th Cong., 1st sess.

Conner, Jane Hollenbeck. *Birthstone of the White House and Capitol.* Virginia Beach, VA: Downing, 2005.

Cook, Frank Gaylord. "Robert Morris." *Atlantic Monthly* 66 (November 1890): 607–618.

Cosentino, Andrew J., and Richard W. Stephenson, comps. *City of Magnificent Distances: The Nation's Capital: A Checklist.* Washington, DC: Geography and Map Division, Library of Congress, 1991.

Cowdrey, Aubrey E. *A City for the Nation: The Army Engineers and the Building of Washington, D.C., 1790–1967.* Washington, DC: US Government Printing Office, 1979.

Cox, William V., comp. *Celebration of the One Hundredth Anniversary of the Establishment of the*

Seat of Government in the District of Columbia. Washington, DC: US Government Printing Office, 1901.

Crew, Harvey W., ed. *Centennial History of the City of Washington, D.C. . . .* Dayton, OH: United Brethren Publishing House, 1892.

Curtis, Richard H. "Home of the Presidents." *Northern Light* 23, no. 4 (November 1992): 6–7.

———. "White House Cornerstone." *Northern Light* 23, no. 4 (November 1992): 4–5.

Cutchin, Janine Basile. "The Quarter Master's House, Seneca, Maryland." Typescript prepared for Vernacular Architecture class, University of Virginia, Spring 1978. Montgomery County Historical Society, Rockville, MD.

Davies, William E. "The Geology and Engineering Structures of the Chesapeake and Ohio Canal: An Engineering Geologist's Descriptions and Drawings." Unpublished draft of 1989. Chesapeake and Ohio Canal Association, Glen Echo, MD.

Davis, William A. *The Acts of Congress in Relation to the District of Columbia, from July 16th, 1790, to March 4th, 1831, Inclusive, and of the Legislatures of Virginia and Maryland.* Washington, DC: William A. Davis, 1831.

Davis, William B., Jr. "The History and Construction of Chain Bridge." Tau Beta Pi thesis, University of Maryland, 1937.

Dennis, Bernard G., Jr., Robert J. Kapsch, Robert J. LoConte, Bruce W. Mattheiss, and Steven M. Pennington, eds. *American Civil Engineering History: The Pioneering Years.* Reston, VA: American Society of Civil Engineers, 2002.

District of Columbia Commissioners. "Memorial of the Commissioners Appointed by the President of the United States" (to Congress, to pass an act authorizing the President to borrow such sums). January 8, 1796, no. 67, 4th Cong., 1st sess. In *American State Papers*, 1:133–134.

———. "Recapitulations of Facts in Support of Their Communication of January 28, 1801." January 11, 1802, no. 150, 7th Cong., 1st sess. In *American State Papers*, 1:254–257.

———. *Records of the Commissioners — Day Books, 1791–1793 and 1796–1800.* Record Group 42. National Archives and Records Administration. Not microfilmed as part of National Archives Microfilm Publication M-371.

———. *Records of the District of Columbia Commissioners and of the Offices Concerned with Public Proceedings, 1797–1867 (RDCC).* Twenty-seven microfilm rolls. National Archives Microfilm Publications Microcopy M-371. Washington, DC: National Archives and Records Administration, 1964. Reproduces most of Record Group 42 in the National Archives. This includes letters sent, letters received, proceedings, and indexes. Proceedings for the period July 31, 1795 to October 24, 1796 are not included as they have been missing since at least 1889 and probably since 1864.

———. "Report of the Commissioners Appointed for Establishing the Temporary and Permanent Seat of the Government of the United States on the State of Their Business and of the Public Property Belonging to the Federal Seat." January 30, 1801, no. 141, 6th Cong., 2nd sess. In *American State Papers*, 1:219–231.

District of Columbia Inhabitants. "Memorial for the Purpose of Erecting a Bridge over the River Potomac." February 4, 1805, House Report no. 191, 8th Cong., 2nd sess. In *American State Papers*, 1:422.

Douglas, Paul H., and William K. Jones. "Sandstone, Canals, and the Smithsonian." *Smithsonian Journal of History* 3, no. 1 (Spring 1968): 41–58.

Dowd, Mary-Jane M., comp. *Records of the Office of Public Buildings and Public Parks of the National Capital.* Record Group 42, Inventory no. 16, National Archives and Records Administration. Washington, DC: US Government Printing Office, 1992.

An Earnest Call Occasioned by the Alarming Pestilential Contagion. Anonymous pamphlet. Philadelphia: Jones, Hoff, and Derrick, 1793.

Easby-Smith, James S. *Georgetown University in the District of Columbia, 1789–1907.* 2 vols. New York: Lewis, 1907.

Eberlein, Harold Donaldson, and Cortland Van Dyke Hubbard. *Historic Houses of George-Town and Washington City.* Richmond, VA: Dietz Press, 1958.

Edwards, Llewellyn. "The Evolution of Early American Bridges." Abstract of paper read at Caxton Hall, Westminster, March 15, 1933, and republished in Committee on History and Heritage of American Civil Engineering, American Society of Civil Engineers, *American Wooden Bridges,* ASCE Historical Publication no. 4 (New York: American Society of Civil Engineers, 1976), 143–168.

Ehrenberg, Ralph E. "Nicholas King: First Surveyor of the City of Washington, 1803–1812." In *Records of the Columbia Historical Society, 1969–1970,* edited by Francis Coleman Rosenberger, 69–70:31–65. Washington, DC: Waverly Press for the Columbia Historical Society, 1971.

Elkins, Stanley, and Eric McKitrick. *The Age of Federalism: The Early American Republic 1788–1800.* New York: Oxford University Press, 1993.

Ellet, Charles, Jr. *Report of a Suspension Bridge across the Potomac for Rail-Road and Common Travel: Addressed to the Mayor and City Council of Georgetown, D.C.* 1st ed. Philadelphia: John C. Clarke, 1852.

———. *Report of a Suspension Bridge across the Potomac for Rail Road and Common Travel: Addressed to the Mayor and City Council of Georgetown, D.C.* 2nd ed. Philadelphia: John C. Clarke, 1854.

Elliot, Jonathan. *Historical Sketches of the Ten Miles Square Forming the District of Columbia; with a Picture of Washington, Describing Objects of General Interest or Curiosity at the Metropolis of the Union.* Washington, DC: Printed for the author, 1830.

Emery, Fred A. "Washington Newspapers." In *Records of the Columbia Historical Society,* 27–28:41–72. Washington, DC: Columbia Historical Society, 1937.

———. "Washington's Historic Bridges." In *Records of the Columbia Historical Society of Washington, D.C.,* 39:49–70. Washington, DC: Columbia Historical Society, 1938.

Ewing, M. C. *Notebooks of 1839–1845.* ARHO-468, Arlington Mansion Library, Arlington National Cemetery.

———. "Report on the Alexandria Canal, to Francis L. Smith, 7th December 1846." *Tiller* 11, no. 3 (Fall 1990): 4–5.

Faehtz, Ernest F. M., and F. W. Pratt, comps. *Washington in Embryo; or, the National Capital from 1791 to 1800.* Washington, DC, 1874.

Fetzer, Kristin. "Seneca Creek Sandstone: Its History and Use as a Building Stone." Typescript prepared for Masonry Conservation Seminar, University of Pennsylvania, March 5, 1997. Montgomery County Historical Society, Rockville, MD.

Finley, James. "A Description of the Patent Chain Bridge; Invented by James Finley; of Fayette County, Pennsylvania. with Data and Remarks, Illustrative of the Power, Cost, Durability, and Comparative Superiority of This Mode of Bridging." *Port Folio,* n.s., 3, no. 6 (June 1810): 441–453.

Fishback, Frederick L. "Washington City: Its Founding and Development." In *Records of the Columbia Historical Society,* 20:194–224. Washington, DC: Columbia Historical Society, 1917.

Fisher, Perry G. *Materials for the Study of Washington: A Selected Annotated Bibliography.* GW Washington Studies no. 1, George Washington University. Washington, DC: George Washington University, 1974.

Fitzpatrick, John C., ed. *The Diaries of George Washington, 1748–1799.* Boston: Houghton Mifflin, 1925.

————, ed. *The Writings of George Washington from the Original Manuscript Sources, 1745–1799.* 39 vols. Washington, DC: US Government Printing Office, 1931–1944.

Fletcher, Robert, and J. P. Snow. "A History of the Development of Wooden Bridges." *Proceedings of the American Society of Civil Engineers* (November 1932). Republished in Committee on History and Heritage of American Civil Engineering, American Society of Civil Engineers, *American Wooden Bridges*, ASCE Historical Publication no. 4. New York: American Society of Civil Engineers, 1976, 29–124.

Flexner, James Thomas. *Washington: The Indispensable Man.* Boston: Little, Brown, 1969.

Force, W. *The Builders Guide: Containing Lists of Prices, and Rules of Measurement, for Carpenters, Bricklayers, Stone Masons, Stone Cutters, Plasterers. . . .* Washington, DC: Peter Force, 1842.

Formwalt, Lee W. *Benjamin Henry Latrobe and the Development of Internal Improvements in the New Republic, 1796–1820.* New York: Arno Press, 1982.

————. "Benjamin Henry Latrobe and the Development of Transportation in the District of Columbia, 1802–1817." In *Records of the Columbia Historical Society of Washington, D.C.,* 50:36–66. Washington, DC: Columbia Historical Society, 1980.

Frary, I. T. *They Built the Capitol.* Richmond: Garrett and Massie, 1940.

Freund, Charles Paul. "From Satan to the Sphinx: The Mason Mysteries of D.C.'s Map." *Washington Post,* November 5, 1995, C-3.

Froncek, Thomas, ed. *An Illustrated History: The City of Washington by the Junior League of Washington.* New York: Alfred A. Knopf, 1977.

Gage, Mary, and James Gage. *The Art of Splitting Stone: Early Rock Quarrying Methods in Pre-Industrial New England, 1630–1825.* Amesbury, MA: Powwow River Books, 2002.

Georgetown Potomac Bridge Company. "Memorial," (opposing the bill "authorizing the Corporation of Georgetown to erect a Bridge over the river Potomac, within the District of Columbia," at Three Sisters), May 1, 1826, in Senate Report 86, 19th Cong., 1st sess.

Gleig, George Robert. *Narrative of the Campaigns of the British Army at Washington, Baltimore, and New Orleans. . . .* London: J. Murray, 1821.

Goode, James. *The Evolution of Washington, D.C.: Historical Selections from the Albert H. Small Washingtoniana Collection at the George Washington University.* Washington, DC: Smithsonian Books, 2015.

Green, Constance McLaughlin. *The Secret City: A History of Race Relations in the Nation's Capital.* Princeton, NJ: Princeton University Press, 1967.

————. *Washington: A History of the Capital, 1800–1950.* 1962. Reprint, Princeton, NJ: Princeton University Press, 1976.

————. *Washington: Village and Capital, 1800–1878.* Princeton, NJ: Princeton University Press, 1962.

Griffin, David J., and Caroline Pegum. *Leinster House: 1744–2000: An Architectural History.* Dublin: Irish Architectural Archive, 2000.

Griffin, Martin I. J. "James Hoban, the Architect and Builder of the White House and the Superintendent of the Building of the Capitol at Washington." *American Catholic Historical Researches* (January 1907): 35–52.

Grigg, Milton L. "Thomas Jefferson and the Development of the National Capital." In Holmes and Heine, *Records of the Columbia Historical Society,* 53–56:81–100.

Griggs, Francis E., ed. *A Biographical Dictionary of American Civil Engineers,* vol. 2. New York: American Society of Civil Engineers, 1991.

————. "Timothy Palmer: The Nestor of American Bridge Builders." *Structures* (2004): 34–35, www.cc.memphis.edu/3121/stuff/general/timothy_palmer.html.

Guidas, John, ed. *The White House: Resources for Research at the Library of Congress.* Washington, DC: US Government Printing Office, 1993.

Gutheim, Frederick. *The Potomac.* New York: Rinehart, 1949.

———. *Worthy of The Nation: The History of Planning for the National Capital.* National Capital Planning Commission Historical Studies. Washington, DC: Smithsonian Institution Press, 1977.

Hafertepe, Kenneth. *America's Castle: The Evolution of the Smithsonian Building and Its Institution, 1840–1878.* Washington, DC: Smithsonian Institution Press, 1984.

Hahmann, Lydia. "How Stiff Is a Curved-Plank Structure." In *Proceedings of the Second International Conference on Construction History,* edited by Malcolm Dunkeld, James W. P. Campbell, Hentie Louw, Michael Tutton, Bill Addis, and Robert Thorne, 2:1501–1516. Exeter, UK: Short Run Press, 2006.

Hahn, Thomas F. *Towpath Guide to the C & O Canal, Section One Georgetown Tidelock to Cumberland.* Shepherdstown, WV: American Canal and Transportation Center, 1997.

Hahn, Thomas Swiftwater, and Emory L. Kemp. *The Alexandria Canal: Its History and Preservation.* Morgantown, WV: Institute for the History of Technology and Industrial Archaeology, 1992.

———. *Cement Mills along the Potomac River.* Monograph Series, vol. 2, no. 1. Morgantown, WV: Institute for the History of Technology and Industrial Archaeology at West Virginia University, 1994.

Hains, Peter C. "Improvement of the Harbors at Washington and Georgetown District of Columbia—Improvement of the Potomac River in the Vicinity of Washington, District of Columbia," July 31, 1883. In *Annual Report of Chief of Engineers to the Secretary of War for the Year 1888,* pt. 1. U.S. House of Representatives, 49th Cong. 1st sess. (1888), Ex. Doc. 1, pt. 2, vol. 2, Appendix 1, 763–794.

———. "Reclamation of the Potomac Flats at Washington, D.C." In *Transactions of the American Society of Civil Engineers,* 31:55–80 (January 1894).

Harbaugh, Thomas. "A Journal, of Accounts, Etc. Thomas Harbaugh, with the Potomac Company and Others, from 1803 to 1833." Unpublished manuscript. Western Maryland Room, Hagerstown Library, Hagerstown, MD.

Harper, Kenton N., comp. *History of the Grand Lodge and of Freemasonry in the District of Columbia with Biographical Appendix.* Washington, DC: R. Beresford, 1911.

Harris, C. M. "The Best Friend I Had on Earth: William Thornton's 'Great Patron,' George Washington." *White House History* 6 (Fall 1999): 360–369.

———. "The Politics of Public Building: William Thornton and President's Square." *White House History* 3 (Spring 1998): 46–59.

———. "Specimens of Genius and Nicknacks: The Early Patent Office and Its Museum." *Prologue* 23, no. 4 (Winter 1991): 406–417.

———. "Washington's 'Federal City,' Jefferson's 'Federal Town.'" *Washington History* 12, no. 1 (Spring–Summer 2000): 49–53.

———. "Washington's Gamble, L'Enfant's Dream: Politics, Design, and the Founding of the National Capital." *William and Mary Quarterly* 56, no. 3 (1999): 527–564.

———. "William Thornton (1759–1828)." On Internet: biography of William Thornton (1759–1828), Prints and Photographs Reading Room, Library of Congress, https://www.loc.gov/rr/print/adecenter/essays/B-Thornton.html, accessed August 31, 2014.

Harris, C. M., and Daniel Preston, eds. *Papers of William Thornton:* Vol. 1, *1781–1802.* Charlottesville: University Press of Virginia, 1995.

Hazleton, George C. *The National Capitol: Its Architecture, Art, and History.* New York: J. F. Taylor, 1914.

Heine, Cornelius W. "The Washington City Canal." In Holmes and Heine, *Records of the Columbia Historical Society*, 53–56:1–27.

Henderson, Jane. "Government Island: Its Forgotten History and Interesting Stone." *Aquia*, October 1980, 81–104.

Herrick, Carole L. *Ambitious Failure: Chain Bridge: The First Bridge across the Potomac River.* Reston, VA: Higher Education Publications, 2012.

Hibben, Henry B. *Navy-Yard, Washington: History from Organization, 1799 to Present Date.* Exec. Doc. 22, 51st Cong., 1st sess. (1890).

Hines, Christian. *Early Recollections of Washington City.* Washington, DC: Chronicle Books, 1866. Reprint, Washington, DC: Junior League of Washington, 1981.

Hirschfeld, Fritz. *George Washington and Slavery: A Documentary Portrayal.* Columbia: University of Missouri Press, 1984.

Historic American Engineering Record. *Rock Creek and Potomac Parkway.* HAER no. DC-20, Library of Congress.

Hodgins, George W. "Aquia Creek Freestone Quarries Used for Early Washington Buildings." Typescript, October, 1953. Office of the Architect of the Capitol, Washington, DC.

Holmes, Oliver W., and Cornelius W. Heine, eds. *Records of the Columbia Historical Society of Washington, D.C., 1953–1956.* Vols. 53–56. Washington, DC: National Republic for the Columbia Historical Society, 1959.

Horne, Robert C. "Bridges across the Potomac." In Holmes and Heine, *Records of the Columbia Historical Society*, 53–56: 249–258.

Horsman, Reginald. *The New Republic: The United States of America, 1789–1815.* New York: Longman, 2000.

Howard, James Q. "The Architects of the Capitol." *International Review* 1 (1874): 736–753.

[Howard, William]. *Report on the Survey of a Canal from the Potomac to Baltimore.* Baltimore: B. Edes, 1828.

Howe, Henry. *Historical Collections of Virginia; Containing a Collection of the Many Interesting Facts, Traditions, Biographical Sketches, Anecdotes etc. etc. Relating to Its History and Antiquities . . . Including a Historical and Descriptive Sketch of the District of Columbia.* Charleston, SC: Wm. R. Babcock, 1846, 1852.

Hughes, Charles B., and George W. Hughes. *Report on the Examination of Canal Routes from the Potomac River to the City of Baltimore, Especially in Relation to the Supply of Water for Their Summit Levels.* Annapolis: William M'Neir, 1837.

Hutchins, Frank, and Cortelle Hutchins. *Washington's Washington.* Washington, DC: Historical Research Service, 1925.

Hutchins, Stilson, and Joseph West Moore. *The National Capital, Past and Present: The Story of Its Settlement, Progress, and Development.* Washington, DC: Post, 1885.

Jackson, Donald, and Dorothy Twohig, eds. *The Diaries of George Washington.* 6 vols. Charlottesville: University Press of Virginia, 1979.

Jackson, Richard P. *The Chronicles of Georgetown, D.C.: From 1751 to 1878.* Washington, DC: R. O. Polkinhorn, 1878.

Jennings, J. L. Sibley, Jr. "Artistry as Design: L'Enfant's Extraordinary City." *Quarterly Journal of the Library of Congress* 36 (1979): 225–278.

Kapsch, Robert J., ed. *Aquia Quarry on Government Island: Report of the Government Island Committee.* Stafford County, VA: Stafford County Government, 2002.

———. "Baltimore and the Maryland Cross-Cut Canal: 1820–1851." In *Baltimore Civil Engineering History*, edited by Bernard G. Dennis Jr. and Matthew C. Fenton IV. Reston, VA: American Society of Civil Engineers, 2004.

———. "Benjamin Wright, Bad Stone, Poor Cement and One Hundred Miles to Go: Building the Monocacy Aqueduct of the Chesapeake and Ohio Canal." In Dennis et al., *American Civil Engineering History*, 439–464.

———. "Benjamin Wright and the Design and Construction of the Monocacy Aqueduct." In *Canal History and Technology Proceedings: Vol. 19, March 18, 2000*. edited by Lance E. Metz, 181–222. Easton, PA: Canal History and Technology Press, National Canal Museum, 2002.

———. "Building the Infrastructure of the New Federal City, 1793–1800." In Rogers et al., *Civil Engineering History*, 74–85.

———. *Canals*. The Norton / Library of Congress Visual Sourcebook in Architecture, Design and Engineering. New York: W. W. Norton, 2004.

———. "The Construction of the Potomac Aqueduct (1833–1841): Pier Construction in Deep Water Conditions." In *Proceedings of the First International Congress on Construction History, Madrid, 20th–24th January 2003*, 1201–1214. Madrid: Instituto Juan de Herrera, Escuela Técnica Superior de Arquitectura, 2003.

———. "George Washington, the Potomac Canal and the Beginning of American Civil Engineering History: Problems and Solutions." In Dennis et al., *American Civil Engineering History*, 129–194.

———. "A Labor History of the Construction and Reconstruction of the White House, 1793–1817." PhD diss., University of Maryland, 1993.

———. *Over the Alleghenies: Historic Canals and Railroads of Pennsylvania*. Morgantown, WV: West Virginia University Press, 2013.

———. "The Potomac Canal: A Construction History." In *Canal History and Technology Proceedings: Vol. 21, March 23, 2002*, edited by Lance E. Metz, 143–235. Easton, PA: Canal History and Technology Press, National Canal Museum, 2002.

———. *The Potomac Canal: George Washington and the Waterway to the West*. Morgantown, WV: West Virginia University Press, 2007.

———. "The Untold Story of Blacks in the White House." *American Visions*, February–March 1995, 8.

Kapsch, Robert J., and Elizabeth Perry Kapsch. *The Monocacy Aqueduct on the Chesapeake and Ohio Canal*. Center for Historic Engineering and Architecture Research. Poolesville, MD: Medley Press, 2005.

Kearney, [J.], [W.] Turnbull, W. M. C. Fairfax, and M. C. Ewing, comps. "Chart of the Head of Navigation of the Potomac River Shewing the Route of the Alexandria Canal, Made in Pursuance of a Resolution of the Alexa. Canal Company, Oct. 1838." Library of Congress.

Kemp, Emory L., ed. *American Bridge Patents: The First Century (1790–1890)*. Morgantown, WV: West Virginia University Press, 2005.

———. "James Finley and the Origins of the Modern Suspension Bridge." In *Essays on the History of Transportation and Technology*, edited by Emory L. Kemp, 142–167. Morgantown, WV: West Virginia University Press, 2014.

Kennon, Donald R., ed. *The United States Capitol: Designing and Decorating a National Icon*. Athens: Ohio University Press for the United States Capitol Historical Society, 2000.

Kilty, William. *The Laws of Maryland*. 2 vols. Annapolis, MD: Frederick Green, 1800.

Kimball, Fiske. "Benjamin Henry Latrobe and the Beginning of Architectural and Engineering Practice in America." *Michigan Technic* 30 (1917): 218–233.

Kimball, Fiske, and Wells Bennett. "The Competition for the Federal Buildings, 1792–1793."

Journal of the American Institute of Architects 7 (1919): no. 1, 8–12; no. 3, 98–102; no. 5, 202–210; no. 8, 355–361.

———. "William Thornton and the Design of the United States Capitol." *Art Studies* 1 (1923): 89–90.

King, Horatio. "The Battle of Bladensburg: Burning of Washington in 1814." *Magazine of American History* 14 (1885): 438–457.

Kite, Elizabeth S. *L'Enfant and Washington, 1791–1792: Published and Unpublished Documents Now Brought Together for the First Time.* Baltimore: Johns Hopkins Press, 1929.

Kohn, David, and Bess Glenn, eds. *Internal Improvement in South Carolina 1817–1828 from the Reports of the Superintendent of Public Works and from Contemporary Pamphlets, Newspaper Clippings, Letters, Editions, and Maps.* Compiled by David Kohn. Washington, DC: Privately printed, 1938.

Knox, Henry. *Instructions to John Vermonnet, regarding fortifications at Annapolis, Maryland and Alexandria, Virginia.* May 12, 1794. In *American State Papers,* vol. 1: 93.

Kuff, Karen R, and James R. Brooks. *Building Stones of Maryland.* Baltimore: Maryland Geological Survey, 1990.

Labelye, Charles. *A Short Account of the Methods made Use of in Laying the Foundations of the Piers of Westminster Bridge.* London: A. Parker, 1751.

Lanier, John J. *The Great American Mason.* New York: Macoy, 1922.

Latrobe, Benjamin. "Dry Docks [At Washington Navy Yard Sufficient for Twelve Frigates with Canal to Little Falls]." December 4, 1802, 7th Cong., 2nd sess., no. 30. Reprinted in *American State Papers: Documents, Legislative and Executive of the Congress of the United States, from the First Session of the First to the Second Session of the Eighteenth Congress, Inclusive,* March 3, 1789 to March 5, 1825, vol. 6, *Naval Affairs,* 104–108. Washington, DC: Gales & Seaton, 1834.

———. *The Journal of Latrobe: Being the Notes and Sketches of an Architect, Naturalist and Traveler in the United States from 1796 to 1820.* New York: D. Appleton, 1905.

———. Letter report on the U.S. Capitol, Enclosure, Benjamin Latrobe to President Thomas Jefferson, April 4, 1803. Reprinted in Van Horne and Formwalt, *Correspondence of Latrobe,* vol. 1, 1784–1804, 268–284.

———. Letter to President Thomas Jefferson, April 4, 1803. Reprinted in Van Horne and Formwalt, *Correspondence of Latrobe,* vol. 1, 1784–1804, 429–433; and in Padover, *Jefferson and the Capital,* 335–340.

———. *Memorial of Benjamin H. Latrobe, Late Surveyor of the Public Buildings, in the City of Washington, in Vindication of His Professional Skills,* January 5, 1819. Washington, DC: E. De Kraft, 1819.

———. *Message from the President of the United States Transmitting a Report of the Surveyor of Public Buildings of the United States in the City of Washington,* December 1, 1808 (Latrobe's sixth annual report, for 1808). Washington, DC: A & G Way, 1808. Reprinted in Van Horne et al., *Correspondence of Latrobe,* vol. 2, 1805–1810, 670–676; in U.S. House, *Documentary History,* 146–152; and in Padover, *Jefferson and the Capital,* 445–452.

———. "Observations on the Foregoing Correspondence [i.e., Correspondence between Capt. William Jones of Philadelphia, and William Jones Esq. Civil Engineer of Calcutta, relative to the principle of Building in India]: by B. H. Latrobe, Surveyor of the Public Buildings of the U. States." *Transactions of the American Philosophical Society* 6 (1809): 384–391.

———. *Opinion on a Project for Removing the Obstructions to a Ship Navigation to Georgetown Col.* Washington, DC: W. Cooper, 1812.

———. *A Private Letter to the Individual Members of Congress on the Subject of Public Buildings of the United States at Washington*, November 28, 1806. Washington, DC: Samuel H. Smith, 1806). Reprinted in Van Horne et al., *Correspondence of Latrobe*, vol. 2, *1805–1810*, 296–316.

———. Report of B. Henry Latrobe on public buildings communicated to the Senate, June 13, 1809, June 12, 1809. Reprinted in U.S. House, *Documentary History*, 155–160.

———. "Report of B. Henry Latrobe to the President of the United States, February 20, 1804, Transmitted to Congress February 22, 1804" (Latrobe's first annual report, for 1803). Reprinted in U.S. House, *Documentary History*, 104–106.

———. "The Report of the Surveyor of the Public Buildings of the United States, December 11, 1809," transmitted to Congress December 16, 1809. Reprinted in U.S. House, *Documentary History*, 157–160.

———. "The Report of the Surveyor of the Public Buildings of the United States, December 28, 1810," transmitted to Congress January 14, 1811. Reprinted in Van Horne et al., *Correspondence of Latrobe*, vol. 2, *1805–1810*, 945–949; and in U.S. House, *Documentary History*, 163–165.

———. "The Report of the Surveyor of the Public Buildings of the United States, at Washington," November 25, 1806, communicated to Congress on December 15, 1806 (Latrobe's fourth annual report, for 1806). Reprinted in U.S. House, *Documentary History*, 120–123; and in Padover, *Jefferson and the Capital*, 373–380.

———. "Report of the Surveyor of the Public Buildings of the United States at Washington, March 23, 1808, submitted to Congress March 25, 1808" (Latrobe's fifth annual report, for 1807). Reprinted in Van Horne et al., *Correspondence of Latrobe*, vol. 2, *1805–1810*, 565–577; in U.S. House, *Documentary History*, 131–137; no. 249, in *American State Papers*, 1:719–724; and in Padover, *Jefferson and the Capital*, 399–413.

———. "Report on the City of Washington Public Buildings," October 27, 1807, 10th Cong., 1st sess., no. 227. Reprinted in Van Horne et al., *Correspondence of Latrobe*, vol. 2, *1805–1810*, 491–494; in *American State Papers*, 1:482–483; and in U.S. House, *Documentary History*, 129–131.

———. "Report on the Freestone Quarries on the Potomac and Rappahannock Rivers," February 19, 1807. *Transactions of the American Philosophical Society* 6 (1809): 283–293. Reprinted in Van Horne et al., *Correspondence of Latrobe*, vol. 2, *1805–1810*, 380–389.

———. Report on the Present State of the Capitol, November 28, 1816. Reprinted in U.S. House, *Documentary History*, 190–193.

———. "Report on the Public Buildings: The Report of the Surveyor of the Public Buildings of the United States at Washington," December 1, 1804 (Latrobe's second annual report, for 1804). In Van Horne and Formwalt, *Correspondence of Latrobe*, vol. 1, *1784–1804*, 577–584; in U.S. House, *Documentary History*, 111–114; and in Padover, *Jefferson and the Capital*, 347–354.

———. "Report on the Public Buildings," December 22, 1805 (Latrobe's third annual report, for 1805). Reprinted in Van Horne et al., *Correspondence of Latrobe*, vol. 2, *1805–1810*, 168–174; and in U.S. House, *Documentary History*, 115–118.

———. Report on the Washington Bridge, to Philip R. Thompson, March 25, 1806. Reprinted in Van Horne et al., *Correspondence of Latrobe*, vol. 2, *1805–1810*, 203–209.

———. "To Samuel Harrison, Editor of the *National Intelligencer*, November 22, 1807" (a description of the House of Representatives, south wing, U.S. Capitol). Reprinted in Van Horne et al., *Correspondence of Latrobe*, vol. 2, *1805–1810*, 499–506; and printed in *National Intelligencer*, November 30, 1807.

———. "To the Chairman of the Committee of the House of Representatives in Congress, to Whom was Referred the Message of the President of the United States of the 22nd of February 1804 Transmitting a Report of the Surveyor of Public Buildings of the 20th of February 1804, February 28, 1803 [*sic*—should read 1804]." Reprinted in U.S. House, *Documentary History*, 107–110.

Latrobe, John H. B. *Memoir of Benjamin Banneker.* Pamphlet. Baltimore: John D. Toy, 1845.

————. *The Capitol and Washington at the Beginning of the Present Century.* Baltimore: W. K. Boyle, 1881.

Law, Thomas. *Observations on the Intended Canal in Washington, 1804.* Washington, DC, 1804. Reprinted in *Records of the Columbia Historical Society of Washington, D. C.*, 8:159–168. Washington, DC: Columbia Historical Society, 1905.

————. *A Reply to Certain Institutions.* Pamphlet. Washington: N.p., 1824.

Lawler, Edward J. "The President's House Revisited." *Pennsylvania Magazine of History and Biography* 129 (October 2005): 371–410.w

Leach, Sara Amy, ed. *Capital IA: Industrial Archeology of Washington, D.C.* Washington, DC: Society for Industrial Archeology, 2001.

Lear, Tobias. *Observations on the River Potomack, Etc.* Baltimore: Samuel T. Chambers, 1940. Reprinted in *Records of the Columbia Historical Society of Washington, D. C.*, 8:117–140. Washington, DC: Columbia Historical Society, 1905. This was a reprint of a 1793 pamphlet originally published in Philadelphia.

Lee-Thorp, Vincent. *Washington Engineered.* Baltimore: Noble House, 2006.

Leish, Kenneth W. *The White House: A History of the Presidents.* New York: Newsweek, 1972.

Library of Congress. *District of Columbia Sesquicentennial of the Establishment of the Permanent Seat of the Government.* Catalog of exhibits in the Library of Congress, April 24, 1950 to April 24, 1951. Washington, DC: US Government Printing Office, 1950.

Littlefield, Douglas R. "Eighteenth-Century Plans to Clear the Potomac River." *Virginia Magazine of History and Biography* 93, no. 3 (July 1985): 291–322.

————. "A History of the Potomac Company and Its Colonial Predecessors, 1748–1828." MA thesis, University of Maryland, 1979.

————. "Maryland Sectionalism and the Development of the Potomac Route to the West, 1768–1826." *Maryland Historian* 14 (Fall–Winter 1983): 31–52.

————. "The Potomac Company: A Misadventure in Financing an Early American Internal Improvement Project." *Business History Review* 58 (1984): 562–585.

Livingood, James Weston. *The Philadelphia-Baltimore Trade Rivalry, 1780–1860.* Harrisburg, PA: Pennsylvania Historical and Museum Commission, 1947.

Lockwood, Mary Smith. *Yesterdays in Washington.* 2 vols. Roslyn, VA: Commonwealth, 1915.

Loftin, T. L. *Contest for a Capital: George Washington, Robert Morris, and Congress, 1783–1791 Contenders.* Washington, DC: Tee Loftin, 1989.

Look before You Leap; or, A Few Hints to Such Artizans, Mechanics, Labourers, Farmers and Husbandmen as Are Desirous of Emigrating to America. . . . London: Printed for W. Roe, 1796.

Loveday, Amos J., Jr. *The Rise and Decline of the American Cut Nail Industry: A Study of the Interrelationships of Technology, Business Organization, and Management Techniques.* Westport, CT: Greenwood Press, 1983.

Lowry, Bates, ed. *The Architecture of Washington, D.C.* Washington, DC: Dunlap Society, 1976.

————. *Building a National Image: Architectural Drawings for the American Democracy, 1789–1912.* Washington, DC: National Building Museum, 1985.

Mackall, Sally Somervell. *Early Days of Washington.* Washington, DC: Neale, 1899.

Maclay, William. *The Diary of William Maclay and Other Notes on Senate Debates.* Edited by Kenneth R. Bowling and Helen E. Veit. Baltimore: Johns Hopkins University Press, 1988.

Mahan, D. H. *An Elementary Course on Civil Engineering for the Use of Cadets of the United States Military Academy.* New York: John Wiley, 1846.

Maryland Geological Survey: vol. 2, *1898.* Baltimore: John Hopkins Press, 1898.

Marolda, Edward. *The Washington Navy Yard: An Illustrated History.* Washington, DC: Naval Historical Center, 1999.

Mason, Frances Norton, ed. *John Norton and Sons, Merchants of London and Virginia: Being the Papers from Their Counting House for the Years 1750 to 1795.* New York: Augustus M. Kelley, 1968.

Mason, John. "Memorial of John Mason ['to erect a bridge over the river Potomac, within the District of Columbia, are such as will unjustly and injuriously affect his interests, and those of the community at large']." May 13, 1826. House Doc. 179, 19th Cong. 1st sess.

Mathews, Catharine Van Cortlandt. *Andrew Ellicott: His Life and Letters.* New York: Grafton Press, 1908.

Matthews, William. *Report on a Survey for the Improvement of the Navigation of Goose Creek, Little River, and Beaver Dam.* Washington, DC, 1832.

McClellan, Phyllis I. *Silent Sentinel on the Potomac: Fort McNair, 1791–1991.* Bowie, MD: Heritage Books, 1993.

McCullough, David. *John Adams.* New York: Simon & Schuster, 2001.

McDonald, Forrest. *Alexander Hamilton; A Biography.* New York: W. W Norton, 1982.

McKee, Harley J. *Introduction to Early American Masonry: Stone, Brick, Mortar and Plaster.* Washington, DC: National Trust for Historic Preservation and Columbia University, 1973.

McLean, A. B. Letter (estimate to repair Washington Bridge), June 6, 1836. In Woodbury, "Letter from Secretary of Treasury."

McNeil, Priscilla W., and Don Hawkins, comps. *Map Showing Tracts of Land in Prince George's County, Maryland, Conveyed for the Federal City and Ownership of the Land on June 28 & 29, 1791 When the First Trust Deeds Were Signed.* February 1991. Library of Congress.

McNeil, William G. "Report of Captain Wm. G. McNeill on the Condition of the Chesapeake and Ohio Canal," December 1, 1833. In U.S. House, *Chesapeake and Ohio Canal,* 141–147.

Meglis, Anne Llewellyn, comp. *A Bibliographic Tour of Washington, D.C.* Washington, DC: D.C. Redevelopment Land Agency, 1974.

Metcalf, Paul. *Waters of Potowmack.* Charlottesville: University of Virginia Press, 2002.

Michler, N. [Nathaniel]. "Report of Brevet Brigadier General N. Michler, in Charge of Public Buildings, Grounds, and Works in the City of Washington, D.C.," September 30, 1869. In *Report of the Chief of Engineers to the Secretary of War, for the Year 1869,* 40th Cong., 2nd sess., vol. 2, pt. 2, Appendix W, 493–529. A separate report of the same title was published in Washington, DC, by the U. S. Government Printing Office in 1869.

Miller, Iris. *Washington in Maps, 1608–2000.* New York: Rizzoli International, 2002.

Miller, John C. *The Federalist Era, 1789–1801.* New York: Harper Torchbooks, 1960.

———. *The Wolf by the Wars: Thomas Jefferson and Slavery.* Charlottesville: University Press of Virginia, 1991.

Mills, Robert. *Guide to the Capitol of the United States, Embracing Every Information Useful to the Visitor, Whether on Business or Pleasure.* Washington, DC, 1834.

Miner, Craig. "The Capitol Workmen: Labor Policy on a Public Project." *Capitol Studies* 3 (Fall 1975): 45–52.

Minutes of the Proceedings of the Committee Appointed on 14 September 1793 to Attend to and Alleviate the Suffering of the Afflicted with the Malignant Fever. Philadelphia: Crissy and Markley, 1848.

Mitchell, Stewart. *New Letters of Abigail Adams, 1788–1801.* Boston: Houghton Mifflin, 1947.

Monkman, Betty C. *The White House, 1792–1992: Image in Architecture.* Washington, DC: American Architectural Foundation, 1992.

Moore, Charles. *Federal and Local Legislation Relating to Canals and Steam Railroads in the District of Columbia, 1802–1903.* Senate Doc. 220, 57th Cong., 2nd sess. (1903).

Moore, Thomas. *Address to the Citizens of Georgetown and Washington on Improving the Navigation of the River Potomac.* Georgetown, DC, 1806.

———. *Ship Navigation to Georgetown.* Georgetown, DC, 1811.

Morale-Vazquez, Rubil. "Imagining Washington: Monuments and Nation Building in the Early Capital." *Washington History* 12, no. 1 (Spring–Summer 2000): 12–29.

Morgan, James Dudley. "Maj. Pierre Charles L'Enfant, the Unhonored and Unrewarded Engineer." In *Records of the Columbia Historical Society, Washington, D. C.,* 2:118–57. Washington, DC: Columbia Historical Society, 1899.

Morgan, Maxine Goff. "A Chronological History of the Alexandria Canal (Part I)." *Arlington Historical Magazine* 3, no. 1 (October 1965): 3–16.

———. "A Chronological History of the Alexandria Canal (Part II)." *Arlington Historical Magazine* 3, no. 2 (October 1966): 3–23.

Morris, Edwin Bateman, comp. *Report of the Commission on the Renovation of the Executive Mansion.* Washington, DC: US Government Printing Office 1952.

Morrison, Alfred J. *The District in the Eighteenth Century: History, Site-Strategy, Real Estate Market, Landscape, etc. as Described by the Earliest Travelers. . . .* Washington, DC: Judd and Detweiler, 1909.

Munroe, Thomas. "Report on the Affairs of the City under His Care," December 20, 1802, 7th Cong., 2nd sess., no. 159. In *American State Papers,* 1:337–339.

Murphy, Jim. *An American Plague: The True and Terrifying Story of the Yellow Fever Epidemic of 1793.* New York: Clarion Books, 2003.

Myer, Donald Beekman. *Bridges and the City of Washington.* 1974. Reprint, Washington, DC: US Government Printing Office, 1983, 1992.

———. *Building the Potomac Aqueduct 1830.* Washington, DC: National Trust for Historic Preservation, 1976.

National Park Service Division of Publications. *Chesapeake and Ohio Canal: A Guide to Chesapeake and Ohio Canal National Historical Park, Maryland, District of Columbia and West Virginia.* Handbook 142. Washington, DC: National Park Service, U.S. Department of the Interior, 1991.

Nelson, Lee. *The Colossus of Philadelphia.* Washington, DC: American Society of Civil Engineers, 1990.

———. "Crafted from Stone." In *Our Changing White House,* edited by Wendell Garrett, 31–55. Boston: Northeastern University Press, 1995.

———. *Nail Chronology as an Aid to Dating Old Buildings.* Technical Leaflet 48. American Association for State and Local History, Nashville, TN.

———. *White House Stone Carving: Builders and Restorers.* Washington, DC: US Government Printing Office 1992.

Newell, Frederick Haynes, ed. *Planning and Building the City of Washington.* Washington, DC: Randell for the Washington Society of Engineers, 1932.

Nute, Grace L. "Washington and the Potomac: Manuscripts of the Minnesota Historical Society, 1764–1796." *American Historical Review* 27 (1923): 497–519.

Oberholtzer, Ellis P. *Robert Morris: Patriot and Financier.* New York: Macmillan, 1903.

[Office of the President of the United States]. *Message from the President of the United States*

Transmitting Plans and Estimates of a Dry Dock, for the Preservation of Our Ships of War, 28 December 1802. Washington, DC: William Duane, 1802.

Osborne, John Ball. "The Removal of the Government to Washington." In *Records of the Columbia Historical Society, Washington D.C.*, 3:136–60. Washington, DC: Columbia Historical Society, 1900.

Owen, Claude W. "Seneca, Once a Commercial Center." Undated typescript. Montgomery County Historical Society, Rockville, MD.

Owen, David Dale. "Report on Aquia Creek and Other Stafford County Freestones," March 30, 1847. In *Report of the Board of Regents of the Smithsonian Institution, Showing the Operations, Expenditures, and Condition of the Institution.* January 6, 1848, U.S. Senate Miscellaneous Report no. 23, 30th Cong. 1st sess., 109–113.

———. "Report on the Sandstones of the Potomac," March 15, 1847. In *Report of the Board of Regents of the Smithsonian Institution, Showing the Operations, Expenditures, and Condition of the Institution,* January 6, 1848, U.S. Senate Miscellaneous Report no. 23, 30th Cong. 1st sess., 36–39.

Owens, Malcolm H. "A History of Tiber Creek, Washington, D.C." Tau Beta Pi thesis, University of Maryland, 1937.

Padgett, James A., ed. "Letters of James Rumsey." *Maryland Historical Magazine* 32 (1937): 10–28, 136–155, 271–285.

Padover, Saul K., ed. *Thomas Jefferson and the National Capital: Containing Notes and Correspondence Exchanged between Jefferson, Washington, L'Enfant, Ellicott, Hallet, Thornton, Latrobe, the Commissioners, and Others, Relating to the Founding, Surveying, Planning, Designing, Constructing, and Administering of the City of Washington, 1783–1818.* Washington, DC: US Government Printing Office 1946.

Parkinson, Richard. *A Tour in America in 1798, 1799, and 1800.* London: J. Harding, 1805.

Partridge, William T. "L'Enfant's Methods and Features of His Plans for the Federal City." In *National Capital Park and Planning Commission Reports and Plans,* 21–33. Washington Region, Supplemental Technical Data to Accompany Annual Report. Washington, DC, 1930.

Passonneau, Joseph R. *Washington through Two Centuries: A History in Maps and Images.* New York: Monacelli Press, 2004.

Patton, Justin S., and Varna G. Boyd. *Archaeological Monitoring Report.* December 2004, U.S. Department of Transportation Headquarters, Southeast Federal Center, Washington, DC.

Pearce, Mrs. John Newton. "The Creation of the President's House." In *Records of the Columbia Historical Society of Washington, D. C.,* edited by Francis Coleman Rosenberger, 63–65:32–48. Washington, DC: Columbia Historical Society, 1966.

Peatross, C. Ford, ed. *Capital Drawings: Architectural Designs for Washington, D.C., from the Library of Congress.* Baltimore: Johns Hopkins University Press in association with the Library of Congress, 2005.

Peter, Armistead, III. *Tudor Place, Design by Dr. William Thornton.* Georgetown: Privately printed, 1970.

Peterson, Arthur G. "The Old Alexandria-Georgetown Canal and Potomac Aqueduct." *Virginia Magazine of History and Biography* 40, no. 4 (October 1932): 307–316.

Phenix, Thomas. *The State Convention on Internal Improvements. . . .* Baltimore: Wm. Ogden Niles, 1825.

Phillips, P. Lee. *The Beginnings of Washington: As Described in Books, Maps and Views.* Washington, DC: W. F. Roberts, 1917.

Pickell, John. *A New Chapter in the Early Life of Washington in Connection with the Narrative History of the Potomac Company.* 1858. Reprint, New York: Burt Franklin, 1970.

Pitch, Anthony. *The Burning of Washington: The British Invasion of 1814.* Annapolis, MD: Naval Institute Press, 1998.

Placzek, Adolf K., ed. *Macmillan Encyclopedia of Architects.* 4 vols. New York: Free Press, 1982.

Pogue, Dennis J. "Interpreting the Dimensions of Daily Life for the Slaves Living at the President's House and at Mount Vernon." *Pennsylvania Magazine of History and Biography* 129, no. 4 (2005): 433–444.

Powell, J. M. *Bring Out Your Dead: The Great Plague of Yellow Fever in Philadelphia in 1793.* Philadelphia: University of Pennsylvania Press, 1991.

Proctor, John Clagett, ed. *Washington: Past and Present: A History.* 5 vols. New York: Lewis Historical, 1930.

Putnam, Raymond S. "The History and Construction of The B Street Canal." Tau Beta Pi thesis, University of Maryland, 1937.

Reiff, Daniel D. *Washington Architecture, 1791–1861: Problems in Development.* Washington, DC.: U.S. Commission of Fine Arts, US Government Printing Office 1971.

Renwick, James, Jr. "Report to the Building Committee of the Smithsonian Institution" (on freestone quarries on the farm of Mr. Peter, at Seneca creek and Bull run, and those on the adjacent lands), March 24, 1847. In *Report of the Board of Regents of the Smithsonian Institution, Showing the Operations, Expenditures, and Condition of the Institution.* January 6, 1848. U.S. Senate Miscellaneous Report no. 23, 30th Cong. 1st sess., 105–107.

Reps, John W. *Monumental Washington: The Planning and Development of the Capital Center.* Princeton, NJ: Princeton University Press, 1967.

———. *Washington on View: The Nation's Capital since 1790.* Chapel Hill: University of North Carolina Press, 1991.

Riley, J. F. *The History of Old Washington.* New York: J. F. Riley, 1902.

Roberdeau, Isaac. "Mathematics and Treatise on Canals." Unpublished manuscript, 1796. Manuscripts Division, Library of Congress.

Rochefoucauld-Liancourt, Francois-Alexandre-Frederic. *Travels through the United States of North America, the Country of the Iroquois and Upper Canada, in the Years 1795, 1796, and 1797.* 2 vols. London: R. Phillips, 1799.

———. "Voyage to Federal City in 1797: The Duc De La Rochefoucauld-Liancourt's Visit in 1797: A New Translation." Translated by David J. Brandenburg and Millicent H. Brandenburg. In *Records of the Columbia Historical Society of Washington 1973–74,* edited by Francis Coleman Rosenberger, 49:35–60. Baltimore: Waverly Press, 1975.

Rogers, Jerry R., Donald Kennon, Robert T. Jaske, and Francis E. Griggs, eds. *Civil Engineering History: Engineers Make History: Proceedings of the First National Symposium on Civil Engineering History, November 10–14, 1996.* New York: American Society of Civil Engineers, 1996.

Royall, Ann. *Sketches of History, Life and Manners, in the United States, by a Traveller.* New Haven, CT: Printed for the author, 1826. Reprint, New York: Johnson Reprint, 1970.

Rumsey, James. Papers of James Rumsey, Manuscripts Division, Library of Congress.

Ryan, William, and Desmond Guinness. *The White House: An Architectural History.* New York: McGraw-Hill, 1980.

Sakolski, Aaron M. "Washington, America's First Boom Town." In *The Great American Land Bubble: The Amazing Story of Land-Grabbing, Speculating, and Booms from Colonial Days to the Present Time,* 147–168. New York: Harper, 1932.

Sanderlin, Walter S. *The Great National Project: A History of the Chesapeake and Ohio Canal.* Baltimore: Johns Hopkins Press, 1946.

———. "The Maryland Canal Project—An Episode in the History of Maryland's Internal Improvements." *Maryland Historical Magazine* 41 (1946): 51–65.

Sayenga, Donald. "James Finley." *Structure* (November 2008), www.structuremag.org
/article.aspx?articleID=804.

Schechter, Stephen L., and Richard B. Bernstein. *Well Begun: Chronicles of the Early National
Period.* Albany: New York State Commissioner on the Bicentennial of the United States
Constitution, 1989.

Schwengel, Frederic D. "The Masons and the Capitol of the United States." *New Age* 73 (March
1965): 40.

Scott, Pamela. "Moving to the Seat of Government: Temporary Inconveniences and Priva-
tions." *Washington History* 12, no. 1 (Spring–Summer 2000): 70–73.

———. *Temple of Liberty: Building the Capitol for a New Nation.* New York: Oxford University
Press, 1995.

———. "'This Vast Empire': The Iconography of the Mall, 1791–1848." In *The Mall in Washington, 1791–1991,* edited by Richard Longstreth. Studies in the History of Art 30, Center for
Advanced Study in the Visual Arts Symposium Papers, 16. Hanover, NH: National Gallery
of Art, Washington, 1991.

Scott, Pamela, and Antoinette J. Lee. *Buildings of the District of Columbia.* New York: Oxford
University Press, 1993.

Seale, William. *The President's House: A History.* 2 vols. Washington, DC: White House Historical Association with the cooperation of the National Geographic Society and Harry N.
Abrams, 1986.

———. *The White House: The History of an American Idea.* Washington, DC: White House Historical Association, 1992.

Semple, George. *A Treatise on Building in Water.* Dublin, 1776.

Shaffer, Donald R. "'We Are Again in the Midst of Trouble': Flooding on the Potomac River
and the Struggle for the Sustainability of the Chesapeake and Ohio Canal, 1828–1996."
Typescript. Contract report prepared for the Chesapeake and Ohio Canal National Historical Park, July 1997, National Park Service, University of Maryland, College Park.

Shomette, Donald G. *Maritime Alexandria: The Rise and Fall of an American Entrepot.* Bowie,
MD.: Heritage Books, 2003.

Simms, Colonel Charles. *Papers of Colonel Charles Simms.* Manuscripts Division, Library of
Congress.

Singleton, Esther. *The Story of the White House.* New York: S. S. McClure, 1907.

Skempton, A. W., M. M. Chrimes, R. C. Cox, P. S. M. Cross-Rudkin, R. W. Rennison, and E. C.
Ruddock. *A Biographical Dictionary of Civil Engineers in Great Britain and Ireland:* Vol. 1, *1500–
1830.* London: Thomas Telford for the Institution of Civil Engineers, 2002.

Smith, Mrs. Samuel Harrison [Margaret Bayard Smith]. *The First Forty Years of Washington
Society.* New York: Charles Scribner's, 1906. Reprint, New York: Frederick Ungar, 1965. Page
references are to the 1965 edition.

Smith, William Loughton. *The Journal of William Loughton Smith.* Proceedings of the Massachusetts Historical Society, 51. Cambridge, MA: University Press, 1917.

Snow, Peter. *When Britain Burned the White House: The 1814 Invasion of Washington.* London:
John Murray, 2013.

Spies, Gregory C. "Major Andrew Ellicott, Esq.: Colonial American Astronomical Surveyor,
Patriot, Cartographer, Legislator, Scientific Instrument Maker, Boundary Commissioner
and Professor of Mathematics." Unpublished paper, available at www.fig.net/pub/fig
_2002/HS2/HS2_spies.pdf.

Spofford, Ainsworth R. *The Founding of Washington City.* Baltimore: J. Murphy, 1881.

Spofford, Alexander R. "Removal of the Government to the District of Columbia in 1800." In Cox, *Celebration of Anniversary*, 243–252.

Spratt, Zack. "Ferries in the District of Columbia." In Holmes and Heine, *Records of the Columbia Historical Society*, 53–56:183–192.

———. "Rock Creek's Bridges." In Holmes and Heine, *Records of the Columbia Historical Society*, 53–56:101–134.

Standiford, Les. *Washington Burning: How a Frenchman's Vision for Our Nation's Capital Survived Congress, the Founding Fathers, and the Invading British Army*. New York: Crown, 2008.

Stapleton, Darwin H., ed. *The Engineering Drawings of Benjamin Henry Latrobe*. New Haven, CT: Published for Maryland Historical Society by Yale University Press, 1980.

Stearns, Elinor, and David N. Yerkes. *William Thornton: A Renaissance Man in the Federal City*. Washington, DC: American Institute of Architects Foundation, 1976.

Stephenson, Richard W. "The Delineation of a Grand Plan." *Quarterly Journal of the Library of Congress* 36 (Summer 1979): 207–224.

———. *"A Plan Whol[l]y New": Pierre Charles L'Enfant's Plan of the City of Washington*. Washington, DC: Library of Congress, 1993.

St. George, Judith. *The White House: Cornerstone of a Nation*. New York: G. P. Putnam's, 1990.

Stillman, Damie. "Six Houses for the President." *Pennsylvania Magazine of History and Biography* 129 (October 2005): 411–431.

Studebaker, Marvin F. "Freestone from Aquia." *Virginia Cavalcade* 9, no. 1 (Summer 1959): 35–41.

Sturge, Joseph. *A Visit to the United States in 1841*. New York: Augustus M. Kelly, 1969.

Sween, Jane C. "Seneca." *Montgomery County Story* 25, no. 1 (1971): 1–10.

Sweig, Donald Mitchell. "Northern Virginia Slavery: A Statistical and Demographic Investigation." PhD diss., College of William and Mary, 1982.

Swift, William H., and Nathan Hale. *Report on the Present State of the Chesapeake and Ohio Canal: The Estimated Cost of Completing It to Cumberland*. Boston: Dutton and Wentworth, 1846.

Taggart, Hugh T. *Old Georgetown (District of Columbia)*. Lancaster, PA: New Era, 1908. Reprinted from *Records of the Columbia Historical Society, Washington, D.C.*, 11:120–224 (Washington, DC: Columbia Historical Society, 1908).

———. "The Presidential Journey, in 1800, from the Old to the New Seat of Government." In *Records of the Columbia Historical Society, Washington D.C.*, 3:180–209. Washington, DC: Columbia Historical Society, 1900.

Taste, Timothy [pseud.]. *The Freaks of Columbia; or, The Removal of the Seat of Government: A Farce*. Washington, DC: N.p., 1808.

[Thornton, William]. *Index to My Private Letter, etc.* Washington, DC, 1807.

Thurston, Robert H. *The Materials of Engineering, Part 1, Non-Metallic Materials*. New York: John Wiley, 1883.

Timoshenko, Stephen P. *History of Strength of Materials*. 1953. Reprint, New York: Dover, 1983.

Tindall, William. *Standard History of the City of Washington: From a Study of Original Sources*. Knoxville, TN: H. W. Crew, 1914.

Tompkins, Sally. *Quest for Grandeur*. Washington, DC: Smithsonian Institution Press, 1993.

[Toms, W. H., and John Devoto]. *The Builder's Dictionary; or, Gentleman and Architect's Companion, Being a Complete and Unabridged Reprint of the Earlier Work Published by A. Bettesworth and C. Hitch*. 2 vols. Washington, DC: Association for Preservation Technology, 1981. Original publication 1734.

Tremain, Mary. *Slavery in the District of Columbia*. New York: G. P. Putnam's, 1892.

[Trimble, Isaac]. *Report of the Engineer Appointed by the Commissioners of the Mayor and City Council of Baltimore on the Subject of the Maryland Canal.* Baltimore: Lucas & Deaver, 1837.

Turnbull, [William]. "Potomac Aqueduct: Captain Turnbull's Report on the Survey and Construction of the Potomac Aqueduct, by Order of the House of Representatives, January 1, 1836." *Civil Engineer and Architect's Journal* 1, no. 1 (October 1837): 147.

[————]. *Report from the Secretary of War in Compliance with a Resolution of the Senate in Reference to the Construction of the Potomac Aqueduct, and the Kyanizing of Timber for the Use of the Same.* U.S. Senate Doc. 178, 26th Cong., 2nd sess. (1841).

————. *Reports on the Construction of the Piers of the Aqueduct of the Alexandria Canal across the Potomac River at Georgetown, District of Columbia, 1835–1840.* Washington, DC: US Government Printing Office 1873.

Tyson, Martha E. *Banneker, the Afric-American Astronomer.* Philadelphia: Friends' Book Association, 1884.

————. *A Sketch of the Life of Benjamin Banneker; from Notes Taken in 1836: Read by J. Saurin Morris before the Maryland Historical Society, October 5th, 1854.* Baltimore: J. D. Toy, 1854.

U.S. Army. "The Record Book of Fort Washington." Manuscript, 1904–1905. National Park Service Museum Resource Center, Lanham, MD.

U.S. Commission on the Renovation of the Executive Director. *Report of the Commission on the Renovation of the Executive Mansion.* Washington, DC: US Government Printing Office, 1952.

U.S. Congress. "An Act for the Relief of the Anacostia Bridge Company," March 3, 1815, 13th Cong., 3rd sess.

————. "An Act for the Relief of the Eastern Branch Bridge Company," March 3, 1815, 13th Cong. 3rd sess.

————. "An Act for the Relief of William Benning [Re: The Anacostia Bridge Company]," April 28, 1828, 20th Cong., 1st sess.

————. "An Act to Incorporate a Company for Opening the Canal in the City of Washington" (charter of the Washington Canal Company), February 16, 1809, 10th Cong., 2nd sess.

————. "An Act to Incorporate a Company to Build a Bridge over the Eastern Branch of [the] Potomac, between Eleventh and Twelfth Streets, East, in the City of Washington" (charter of the Navy Yard Bridge Company), February 24, 1819, 15th Cong., 2nd sess.

————. *Laws of the United States of America, from the 4th of March, 1789, to the 4th of March, 1815.* 5 vols. Philadelphia: John Bioren and W. John Duane; Washington: R.C. Wrightman, 1815.

U.S. General Accounting Office. General Accounting Office Accounts of the Commissioners of the City of Washington, 1794–1802. Record Group 217, 7 boxes. National Archives and Records Administration.

U.S. Geological Survey. *Building Stones of Our Nation's Capital.* U.S. Department of the Interior Geological Survey, USGS: INF-74-35. Washington, DC: US Government Printing Office 1975.

U.S. House of Representatives. *Alexandria Canal.* January 31, 1831, Report no. 71, 21st Cong., 2nd sess.

————. *Alexandria Canal Company.* March 30, 1832, 22nd Cong., 1st sess.

————. *Annual Report of the President and Directors of the Washington Canal Company.* February 25, 1824. Washington, DC: Gales & Seaton, 1825.

————. *Annual Report of the President and Directors of the Washington Canal Company.* March 10, 1826, Doc. 127, 19th Cong. 1st sess. Washington, DC: Gales & Seaton, 1826.

————. *Annual Report of the Washington Canal Company.* April 21, 1824. Washington, DC: Gales & Seaton, 1824.

———. *Aqueduct Bridge — Battery Cove, Hearings Before the Committee on Military Affairs,* 69th Cong., 1st sess. (1926).

———. *Canal Company — Alexandria, Report: The Committee on the District of Columbia Have under Consideration the Memorial from the President and Directors of the Alexandria Canal Company.* May 30, 1834, Report no. 498, 23rd Cong., 1st sess.

———. "Capture of the City of Washington," communicated to the House of Representatives on November 29, 1814. 13th Cong., 3rd sess., no. 137. In *American State Papers,* 1:524–599.

———. *Chesapeake and Ohio Canal.* April 17, 1834, House Report no. 414, 23rd Cong., 1st sess.

———. *Chesapeake and Ohio Canal, Letter from the Secretary of War: Report of Captain William Gibbs McNeil, Top. Engrs.* Doc. 38, January 14, 1834, 23rd Cong., 1st sess.

———. *Chesapeake and Ohio Canal.* House Report no. 47, January 22, 1828, 20th Cong. 1st sess.

———. *Chesapeake and Ohio Canal.* Report no. 90, 19th Cong., 2nd sess., January 30, 1827.

———. *Development of the United States Capital: Addresses Delivered in the Auditorium of the United States Chamber of Commerce Building, Washington, D.C., at Meetings Held to Discuss the Development of the National Capital, April 25–26, 1929.* House Doc. 35, 71st Cong., 1st sess. (1930).

———. *Documentary History of the Construction and Development of the United States Capitol Building and Grounds (DH).* Report no. 646, 58th Cong. 2nd sess. (1904).

———. *Drawings Accompanying the Report of Captain Turnbull, on the Survey and Construction of the Alexandria Aqueduct, Made to the House of Representatives, 2d of July, 1838.* Washington, DC: US Government Printing Office, 1838.

———. *Final Report from the Secretary of the Treasury on the Subject-Matter of the 3d Section of 'An Act Authorizing the Construction of a Bridge across the Potomac,' etc.* (Subtitle: Bridge across the Potomac — O. H. Dibble), December 29, 1834, Doc. 51, 23rd Cong., 2nd sess.

———. *Letter from the Secretary of the Interior, Transmitting Reports, Plans, and Cost Estimates for Permanent Bridges across the Potomac, and for the Improvement of the River.* February 7, 1857, Ex. Doc. 68, 34th Cong., 3rd sess.

———. *Letter from the Secretary of War, Transmitting . . . a Report and Plans of the Survey of a Route for a Canal from the City of Baltimore to the Contemplated Chesapeake and Ohio Canal.* Doc. 58, 20th Cong., 1st sess., January 14, 1828.

———. *Letter from the Secretary of War Transmitting a Report, Map, and Estimate of the Chesapeake and Ohio Canal to Alexandria, in the District of Columbia.* 20th Cong., 1st sess., Report no. 254, April 21, 1828. Washington, DC: Gales & Seaton, 1828.

———. *Memorial of the Central Committee of the Chesapeake and Ohio Canal Convention and of the Commissioners Appointed by the States of Virginia and Maryland and Met by the United States to Open Books for Subscription of Stock to Said Company.* Doc. 151, 19th Cong., 1st sess., April 3, 1826.

———. *Message from the President of the United States Transmitting a Report from the Secretary of War . . . Internal Improvements concerning the Proposed Chesapeake and Ohio Canal.* 19th Cong., 2nd sess., Doc. 10, December 7, 1826.

———. *Potomac Aqueduct. 1838. Captain Turnbull's Report on the Survey and Construction of the Alexandria Aqueduct. 1838.* Doc. 459, 25th Cong., 2nd sess.

———. *Report of Committee on Bill Authorizing Loan for the Use of the city of Washington* (report of Commissioner Alexander White), March 11, 1796, 4th Cong., 1st sess., no. 78. In *American State Papers,* 1:142–144.

———. *Report of Committee on Expenditure of Money Made by the Commissioners of the City* of Washington, the Disposition of Public Property and Generally into the Transactions of the Commissioners, February 27, 1801, 6th Cong., 2nd sess., no. 145. In *American State Papers*, 1:243–246.

————. *Report of Committee on Memorial of the Commissioners Appointed for Establishing the Seat of Government at Washington.* January 25, 1796, 4th Cong., 1st sess., no. 70. In *American State Papers*, 1:136–137.

————. *Report of the Committee on the Circumstances Which Produced the Deficit for the Public Buildings and How Far It May Be Consistent with the Public Interest to Abolish the Office of Surveyor of Public Buildings.* April 21, 1808, no. 255, 10th Cong., 1st sess. In *American State Papers*, 1:955–956.

————. *Report of the Committee on the Disputes between the Commissioners of the City of Washington and Other Persons Who May Conceive Themselves Injured by the Several Alterations Made in the Plan of the City.* April 8, 1802, 7th Cong., 1st. sess., no. 157. In *American State Papers*, 1:330–336.

————. *Report of the Committee on the Petition of Sundry Citizens of the District of Columbia Praying that Congress Will Pass an Act, Incorporating a Company for the Purpose of Opening a Canal to Unite the Waters of the Potomac and the Eastern Branch, through Tyber Creek and the Low Grounds at the Foot of Capitol Hill.* February 11, 1802, 7th Cong., 1st sess., no. 152. In *American State Papers*, 1:258–259.

————. *Report of the Committee on the Petition of Sundry Inhabitants on a Bridge across the Potomac at Washington.* January 21, 1806, no. 200, 9th Cong., 1st sess. In *American State Papers*, 1:437–439.

————. *Report of the District of Columbia to Whom Were Referred Sundry Memorials from the Inhabitants of Pennsylvania, Maryland, and Virginia, Praying the Aid of the Federal Government towards the Improvement of the Navigation of the River Potomac.* Report no. 111, May 3, 1822, 17th Cong., 1st sess.

————. *Report of the President and Directors of the Washington Canal Company, of the Amount of Their Expenditure, and the Clear Profit Thereof; Made in Pursuance of a Requisition of Their Charter, January 31, 1817.* Doc. 63. Washington, DC, 1817.

————. *Report on Bridge across the Potomac River at Washington.* February 4, 1805, 8th Cong., 2nd sess., no. 191. In *American State Papers*, 1:422.

————. *Report Recommending the Offices of Two of the Commissioners of the City Ought to Be Abolished and to Be Vested in One Only.* February 12, 1802, 7th Cong., 1st sess., no. 153. In *American State Papers*, 1:260.

————. *Survey of the Anacostia River. Letter from the Secretary of the Treasury, Transmitting an Estimate from the Secretary of War of Appropriation for a Survey of Anacostia River, District of Columbia.* January 14, 1888, Ex. Doc. 97, 50th Congress, 1st session.

U.S. Office of the Chief of Engineers, U.S. Army. *Laws of the United States Relating to the Construction of Bridges over Navigable Waters of the United States from March 2, 1805, to March 3, 1887.* Compiled by John G. Parke. Washington, DC: US Government Printing Office, 1887.

U.S. Secretary of Treasury. *"Expenditures in the District of Columbia: Letter from the Secretary of the Treasury in Answer to a Resolution of the House of February 11, Transmitting a Statement of the Expenditures for Public and Private Purposes in the District of Columbia, from the Establishment of the Seat of Government to the 31st of December, 1869.* House Ex. Doc. 156, 41st Cong., 2nd sess. (1870).

————. *Expenditures of Money in the District of Columbia: Letter from the Secretary of the Treasury, Transmitting a Statement of the Money Annually Expended within the District of Columbia since 1790.* 31st Cong., 1st sess., House Ex. Doc. 83 (1849).

————. *Report from the Secretary of the Treasury, Transmitting in Compliance with a Resolution of the Senate, a Statement Showing the Appropriations to Be Expended in the District of Columbia since the Location of the Seat of Government Therein.* Senate Doc. 600, 26th Cong., 1st sess. Washington, DC: Blair and Rives, 1840.

————. *Report of the Secretary of the Treasury on the Subject of Public Roads and Canals. . . .* (Gallatin report on roads and canals). April 4, 1808, 10th Cong., 1st sess., no. 250. In *American State Papers*, 1:724–921. Reprint of 1808 work, New York: Augustus M. Kelley, 1968.

————. *Report on the Debt of the City of Washington.* January 15, 1802, 7th Cong., 1st sess., no. 151. In *American State Papers*, 1:258.

U.S. Senate. *Documents* (of the Chesapeake and Ohio Canal). Senate Report 610, 26th Cong., 1st sess., July 11, 1840.

————. *Papers Relating to the City of Washington: Printed for the Senate Committee on the District of Columbia.* Washington, DC: US Government Printing Office 1900.

————. *Report from the Secretary of War in Compliance with a Resolution of the Senate in Reference to the Construction of the Potomac Aqueduct, and the Kyanizing of Timber for the Use of the Same, 1841.* Doc. 178, 26th Cong., 2nd sess. (1841).

————. *Report* (on Senate bill 310, "To incorporate the Georgetown and Washington Canal and Sewerage Company"). July 15, 1868, Rep. Com. no. 167, 40th Cong., 2nd sess.

————. *Resolutions of the Legislature of Maryland in Favor of Making the Bridges over the Anacostia or Eastern Branch of the Potomac River Free of Toll.* April 5, 1844, Senate Doc. 271, 28th Cong., 1st sess.

Van Dyne, Larry. "If These Stones Could Talk." *Washingtonian Magazine*, October 2002, 50.

Van Horne, John C., Jeffrey A. Cohen, Darwin H. Stapleton, William B. Forbush III, and Tina H. Sheller, eds. *The Correspondence and Miscellaneous Papers of Benjamin Henry Latrobe:* Vol. 3, *1811–1820 (CL3).* New Haven, CT: Yale University Press for the Maryland Historical Society, 1988.

Van Horne, John C., Jeffrey A. Cohen, Darwin H. Stapleton, Lee W. Formwalt, William B. Forbush III, and Tina H. Sheller, eds. *The Correspondence and Miscellaneous Papers of Benjamin Henry Latrobe:* Vol. 2, *1805–1810 (CL2).* New Haven, CT: Yale University Press for the Maryland Historical Society, 1986.

Van Horne, John C., and Lee W. Formwalt, eds. *The Correspondence and Miscellaneous Papers of Benjamin Henry Latrobe:* Vol. 1, *1784–1804 (CL1).* New Haven, CT: Yale University Press for the Maryland Historical Society, 1984.

Verner, Coolie. "Surveying and Mapping the New Federal City." *Imago Mundi* 23 (1969): 59–72.

Virginia House of Delegates. Report of the Falls Bridge Turnpike Company, December 6, 1820. In *Journal of the House of Delegates*, xviii–xix. Richmond, VA: Thomas Ritchie, 1820.

Von Gerstner, Franz Anton Ritter. *Die innern Communicationen der Vereinigten Staaten von Nordamerica.* 1842. Translated by David J. Diephouse and John C. Decker as *Early American Railroads and Canals* and edited by Frederick C. Gamst. Stanford, CA: Stanford University Press, 1997.

Wagner, Frederick. *Robert Morris: American Patriot.* New York: Dodd, Mead, 1976.

Walker, George. *A Description of the Situation and Plan of the City of Washington.* Pamphlet. London: George Walker, 1792.

Walston, Mark. "Seneca Stone: Building Block of the Nation's Capital." *Maryland Magazine*, Winter 1985, 39–42.

Ward, George Washington. *The Early Development of the Chesapeake and Ohio Canal Project.* Johns Hopkins University Studies in Historical and Political Science, Series 17, Nos. 9–10–11. New York: Johnson Reprint, 1973. Original publication 1899.

Warden, D. B. *A Chorographical and Statistical Description of the District of Columbia, the Seat of the General Government of the United States.* Paris: Smith, 1816.

Washington Canal Company. *Report of the President and Directors of the Washington Canal Company.* Washington, DC, January 31, 1817.

————. *To the Senate and House of Representatives of the United States of America* (annual report). Washington, DC, April 21, 1824.

————. *To the Senate and House of Representatives of the United States of America* (annual report). Washington, DC, February 24, 1825.

————. *To the Senate and House of Representatives of the United States of America* (annual report). Washington, DC, March 10, 1826.

Watterston, George. *A Picture of Washington: Giving a Description of All the Public Buildings, Grounds, etc.* Washington, DC: William M. Morrison, 1840.

Weld, Isaac, Jr. *Travels through the States of North America and the Provinces of Upper and Lower Canada during the Years 1795, 1796, and 1797.* London: John Stockdale, 1799.

Wennersten, John R. *The Historic Waterfront of Washington, D.C.* Charleston, SC: History Press, 2014.

Wiebenson, Dora. "The Domes of the Halle au Blé in Paris." *Art Bulletin* 55, no. 2 (June 1973): 262–279.

Woodbury, Levi. "Letter from the Secretary of the Treasury: Bridge across the Potomac River." In U.S. House of Representatives, Doc. 281, 24th Cong. 1st sess., June 7, 1836.

Wright, Carroll D. "The Economic Development of the District of Columbia." *Proceedings of the Washington Academy of Sciences* 1 (1899): 161–187.

Wright, G. Frederick, and Frederick Bennett Wright, eds. "Washington's Canoe Trip down the Potomac Related in a Letter to Colonel Innes." *Records of the Past* 9 (1910): 74–79.

Young, James Sterling. *The Washington Community, 1800–1828.* New York: Columbia University Press, 1966.

ROBERT J. KAPSCH received his BS in engineering from Rutgers University; an MS in management and an MA in American Studies, both from George Washington University; a PhD in architecture and engineering from Catholic University, and a PhD in American Studies from the University of Maryland. He is a researcher and an author at the Center for Historic Engineering and Architecture.

Previously he served as the National Park Service Senior Scholar in Historic Architecture and Engineering (2000–2005) and as project engineer for the Chesapeake and Ohio Canal National Historical Park (1995–2000). Before that, for fifteen years (1980–1995) he was the Chief of the Historic American Buildings Survey / Historic American Engineering Record, and in that post he was able to double the size of those collections in the Library of Congress.

Kapsch is the recipient of the American Society of Civil Engineers History and Heritage award (2016); the Distinguished Service Medal, the highest award of the US Department of Interior, for his work at the National Park Service; and many other awards and honors. He is the author of numerous articles on canals and engineering history as well as books, including *Canals* (W. W. Norton, 2004); *The Potomac Canal: George Washington and the Waterway to the West* (West Virginia University Press, 2007) (recipient of *Foreword* magazine's silver medal for Best Book of the Year for history); *Historic Canals and Waterways of South Carolina* (University of South Carolina Press, 2010); and *Over the Alleghenies: Early Canals and Railroads of Pennsylvania* (West Virginia University Press, 2013) (2014 Gold Medal, Independent Publishers Book Awards, Mid-Atlantic Nonfiction Regional Award). He also coauthored, with his wife, Elizabeth Perry Kapsch, *The Monocacy Aqueduct on the Chesapeake and Ohio Canal* (Medley Press, 2005).

He is a member of numerous American and European organizations on the history of engineering and architecture, including the American Institute of Architects, the American Society of Civil Engineers, the Society for Architectural History, the Society for Industrial Archeology, the International Committee for the Conservation of the Industrial Heritage, and others. Kapsch and his wife live on a farm in Poolesville, Maryland, with three horses, four donkeys, five dogs, seven cats, and ten chickens.